EAI/Springer Innovations in Communication and Computing

Series Editor

Imrich Chlamtac, European Alliance for Innovation, Ghent, Belgium

The impact of information technologies is creating a new world yet not fully understood. The extent and speed of economic, life style and social changes already perceived in everyday life is hard to estimate without understanding the technological driving forces behind it. This series presents contributed volumes featuring the latest research and development in the various information engineering technologies that play a key role in this process. The range of topics, focusing primarily on communications and computing engineering include, but are not limited to, wireless networks; mobile communication; design and learning; gaming; interaction; e-health and pervasive healthcare; energy management; smart grids; internet of things; cognitive radio networks; computation; cloud computing; ubiquitous connectivity, and in mode general smart living, smart cities, Internet of Things and more. The series publishes a combination of expanded papers selected from hosted and sponsored European Alliance for Innovation (EAI) conferences that present cutting edge, global research as well as provide new perspectives on traditional related engineering fields. This content, complemented with open calls for contribution of book titles and individual chapters, together maintain Springer's and EAI's high standards of academic excellence. The audience for the books consists of researchers, industry professionals, advanced level students as well as practitioners in related fields of activity include information and communication specialists, security experts, economists, urban planners, doctors, and in general representatives in all those walks of life affected ad contributing to the information revolution.

Indexing: This series is indexed in Scopus, Ei Compendex, and zbMATH.

About EAI - EAI is a grassroots member organization initiated through cooperation between businesses, public, private and government organizations to address the global challenges of Europe's future competitiveness and link the European Research community with its counterparts around the globe. EAI reaches out to hundreds of thousands of individual subscribers on all continents and collaborates with an institutional member base including Fortune 500 companies, government organizations, and educational institutions, provide a free research and innovation platform. Through its open free membership model EAI promotes a new research and innovation culture based on collaboration, connectivity and recognition of excellence by community.

Michal Balog • Angelina Iakovets
Stella Hrehova
Editors

EAI International Conference on Automation and Control in Theory and Practice

Springer

EAI
RESEARCH MEETS INNOVATION

Editors
Michal Balog
Faculty of Manufacturing Technologies with the seat in Prešov, Department of Industrial Engineering and Informatics
Technical University of Košice
Prešov, Slovakia

Angelina Iakovets
Faculty of Manufacturing Technologies with the seat in Prešov, Department of Industrial Engineering and Informatics
Technical University of Košice
Prešov, Slovakia

Stella Hrehova
Faculty of Manufacturing Technologies with the seat in Prešov, Department of Industrial Engineering and Informatics
Technical University of Košice
Prešov, Slovakia

ISSN 2522-8595 ISSN 2522-8609 (electronic)
EAI/Springer Innovations in Communication and Computing
ISBN 978-3-031-31966-2 ISBN 978-3-031-31967-9 (eBook)
https://doi.org/10.1007/978-3-031-31967-9

This Springer imprint is published by the registered company Springer Nature Switzerland AG
The registered company address is: Gewerbestrasse 11, 6330 Cham, Switzerland

Preface

We are pleased to present the proceedings of the 15th year of the international conference ARTEP 2023 held for the first time under the auspices of the European Alliance for Innovation (EAI). The conference's organizer was the Department of Industrial Engineering and Informatics, Faculty of Manufacturing Technologies with the seat in Prešov of the Technical University of Košice. The conference was focused on automation and management in theory and practice, especially on Industry 4.0 and its applications.

The conference mainly attracted researchers and experts from the European Union who use and develop Industry 4.0 technology and its implementation in practice. The central theme of the ARTEP 2023 conference was the application of intelligent systems in industrial processes and the creation of added value for the Smart Industry in product and process innovations.

The introduction of the conference was dedicated to manufacturing companies from Slovakia, and the subsequent Workshop was carried out in the spirit of Industry 4.0 in the conditions of manufacturing companies. The Workshop aimed to map the situation of Industry 4.0, mainly in selected companies that cooperate with the Department of Industrial Engineering and Informatics in solving specific topics in the field of intelligent systems.

The following ARTEP 2023 technical program consisted of 27 full papers and 3 invited papers in the framework of personal presentations at the main conference circuits.

Coordination by the steering committee – Imrich Chlamtac and Michal Balog – was essential for the conference's success. We sincerely appreciate their continued support and guidance. Working with such an excellent team of the organizing committee was a great pleasure; their hard work in organizing and supporting the conference was top-notch. In particular, the Technical Program Committee was led by TPC co-chairs, Prof. Dr. Ján Pitel and Prof. Dr. Alena Galajdová, who completed the process of mutual evaluation of technical documents and created a high-quality technical program. We would also like to thank Conference Manager Veronika Kissová, Angelina Iakovets and Stella Hrehová, for their help and support with the conference preparation. At the same time, I thank each author, who sent his contribution to the ARTEP 2023 conference and Workshop.

We strongly believe that the ARTEP 2023 conference provides a good ground for all researchers, developers, and practitioners to discuss all aspects of science and technology relevant to computerization, digitization, and Industry 4.0 and to build strong partnerships. We also expect the next ARTEP 2024 conference to be as successful and stimulating as the contributions presented in this collection suggest.

Košice, Prešov, Slovakia

Michal Balog
Angelina Iakovets
Stella Hrehova

Conference Organization

Steering Committee

Imrich Chlamtac	University of Trento, Italy
Michal Balog	Technical University of Košice, Slovakia

Organizing Committee

Honorary Chair

Dr. h. c. mult. prof. Ing. Jozef Zajac, CSc Dean of the Faculty of Manufacturing Technologies with seat in Prešov, Technical University in Košice

General Chair

Michal Balog Technical University of Košice, Slovakia

General Co-Chairs

Dobrotvorskiy Sergey	National Technical University Kharkiv Polytechnic Institute, Ukraine
José Machado	Mechanical Engineering department of University Minho, Portugal

TPC Chair and Co-Chair

Ján Piteľ	Technical University of Košice, Slovakia
Alena Galajdová	Technical University of Košice, Slovakia
Stella Hrehová	Technical University of Košice, Slovakia
Alessandro Ruggiero	University of Salerno, Italy

Sponsorship and Exhibit Chair

Michal Balog Technical University of Košice, Slovakia

Local Chair

Michal Balog	Technical University of Košice, Slovakia

Workshops Chair

Stella Hrehová	Technical University of Košice, Slovakia
Jana Mižáková	Technical University of Košice, Slovakia
Marek Vagaš	Technical University of Košice, Slovakia

Publicity & Social Media Chair

Róbert Rákay	Technical University of Košice, Slovakia
Yevheniia Basova	National Technical University Kharkiv Polytechnic Institute, Ukraine

Publications Chair

Angelina Iakovets	Technical University of Košice, Slovakia
Monika Trojanová	Technical University of Košice, Slovakia
Zuzana Šoltysová	Technical University of Košice, Slovakia

Web Chair

Jana Mižáková	Technical University of Košice, Slovakia
Stella Hrehová	Technical University of Košice, Slovakia

Panels Chair

Dagmar Cagáňová	Comenius University in Bratislava, Slovakia

Technical Program Committee

Ján Piteľ	Technical University of Košice, Faculty of Manufacturing Technologies with the seat in Prešov, Slovakia
Alena Galajdová	Technical University of Košice, Faculty of Mechanical Engineering, Department of Industrial Automation and Mechatronics, Slovakia
Alessandro Ruggiero	Faculty of Engineering, Department of Mechanical Engineering, University of Salerno, Italy
Alexander Hošovský	Technical University of Košice, Faculty of Manufacturing Technologies with the seat in Prešov, Slovakia
Tibor Krenický	Technical University of Košice, Faculty of Manufacturing Technologies with the seat in Prešov, Slovakia
Vladimir Modrák	Technical University of Košice, Faculty of Manufacturing Technologies with the seat in Prešov, Slovakia

Michal Hatala	Technical University of Košice, Faculty of Manufacturing Technologies with the seat in Prešov, Slovakia
Svetlana Radchenko	Technical University of Košice, Faculty of Manufacturing Technologies with the seat in Prešov, Slovakia
Jozef Jurko	Technical University of Košice, Faculty of Manufacturing Technologies with the seat in Prešov, Slovakia
Darina Duplaková	Technical University of Košice, Faculty of Manufacturing Technologies with the seat in Prešov, Slovakia
Kamil Židek	Technical University of Košice, Faculty of Manufacturing Technologies with the seat in Prešov, Slovakia
Denisa Olekšaková	Technical University of Košice, Faculty of Manufacturing Technologies with the seat in Prešov, Slovakia
Marek Kočiško	Technical University of Košice, Faculty of Manufacturing Technologies with the seat in Prešov, Slovakia
Peter Trebuňa	Technical University of Košice, Faculty of Mechanical Engineering, Slovakia
Miriam Pekarčíková	Technical University of Košice, Faculty of Mechanical Engineering, Slovakia
Iveta Zolotová	Technical University of Košice, Faculty of Electrical Engineering and Informatics, Slovakia
Peter Papcun	Technical University of Košice, Faculty of Electrical Engineering and Informatics, Slovakia
Martin Straka	Technical University of Košice Faculty of Mining, Ecology, Process Control and Geotechnologies, Slovakia
Dagmar Cagaňová	Comenius University in Bratislava, Faculty of Management, Slovakia
Natalia Horňaková	Comenius University in Bratislava, Faculty of Management, Slovakia
Miloš Čambal	Slovak Technical University of Bratislava, Faculty of Materials Science Technology in Trnava, Slovakia
Pavel Važan	Slovak Technical University of Bratislava, Faculty of Materials Science Technology in Trnava, Slovakia
Pavol Tanuška	Slovak Technical University of Bratislava, Faculty of Materials Science Technology in Trnava, Slovakia
Vitalii Ivanov	Sumy State University, Faculty of Technical Systems and Energy Efficient Technologies, Ukraine
Zuzana Ságová	Department of Automation and Production Systems, University of Žilina, Slovakia
Ivan Zajacko	Department of Automation and Production Systems, University of Žilina, Slovakia

Contents

About the Editors

Michal Balog is head of the Department of Industrial Engineering and Informatics at the Faculty of Manufacturing Technologies based in Presov of Technical University in Košice. His publishing activities are introduced by more than 100 domestic and foreign publications. He works as a lecturer at local universities in Poland and the Czech Republic. He is the organizer of many domestic and international conferences, acts as a tutor on PhD study programs and has participated in numerous Slovak and international projects as a team member and as the project head. Among his most important project activities is the "Support and expansion of the CVD Plus Transport Research Center" project from 2013 to 2020, where he was actively involved as a researcher.

Angelina Iakovets in 2014 has graduated from the Kherson National Technical University, where she got a master's degree in Consolidated Information Analysis. In 2022, she has graduated from the Technical University o Košice, where she successfully received the title of PhD. Currently, she is a researcher at the Faculty of Manufacturing Technology with a seat in Prešov of Technical University in Kosice. She actively takes part in faculty projects and writes scientific articles. Her research area focuses on using mobile applications and solutions and entrepreneurial information systems and automation. She is a member of the EAI research group.

Stella Hrehova graduated from the Mechanical Engineering Faculty of the Technical University in Kosice. She received her doctorate in the field of engineering technology at the Faculty of Manufacturing Technologies with a seat in Prešov, of the Technical University in Košice in 2004. In 2011, she received the title of BSc in computer science at the P.J. Šafarik University in Košice. She has 25 years of teaching experience. Her scientific focus is on data processing, optimization and augmented reality. She has published numerous articles at scientific conferences and in journals. She is a member of the EAI research group.

Part I
Wire Electric, Vibration, Diagnostics, Laser, Technical Diagnostics

Application of Laser Doppler Vibrometer in Wire Electrical Discharge Machinery Diagnostics

Tibor Krenicky, Juraj Ruzbarsky, Monika Trojanova, Zuzana Murcinkova, and Jozef Mascenik

1 Introduction

Increased vibrations of machine tools cause various negative impacts on the given machine or technological process but also have a negative impact on the working environment [1]. Vibrodiagnostics determines the technical condition of the machine tool and its causes through the evaluation of operational vibrations. Vibrations arise on the principle of oscillations, which are formed due to the dynamic effects of mechanical parts of machines (transmissions in the gearbox, wear of bearings, etc.), by which they react to the action of internal and external forces [2]. This can result in reduced efficiency, defects, poor workpiece quality, unplanned downtime, and maintenance, which also increases operational costs [3]. Currently, monitoring systems are increasingly being used to analyze parameters, including vibrations, to improve production control and quality in accordance with the implementation of Industry 4.0 strategy principles.

Electro-spark or electro-erosive machining (EDM – electric discharge machining) is a technology in which material is removed using the interaction of a pulsed electric discharge between the tool and the workpiece itself. The process of wire electro-erosive machining (Wire EDM – WEDM) is designed in such a way that a wire is unwound from a coil of wire, which passes through the series of guide rollers. Brass wire is most often used for this purpose. Subsequently, the wire passes through the guide part, which serves to ensure that the wire has the correct position during machining [4]. The wire is subsequently wound on the collecting coil – Fig. 1. Connecting a wire to an electrical power source causes the wire to act as the

T. Krenicky (✉) · J. Ruzbarsky · M. Trojanova · Z. Murcinkova · J. Mascenik
Faculty of Manufacturing Technologies with a seat in Prešov, Technical University of Košice, Prešov, Slovakia
e-mail: tibor.krenicky@tuke.sk

M. Balog et al. (eds.), *EAI International Conference on Automation and Control in Theory and Practice*, EAI/Springer Innovations in Communication and Computing, https://doi.org/10.1007/978-3-031-31967-9_1

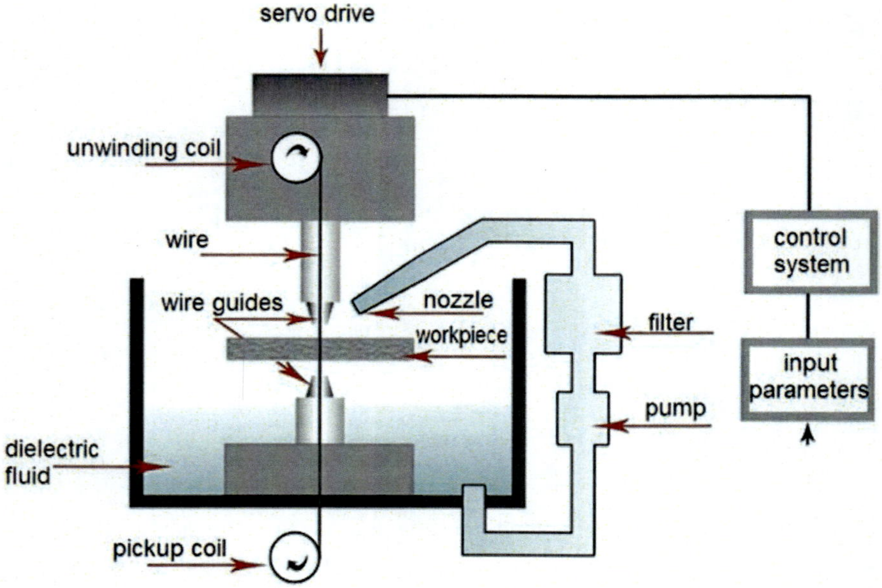

Fig. 1 Schematic of an WEDM cutter

cathode and the workpiece as the anode. When the wire gets close to the workpiece, a spark discharge occurs, causing the material to vaporize from the workpiece and the wire. This discharge is also supported by dielectric flushing, which helps to cool the machining area and remove the debris material. The machined area has larger dimensions compared to the diameter of the wire itself (0.02–0.3 mm), which is the results of the technology using a spark gap. WEDM technology enables machining accuracy of typically 0.005–0.01 mm with a roughness of the machined surface Ra of 0.8 to 3.2. It is often used to create holes with a complex shapes in hard-to-machine materials, e.g., for cutting tools, templates, and hard materials such as sintered carbides [5].

The vibrations that most affect the resulting material cut are presumed to be localized in the area of the wire, where the discharge occurs between the given workpiece and the wire, while the discharge itself is one of the factors for the generation of vibrations [6, 7]. Another common factor of increased vibrations is neglected maintenance and incorrect replacement of machine parts such as wire routing or other influences [8]. Monitoring systems such as a laser vibrometer, despite their higher price compared to conventional sensors, help to analyze the problem of vibrations without disrupting the production process, which is one of the key prerequisites for the possibility of applicability in complex technologies such as WEDM, since it is not possible here to measure vibrations with contact methods, e.g., on a thin unrolled wire. By analyzing and predicting operating parameters, it is possible to plan maintenance at the right time in order to prevent unwanted downtime, improve the efficiency and quality of machining, and also extend the service life of the machine tool [9].

The usefulness of vibrodiagnostic methods in determining the technical condition of technological devices, as well as monitoring technological processes, has been demonstrated by many examples. However, due to the difficult measurement conditions of WEDM technology, research works in this area are rare, and each technical system is unique, especially its characteristics before and after the maintenance performed. This uniqueness is the main motivation of the presented research.

This research aims to study the usability of vibrodiagnostic methods in the diagnosis of vibrations on an electro-erosive cutter using a laser vibrometer, with a focus on the analysis of the state of the technological equipment before planned maintenance and after its execution, with a description of the acquisition and evaluation of the measured data [10].

2 Experimental Part

The principle of the EDM cutter mechanism is that the wire unwound from the coil passes through mechanical rollers that are guided into the guide tube, which ensures that the wire is pushed into the guide parts of the machine. The experimental part was carried out on an electro-erosive cutter by the Fanuc Company – Table 1. Operating vibrations were measured using a portable noncontact Doppler vibrometer PDV-100 made by Polytec was used as a measurement system with a frequency range of 0.5 Hz to 22 kHz and a resolution of 0.02 μm/s, providing analog signal output as documented in Fig. 2.

To investigate the vibration, several measured nodes were selected on the wire and individual parts of the mechanism. Since the given cutter contains a guide tube,

Table 1 Technological measurement parameters

Parameter	Value
Laser source (nm)	HeNe, 633
Target distance (mm)	510
Frequency range (Hz)	0.5–22 k
Velocity range (mm/s/V)	125
Velocity resolution (μm $s^{-1}/\sqrt{Hz}$)	<0.1
Output (V)	velocity ± 4, 24-bit DAC
xyzuv range (mm)	370 × 270 × 255 × 120 × 120
Wire (mm)	brass, 0.25
Feed rate (m min^{-1})	5
Workpiece weight (kg)	21.2
Machine weight (kg)	1200

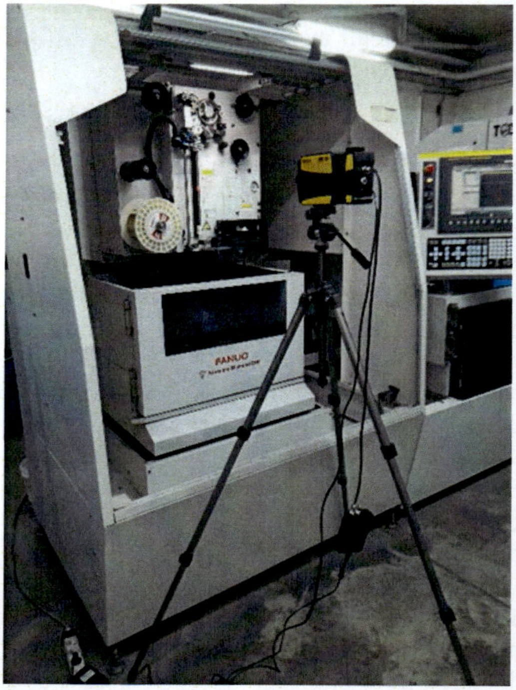

Fig. 2 Location of the vibrometer during the experiments

	Guide tube for wire
	Plastic rectifier
	Wire above the wheel - position 1cm
	Wheel
	Wire under the wheel - position 1 cm
	Wire under the wheel - 10 cm
	Wire under the wheel - 10 cm to the right
	Feeder (top edge diagonally from above)
	Wire visualization

Fig. 3 Measuring points on the WEDM machine

the measurement points were also determined on it, as marked in Fig. 3. Other points were fixed at places like a plastic rectifier, guide wheel, and brass wire.

A repeated measurement was also carried out on the electro-erosive cutter, which was carried out after maintenance. The repeated measurement was carried out three months after the first measurement, after the maintenance of the machine itself. Wire guides, bearings, and other mechanical parts were changed, which underwent cleaning or total replacement. The measured points have not changed. Figure 4 presents the locations where maintenance was performed.

2.1 Signal Analysis

The parameters of the total oscillation and the time courses of the signals are evaluated in the time domain, while the measured values of the total oscillation are compared with the previous measurement when the machine was working without a fault. Critical values are set for further comparison. This method provides initial overview information about total vibration values, displays the time sample of vibrations, and serves as a basis for the use of the subsequent frequency analysis method [11]. The advantage of this method is the speed of evaluation, and the disadvantage is that signals with a small amplitude are lost in vibration noise, which does not allow for locating the cause, place of origin, and occurrence of vibrations. However, if the application of frequency analysis is chosen correctly, it allows the detection of emerging disturbances in individual parts of the object using amplitude and phase spectra. This type of analysis is based on Discrete Fourier Transform and Fast Fourier Transform (FFT). The signals were processed using an FFT analyzer

Fig. 4 Location of the replaced parts

Polytec VibSoft-20, in which time and frequency spectra were plotted from the measured signals.

2.2 *Vibration Spectra Before and After Maintenance*

The frequency spectrum was set to two ranges. The range from 0 to 1000 Hz was set for the rectifier tube, and the range from 0 Hz to 20 kHz was used for the other parts of the measured points.

Wire Guide Tube: Top Part

At the measured point of the guide tube (upper part of the tube) (Fig. 5 top), the frequency spectrum showed several frequency peaks and groups of peaks. The most prominent frequency peak in the spectrum can be seen at the beginning of the measured spectrum, at approximately 10 Hz. From this most prominent, there are also hints of harmonics. The second and third frequency peaks are lower by about 70%. Another significant frequency peak is in the 50 Hz region, where an increase of about half was noted. Around it, you can also see a smaller cluster of frequency peaks from 55 Hz to 100 Hz and from 150 Hz to 200 Hz. During the repeated measurement after maintenance (Fig. 5 bottom), a decrease in noise can be seen in the frequency spectrum, but also a decrease or reduction of clusters of frequency peaks. The most prominent peak was also recorded in the 10 Hz region, similarly to before maintenance. However, they were only lower than this most prominent one. The second frequency peak was about 60% lower, and the third and fourth were about 75% lower. The increase was noted at the fifth frequency peak, representing approximately 50%. Clusters of frequency peaks can be seen in the region of 50–90 Hz.

Wire guide tube: top part.

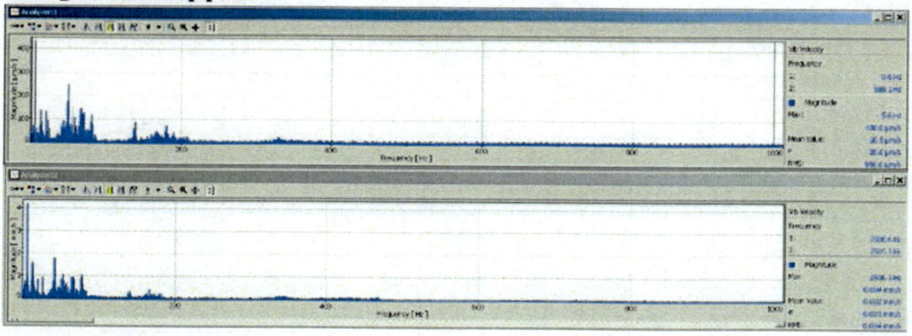

Fig. 5 Frequency spectrum measured on the wire guide tube top part before (top) and after (bottom) maintenance

Wire Guide Tube: Middle Part

At the second measurement point of the guide tube (middle part), the frequency spectrum shows one significant frequency peak in the region of 140 Hz and several clusters of peaks of lower values by approximately 75–80% in the bands around 50–100 Hz and 150–200 Hz (Fig. 6 top). When repeated measurement after maintenance, it is possible to observe a greater representation of frequency peaks in the frequency spectrum (Fig. 6 bottom) compared to the state before maintenance. At low frequencies, it is possible to observe the appearance of the second-highest frequency peak reaching approximately 40% of the maximum peak value. From this second-highest peak, it is possible to register a series of harmonic multiples with decreasing peaks of approximately 50% and 75%. From 50 Hz to 100 Hz, clusters of falling peaks can be seen, about 70–80% lower than that peak. Another cluster of peaks is present in the 150–200 Hz region.

Wire Guide Tube: Bottom Part

In the measured point located on the lower part of the guide tube (Fig. 7 top), several clusters of peaks can be seen, especially in the regions from 45 Hz to 100 Hz and from 150 Hz to 180 Hz. Significant and distinct frequency peaks are visible at the beginning of the measured spectrum. In the repeated measurement (Fig. 7 bottom), it can be seen that the clusters of frequency peaks and the noise in the frequency spectrum have been significantly reduced. Several frequency peaks can be seen at the beginning of this spectrum, the most prominent being three harmonics with gradually decreasing amplitudes. The decline of these peaks is approximately 40%, 60%, and 80% from the maximum value. From around 50 Hz to 100 Hz and 150 Hz to 180 Hz, clusters of spikes are seen that are significantly lower as compared to the premaintenance spectrum.

Plastic Rectifier

When measured on a plastic rectifier (Fig. 8 top), the frequency spectrum showed two dominant frequency peaks and several clusters of peaks. There is also an increased level of broad-spectral noise in the spectrum, which somewhat blocks the identification of less prominent frequency peaks. Dominant peaks include those in

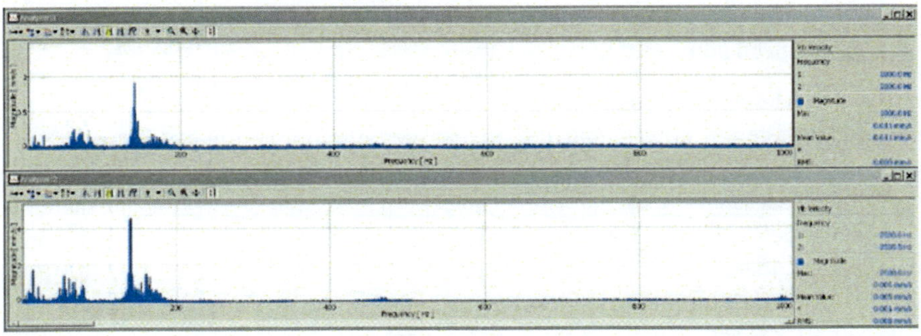

Fig. 6 Frequency spectrum measured on the wire guide tube middle part before (top) and after (bottom) maintenance

Wire guide tube: bottom part.

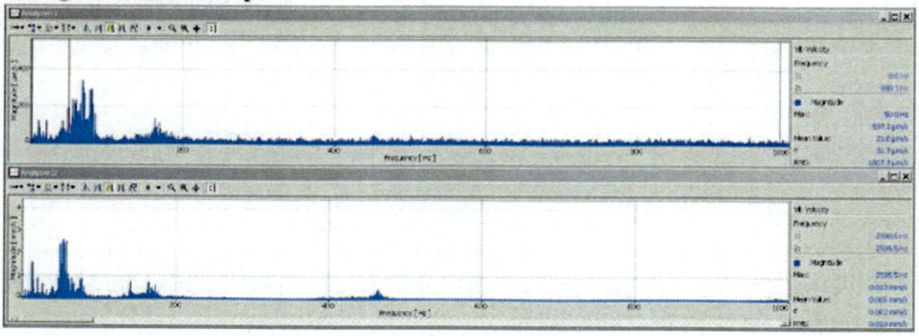

Fig. 7 Frequency spectrum measured on the wire guide tube bottom part before (top) and after (bottom) maintenance

Plastic rectifier.

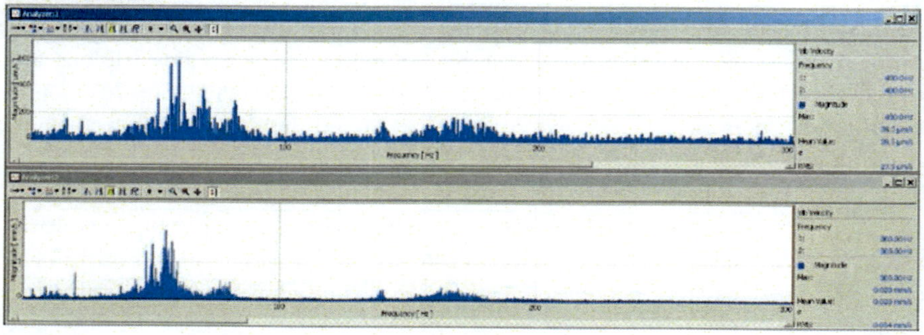

Fig. 8 Frequency spectrum measured on the plastic rectifier before (top) and after (bottom) maintenance

the range of 55–60 Hz. Other significant frequency peaks are found around 10 Hz, 20 Hz, and 50 Hz, with levels ranging from 50% to 70% of the maximum. The noise level is significantly reduced in the repeated measurement after maintenance (Fig. 8 bottom). Dominant peaks include those in the range of 52–55 Hz. Clusters of multiple spikes are also in the 55–60 Hz region. The peak in the 20 Hz region is more pronounced compared to the spectrum measured before maintenance.

Wire 1 cm Above the Wheel

One place where it is not a simple task to measure vibrations from a safe distance is a 0.25 mm thick wire. The laser beam had to be precisely focused.

At the measured point on the wire located 1 cm above the wheel (Fig. 9), there is a distinct structure consisting of a series of harmonic peaks. In the spectrum before maintenance, there is an increased noise level and clusters of frequency peaks (top picture). These clusters are evident at the beginning of the measured spectrum in the region up to 500 Hz. The first significant frequency peak is located in the region of

Wire 1 cm above the wheel.

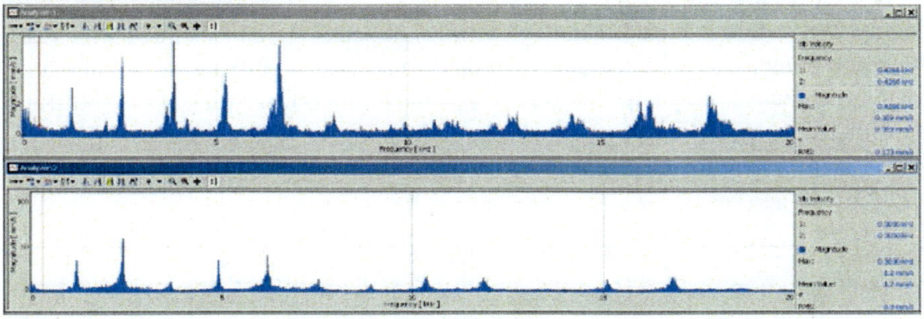

Fig. 9 Frequency spectrum measured on the wire 1 cm above the wheel before (top) and after (bottom) maintenance

Guide wheel.

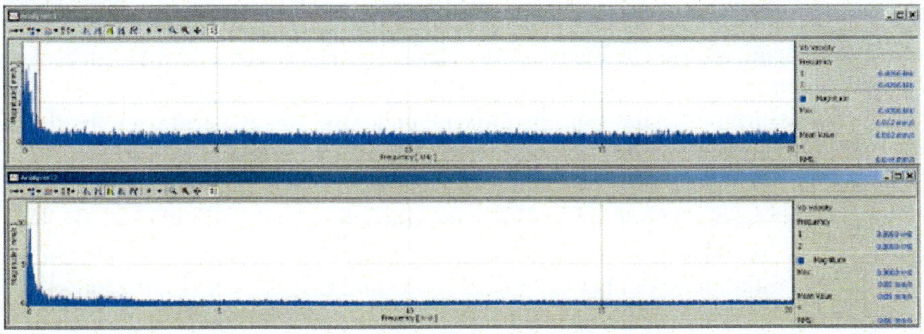

Fig. 10 Frequency spectrum measured on the guide wheel before (top) and after (bottom) maintenance

1.3 kHz, followed by four harmonic multiples with higher amplitudes. Other harmonics are significantly smaller but present in the entire measured range.

In the repeated measurement after maintenance (Fig. 9 bottom), the noise is less pronounced, and clusters of peaks are also suppressed. Several harmonic peaks were also recorded in this spectrum. Of these, they more significantly copy the frequencies of the system before the maintenance, but after the maintenance, there was a reduction in the frequencies of the less prominent peaks.

Guide Wheel

When measuring on the guide wheel, which serves to correct the path of the wire, the spectrum showed significant noise. Before maintenance, more significant values were present in low frequencies up to 0.5 kHz (Fig. 10 top). In the measurement after maintenance (Fig. 10 bottom), a steeper noise drop was recorded with a significant peak in the region of 0.2 kHz.

Wire 1 cm Below the Wheel

Significantly more intense harmonic multiples of peaks were measured at the measurement point located on the wire 1 cm below the wheel compared to the measurement on the wire 1 cm above the wheel. The most prominent peaks in the spectrum before maintenance (Fig. 11 top) were recorded in the range of 2.5 kHz and 2.8 kHz, while the nature of the spectrum recorded after the maintenance (Fig. 11 bottom) is similar.

Wire 10 cm Below the Wheel

Vibrations were measured from two positions on the wire 10 cm below the wheel. The spectrum from the frontal measurement before maintenance (Fig. 12 top) showed the most pronounced frequency peak in the region of approximately 0.3 kHz. Numerous lower harmonic peaks are also present in the spectrum, mostly 70 to 80% lower, while some are extended to the sides.

In the measured spectrum measured from the frontal position after maintenance (Fig. 12 bottom), the character of the generated frequency peaks is similar to that before maintenance, except that the maximum frequency peak is located in the 2.5 kHz region.

When measuring the vibrations on the wire under the 10 cm wheel from the right side, before the maintenance, several more pronounced harmonic frequency peaks were measured (Fig. 13 top); the dominant peak in this spectrum was in the region of 0.4 kHz, the second harmonic of which was lower by two thirds. Other significant peaks were recorded in the 6 kHz, 8 kHz, and 10 kHz. If we compare this spectrum with the spectrum measured from the frontal position (Fig. 12 above), it is possible to see an increase in these frequency peaks by up to half. During the measurement repeated after the maintenance (Fig. 13 bottom), a spectrum of frequency peaks was also recorded, where the peak in the 0.4 kHz region was dominant, identical to the measurement before the maintenance. A proportional decrease of peaks with higher frequencies is also evident in the spectrum.

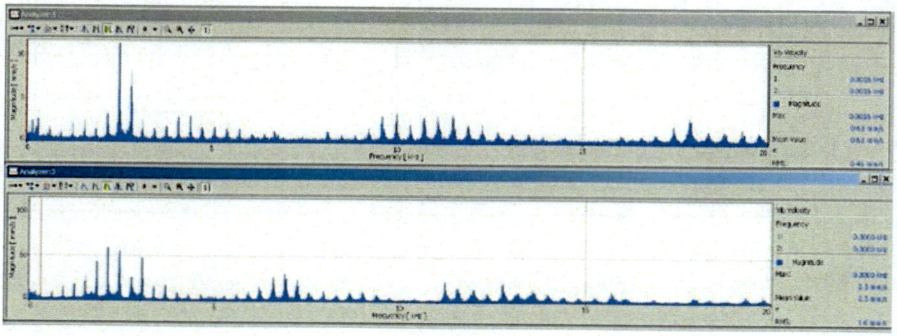

Fig. 11 Frequency spectrum measured on the wire 1 cm below the guide wheel before (top) and after (bottom) maintenance

Wire 10 cm below the wheel.

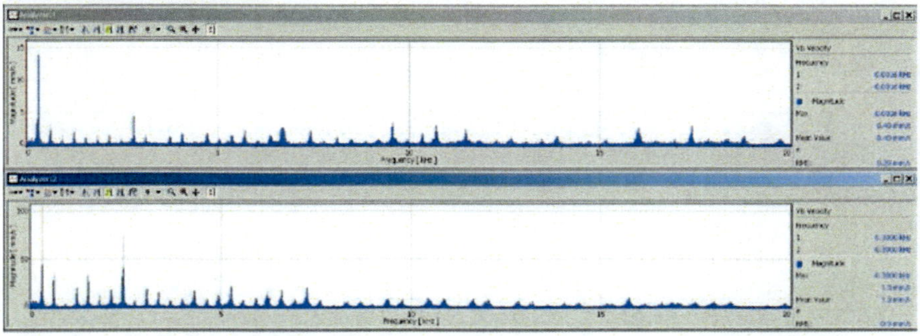

Fig. 12 Frequency spectrum measured frontally on the wire 10 cm below the guide wheel before (top) and after (bottom) maintenance

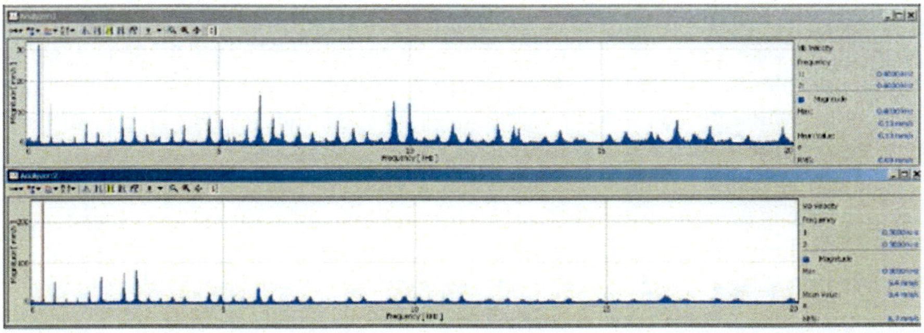

Fig. 13 Frequency spectrum measured horizontally from the side, perpendicularly to the frontal view, on the wire 10 cm below the guide wheel before (top) and after (bottom) maintenance

Feeder Upper Edge

Considerable noise and extended regions of several frequency peaks are seen at this measured point, with the most prominent clusters of these peaks being in the 1.5–5 kHz region (Fig. 14 top). The dominant peaks in this spectrum are around the 0.1–0.2 kHz region. In the repeated measurement after maintenance (Fig. 14 bottom), the clusters of frequency peaks were reduced. In the region from 0.1 kHz to 0.2 kHz, dominant peaks are formed similarly to the case before maintenance in the region 0.1–0.2 kHz.

3 Discussion

From the measured values on selected parts of the Fanuc WEDM system, it is evident that before the maintenance itself, it is possible to observe more noise in the frequency spectra than it was during the repeated measurement after the maintenance.

Feeder upper edge.

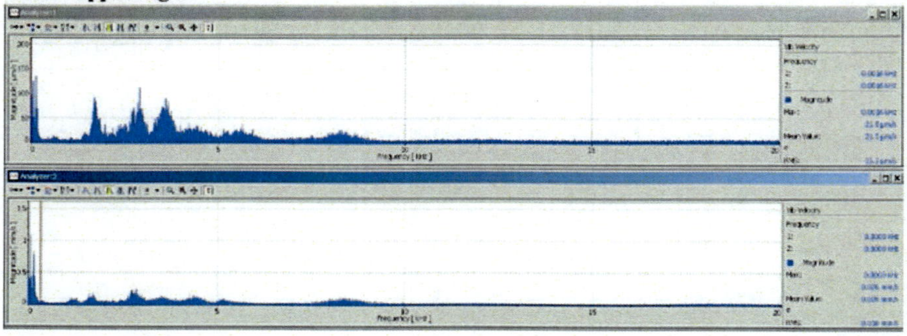

Fig. 14 Frequency spectrum measured on the upper edge of the feeder diagonally from above before (top) and after (bottom) maintenance

During the measurement, problem areas for vibration recording were noted. One problematic aspect was the passage of the beam through water and plexiglass. Another problematic aspect was the flushing, removing unwanted material from the work area, which caused the water and solid waste particles to swirl, causing an increase in noise.

The results of the performed measurements show important diagnostic data that can help both in the setting of effective machining parameters [4, 11] and as feedback for manufacturers or assemblers of individual parts of the technological system, or its spare parts [12, 13].

4 Conclusions

Vibrations are an accompanying phenomenon of the transmission of force effects, which mean effects with an undesirable effect on the operation and condition of technological equipment, e.g., during machining processes. As part of this research, vibrations on a Fanuc EDM cutter before and after maintenance were investigated, that is, on a precision machining device where an increased level of vibration-causing parameters would mean a reduction in production quality. The Polytec PDV-100 laser vibrometer was used as a measuring device for the vibration investigation. Vibrations were measured at locations such as the guide tube for the wire, on the 0.25 mm diameter brass wire itself, and at other selected nodes.

It has been demonstrated that with the given method, it is possible to record vibrations in different places of the system. The measured results found that after the maintenance was carried out, the level of vibration peaks increased several times. The cause of this condition is the initial run-in for settling in the new parts that have been replaced. These vibrations are mostly of low frequency, affecting the state of the machine or on machine parts and, above all, on the machined surface of workpieces. In order to reduce vibrations, it would be advisable to set the given technology in such a way (input parameters: flushing speed, time of switching on the discharge, and time of switching off the discharge on the wire itself) that potential sources of vibration are as small as possible. These experimental changes would have to undergo further investigation in order to confirm these suggestions and then put them into practice [12]. Nevertheless, it can be concluded that the presented method enables the acquisition of usable signals characterizing the process and machinery state.

Acknowledgments This work was supported by the Slovak Ministry of Education within project VEGA No. 1/0823/21 and by the Slovak Research and Development Agency under contract No. APVV-18-0316.

References

1. Zhang, G., Huang, H., Zhang, Z., et al.: Study on the effect of three dimensional wire vibration on WEDM based on a novel thermophysical model. Int. J. Adv. Manuf. Technol. **100**, 2089–2101 (2019). https://doi.org/10.1007/s00170-018-2796-x
2. Di, S., Chu, X., Wei, D., Wang, Z., Chi, G., Liu, Y.: Analysis of kerf width in micro-WEDM. Int. J. Mach. Tools Manuf. **49**(10), 788–792 (2009). ISSN 0890-6955. https://doi.org/10.1016/j.ijmachtools.2009.04.006.
3. Unune, D.R., Mali, H.S.: Experimental investigation on low-frequency vibration assisted micro-WEDM of Inconel 718. Eng. Sci. Technol. J. **20**(1), 222–231 (2017). ISSN 2215-0986. https://doi.org/10.1016/j.jestch.2016.06.010.
4. Straka, Ľ. Čorný, I.: Identification of geometric errors of circular profiles at WEDM caused by the wire tool electrode vibrations and their reduction with support of acoustic emission method. Eng. Fail. Anal. **134**, 106040 (2022). ISSN 1350-6307. https://doi.org/10.1016/j.engfailanal.2022.106040.
5. Fan, S.Y., Zhang, Q.J., Chen, H.W.: Dynamic characteristics analysis and simulations of transverse vibration of axially moving electrode wire in WEDM. Adv. Mater. Res. **328–330**, 1792–1796 (2011). https://doi.org/10.4028/www.scientific.net/amr.328-330.1792
6. Sun, P.: Effect of the vibration on machining gap of WEDM. Appl. Mech. Mater. **395–396**, 1053–1056 (2013). https://doi.org/10.4028/www.scientific.net/amm.395-396.1053
7. Li, C.J., Guo, Y.F., Bai, J.C., Lu, Z.S., Chen, C.: Study on modeling and simulation of wire vibration in WEDM. Adv. Mater. Res. **472–475**, 2013–2017 (2012). https://doi.org/10.4028/www.scientific.net/amr.472-475.2013
8. Habib, S., Okada, A.: Experimental investigation on wire vibration during fine wire electrical discharge machining process. Int. J. Adv. Manuf. Technol. **84**, 2265–2276 (2016). https://doi.org/10.1007/s00170-015-7818-3
9. Manufacturing Guide: Wire EDM. https://www.manufacturingguide.com/en/wire-edm. Last accessed 2020/10/03.
10. Prime: Working of electric dicharge machining. http://www.perime.com/working-of-electric-discharge-machining. Last accessed 2020/10/03.
11. Straka, Ľ., Čorný, I.: Adjustment of wire vibrations in order to improve geometric accuracy and surface roughness at WEDM. Appl. Sci. **11**(11), 4734 (2021). https://doi.org/10.3390/app11114734
12. Almeida, S.T., Mo, J.P.T., Bil, C., et al.: Servo control strategies for vibration-control in robotic wire EDM machining. Archiv. Comput. Methods Eng. **29**, 113–127 (2022). https://doi.org/10.1007/s11831-021-09570-1
13. Wang, P.W., Yang, C.H., Yan, M.T.: High stable mechanism design for micro-wire transport system in WEDM. Adv. Mater. Res. **189–193**, 1366–1369 (2011). https://doi.org/10.4028/www.scientific.net/amr.189-193.1366

Obstacle-Resistant Wireless Strain Gauge Complex for Automated Monitoring of the Steel Structures Condition

Nikolay Sergienko, Serhii Hubskyi, Natalia Pavlova, Olha Turchyn, Oleksandr Hasiuk, and Kamil Židek

1 Introduction

Industrial buildings belong to objects of increased danger. People's lives depend on the safety of their operation. One of the elements that limit the service life of industrial buildings and affect the safety of their operation is steel structure. Also, steel structure cost is more than two-third of the building total cost. For example, metallurgical cranes operating in heavy and very heavy operation modes have a standard service life of 12–16 years. Even a single thermal effect, exceeding the normative, on the bottom flange of the main beam of the crane causes acceleration of degradation processes in the metal, which can reduce the service life of steel structures by several times. Reducing fatigue loads at the design stage of steel structures [1] is not a solution to the problem for existing structures. Fatigue cracks are unacceptable operational defects [2]. Optimal drives control of crane [3] can only partially reduce fatigue processes in steel structures. Therefore, preventing the occurrence of operational defects (cracks, plastic deformations [4]) in the steel structure of industrial buildings by monitoring is an urgent scientific and practical task [5].

Special attention should be paid to the steel structure of industrial buildings that have completed their service life. One of the steps to determine their operational suitability and further extension of the service life is using destructive and/or nondestructive testing (NDT) methods. First, however, it is important to prevent the occurrence of operational defects, to detect them at the nascent stage, that is, the

N. Sergienko · S. Hubskyi · N. Pavlova · O. Turchyn · O. Hasiuk
National Technical University "Kharkiv Polytechnic Institute", Kharkiv, Ukraine

K. Židek (✉)
Faculty of Manufacturing Technologies with the seat in Prešov,
Technical University of Košice, Prešov, Slovakia

M. Balog et al. (eds.), *EAI International Conference on Automation and Control in Theory and Practice*, EAI/Springer Innovations in Communication and Computing, https://doi.org/10.1007/978-3-031-31967-9_2

formation of a stress–strain state (SSS), which can lead to a defect (that is, failure to reach the "bifurcation" point). NDT methods can solve such a problem.

Existing methods of nondestructive testing ("active," e.g., magnetic-coercive, ultrasonic, and "passive": acoustic-emission, magnetic metal memory, strain gauge, etc.) have scientifically sound evidence bases with known disadvantages. Existing studies allow the use of the strain gauge method of NDT in the steel structure condition diagnosing practice based on the "risk analysis" factor [6].

2 Literature Review

Strain gauge stations (SGS) with data transmission via wires have become the most widely used for monitoring the SSS of industrial steel structures based on the NDT strain gauge-metric method. SGS is a multi-channel station, assembled from separate modules – an analog-to-digital converter (ADC), power supplies, amplifiers, and cables that transmit information. Sensors are connected using a bridge scheme (Wheatstone Bridge). One of the sensors is active, the other is compensation, and the other two are ballast. The active sensor measures deformations in the metal structure, the compensating sensor is used to compensate for changes in the resistance of the working sensor when the temperature changes. To register and process the received information from the sensors, a program, and specialized software is used, which provides a recording of information from each channel, filtering of noise and interference in the recorded signal, processing of the received signal with a selection from the entire spectrum of amplitudes and averages values. This SGS with data transmission via wires has the following advantages: increased accuracy of measurements due to the possibility of measurements in several ranges; the direct output of the received information to the computer in real time; cable transmission of an amplified signal from converters, which allows the use of cables longer than 25 m; the possible number of simultaneously used channels is up to 32. However, this tensometric system has a significant drawback – data transmission occurs via wires. That is, each channel needs high-quality wires. The longer the wires, the more noise appears in the signal, and the weaker it becomes. As a result, it is necessary to use expensive high-quality amplifiers and filters, which greatly complicates the equipment for strain gauge control (SGC), its cost, and the complexity of use.

To solve the problems associated with the appearance of noise and signal attenuation, SGC with wireless data transmission is used [7], which corresponds to the concept of Industry 4.0 [8].

In work [9], it is proposed to use a wireless SGS based on smart wireless sensor networks [10] based on the IRIS Mote platform for monitoring steel structure stress–strain state. The wireless sensor itself consists of strain gauges (Wheatstone Bridge with the possibility of coarse and fine adjustment) and a deformation board, an IRIS loop (has a wireless data transfer rate of 250 kbps, RAM 256 kB) with software. This system has an ADC with a 10-bit resolution, which is not enough for high-quality SGC of the steel structure. Therefore, the basic scheme of the wireless strain sensor is implemented on two amplifiers with an adjustable gain. This makes it difficult to conduct SGC of the steel structure [11].

In work [12], it is proposed to use SGC with wireless data transmission based on low-power microprocessors of the Intel WISP platform for monitoring steel structure stress–strain state. The WISP device constantly collects radio frequency energy (using RFID technology) and stores it in a capacitor while the microcontroller is in sleep mode. When enough energy is accumulated, the voltage controller sends a signal to the microcontroller and activates it. The microcontroller executes the programmed algorithm, which consists of activating the power supply of the external sensor and converting its output from analog-to-digital through the built-in 10-bit ADC. Next, after the ADC, the signal is digitally sent to the central unit via the radio channel. Several external sensors can be connected and processed by one WISP device. The WISP device provides power to the Wheatstone Bridge amplifier and is named WISPs/g. WISP's average power consumption when active is 1.08 mW [12]. This SGS using RFID technology and the Intel WISP platform has a significant drawback. The accumulated energy for wireless data transmission under dynamic loads of the controlled steel structure will not be enough for continuous data recording and transmission. This station allows only periodic monitoring of steel structure stress–strain state. Also, the power of the signal transmitted over the radio channel is very small. Therefore, the signal transmission distance will be small, only a few meters. And in the presence of industrial obstacles, despite the digital form of information transmission, this method may not transmit information at all via the radio channel.

In work [13], it is proposed to use for monitoring steel structure stress–strain state SGC with wireless data transmission using RFID technology and wireless sensor networks (WSN) [14]. A WSN is a semi-passive wireless strain sensor that uses a strain gauge (like a wired sensor) but can be passively polled, like an RFID tag. Similar to the RFID system, a polling device emits an electromagnetic wave that affects the antenna of the wireless strain-resistive sensor module, which excites it. The strain gauge changes its resistance according to the applied strain and drives an oscillating circuit that modulates the electromagnetic wave scattered by the sensor antenna. The modulation frequency depends on the applied strain and can be easily measured with a spectrum analyzer or frequency meter after the polling unit receives the modulated back-propagated wave back. The tag-sensor uses a battery to power the oscillating circuit, which is kept switched off at all times except for a short time interval during polling and measurement. For this reason, the battery's service life can be very long. Sensors of WSN networks remain in operation for several decades with sufficiently low maintenance costs [13]. This SGC using RFID technology and wireless sensor networks has the following limitations in use: a small distance (only a few meters) between the surveyed unit and the sensors; the polling unit can simultaneously poll several sensors, but increasing their number affects the speed of signal transmission.

The only thing for all approaches is that the quality control of steel structures and other objects must be carried out by comparing the planned theoretically or experimentally determined quality indicator with its valid value at the current moment. At the same time, it is necessary to ensure high accuracy of the experimental data of the study of the object's current state with the transmission of signals over a long distance with minimal errors. Therefore, quality control consists of checking the state and quality indicators to detect deviations from the expected values. If a deviation is detected, it is necessary to find the level of deviation, and the reason for its occurrence, and after correcting the process, again check the compliance of the adjusted quality indicators with their planned values. It is through such a continuous cycle that it is possible to control the object's condition and check whether it has the necessary quality for further use.

The scientific basis of modern control of the state of the steel structure is mathematical and experimental methods using modern information technologies and programs.

When assessing the quality and state of the object, it is necessary to choose methods that provide reliable information, are simple and understandable, and do not require significant resources.

3 Material and Methods

3.1 *Changes in the Characteristics of the Material of Metal Structures*

A feature of the behavior of metals undergoing destruction is the indispensable presence of micro- or macrodeformation before destruction. Depending on the structural state, the type of loading, and the asymmetry of the cycle, the endurance limit of metals and alloys can be higher or lower than the physical yield strength in its value. In the case when it is below the physical yield strength (the most common case), cyclic deformation begins with the stage of cyclic microyielding. Sometimes the presence of this stage is associated with the phenomenon of delayed fatigue yielding, by analogy with the effect of delayed yielding during static deformation.

After reaching a certain number of cycles (corresponding to the end of the stage of cyclic microyielding), the beginning of the opening of the hysteresis loop and a decrease in the effective stress σ_a are observed. There is a process of macroscopic cyclic softening. This behavior is typical for materials with a physical yield strength and tested for fatigue below the static yield strength.

As an example, we can present the curves of cyclic deformation of carbon and alloy steels (Fig. 1). From a comparison of the static and cyclic deformation curves, the following conclusion can be drawn whether the material is cyclically hardening (if the cyclic deformation curve passes above the static deformation curve) or it is cyclically weakening (the cyclic deformation curve passes below the static deformation curve).

The indicated voltage changes should be considered in the theoretical studies of the steel structure SSS, setting the allowable stresses.

During the operation of the steel structure, it is possible to measure deformations in the most loaded places with the help of strain gauges. In this case, the number of loading cycles should be considered, i.e., carry out dynamic monitoring.

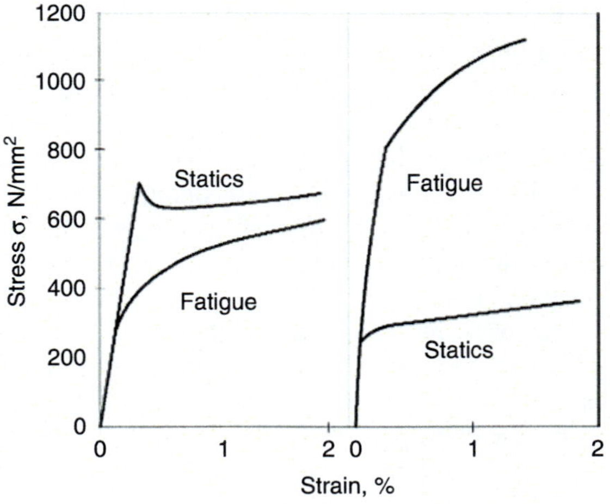

Fig. 1 Curves of steel cyclic deformation: (**a**) carbon steel; (**b**) alloy steel

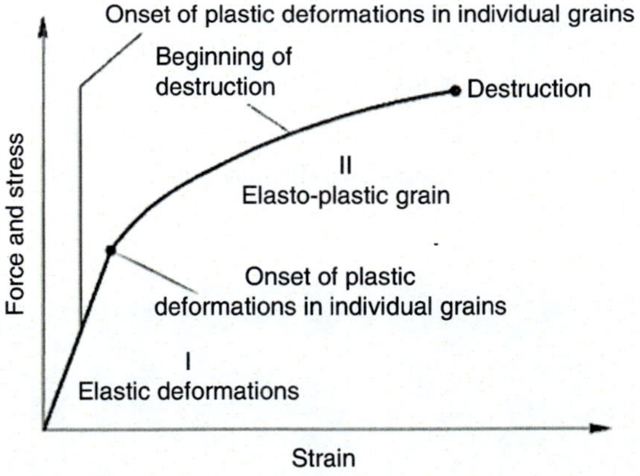

Fig. 2 Graphical relationship between forces and deformations

The integral characteristic of deformation is presented as a graph in Fig. 2, where the elastic (I) and plastic (II) deformation zones are indicated. With a certain level of adjustment of the assumption, the deformation zone II is described by the linear Hooke's law. Such a representation is accepted for studying the SSS in the construction of lifting and transporting machines.

3.2 Theoretical Assessment of Steel Structure Stress–Strain State

It is possible to carry out a theoretical assessment of the steel structure performance by determining its SSS; this requires the development of its 3D model – Fig. 3 [15].

Increasing requirements for the steel structures' capacity necessitates new models of object creation that can withstand not only the force load but also thermal deformation, which in some cases causes plastic deformation of the metal.

The main idea of the finite element method (FEM) is that a continuous physical quantity (temperature, pressure, displacement, stress) is approximated by a discrete model built on a set of piecewise continuous functions defined on a finite number of subdomains. To build a finite element model, linear, quadratic, and cubic elements can be used. A polynomial is often used as an element's piecewise continuous function [16].

The basic FEM equation for the theory of elasticity follows from the potential energy theorem. Writing the expression for the total potential energy of the system in a discrete form, differentiating it, and equating it to zero in order to find the extremum of the potential energy, the basic FEM equation is determined in the form of the following system of linear equations:

$$[K]\{U\} = \{F\} \tag{1}$$

where $[K]$ – system stiffness matrix obtained by summing the stiffness matrices of individual elements,

$\{F\}$ – system load vector,
$\{U\}$ – vector of unknown displacements.

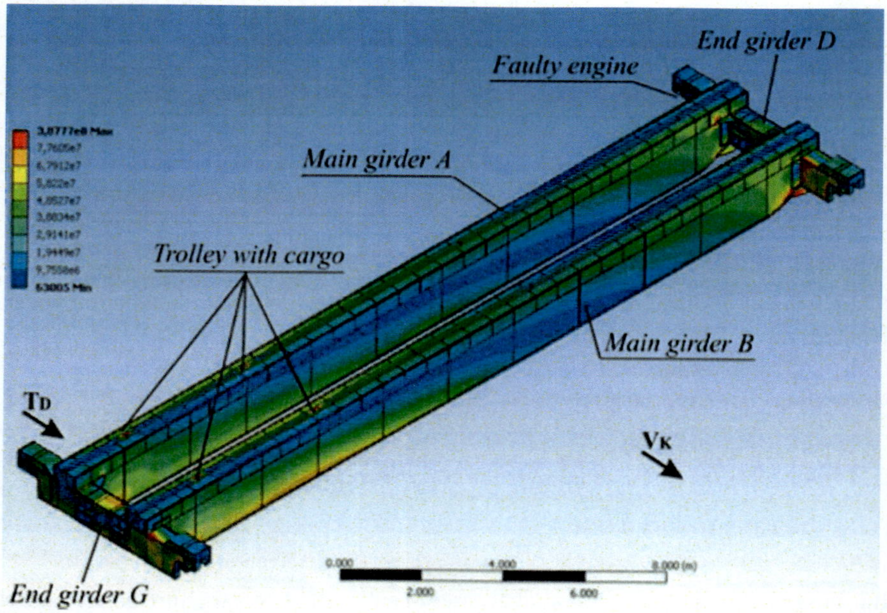

Fig. 3 Steel structure stress–strain state in case of skewing loads

After taking into account the boundary conditions (embedding of the part and external load), the system of equations (1) is solved by the modified Gauss method. The displacements in the nodes of the model established after the solution of system (1) are used to calculate the deformations

$$\{\varepsilon\} = [B]\{U\} \tag{2}$$

and stresses in the elements:

$$\{\sigma\} = [D]\{\varepsilon\} - [D]\{\varepsilon 0\} \tag{3}$$

The values of strains and stresses in the nodes can be obtained in several ways, the most accurate of which is the conjugate approximation method. Knowing the nodal values of physical quantities, interpolation polynomials can be used to calculate their values at any point of an element and construct equal-value zones for any constituent of the strain or stress tensor.

An example of modeling the bridge crane under skewing loads SSS is shown in Fig. 3 [15].

When assessing the technical condition of an object, for example, a crane steel structure, with the subsequent determination of its operational suitability, it is necessary to distinguish the following factors: static and fatigue strength, crack resistance, stability and rigidity.

To assess each factor, experts and NDT specialists have to take into account and investigate the following indicators: safety factors and metal structure (for static strength); characteristic number (for fatigue strength); consideration of local and general deformations (for deformability); presence and growth rate of cracks (for crack resistance).

In this case, the design of the crane must be taken into account, and a differential approach must be provided for each type of crane [17].

When determining the residual static strength, it is also necessary to carry out practical calculations, similar to those performed when designing crane steel structures in accordance with regulatory documentation.

The main difference between these calculations is that they must take into account the actual weakening of the steel structure, corrosion, wear, technological cuts in the elements, etc.

The computer program "Crane" has been developed, which allows to carry out strength calculations of the bridge cranes steel structure with actual geometric parameters [18]. The program "Crane" is based on the algorithm for calculating bridge cranes steel structure by limit states:

– the condition of the first limit state is characterized by the exhaustion of the load-bearing capacity of the material according to static characteristics

$$\sigma_{\max} \leq R \bullet \gamma \tag{4}$$

where σ_i – maximum design stress (stress intensity),

R – design material resistance,
γ – coefficient of working conditions of the calculated element.
the condition of the second limit state is characterized by the exhaustion of the load-bearing capacity of the material in terms of resistance to repeatedly acting loads

$$\sigma_{\max} \leq R^0 \bullet \gamma_c \quad (5)$$

where R^0 – design fatigue resistance, determined by taking into account the nature of the action of the cyclic load.

the third limit state is characterized by the occurrence of deformations that prevent the normal operation of the crane due to a decrease in control accuracy or the occurrence of involuntary movement of the trolley or leads to the occurrence of force factors unacceptable by the criteria of the first or second limit state

$$\overline{y}_{st} \leq \left[\overline{y}_{st}\right]; t_{\text{dump}} \leq \left[t_{\text{dump}}\right] \quad (6)$$

As the initial data (in addition to geometric parameters and weight characteristics of the steel structure units) in the calculations take into account: the group classification of the crane, the limit value of design resistance, the safety factor depending on the load mode, dynamic load factors, distortions during crane movement from sudden failures of electric motors of transmissions, etc.

Dependencies (1)–(6) are the basis for a reasoned choice of safety factors that affect the determination of allowable stresses in steel structures.

3.3 *Wireless Monitoring Complex SGC for Steel Structure*

The block diagram of the obstacle-resistant wireless monitoring complex SGC for steel structure consists of the central unit (Fig. 4a) and sensors (block diagram – Fig. 4b).

ADC with an amplifier (Fig. 4a) is implemented on a two-channel module with an HX711 microcircuit. This module has an ADC resolution of 24, and an amplification factor of 128 is used with a measurement frequency of 80 Hz.

The wireless sensor and the central unit microcontroller are assembled on the 8-bit processor of the Atmega 328, which is part of the Arduino NANO module.

The transceiver of the wireless sensor and the central unit is assembled on the NRF24L01 module, and the data transfer rate is 0.25 Mbps. It is possible to implement a network on one channel with seven modules (one receiver and six transmitters).

The Wheatstone Bridge is implemented according to the half-bridge scheme – Fig. 5a. The appearance of the wireless sensor of the monitoring system of the steel

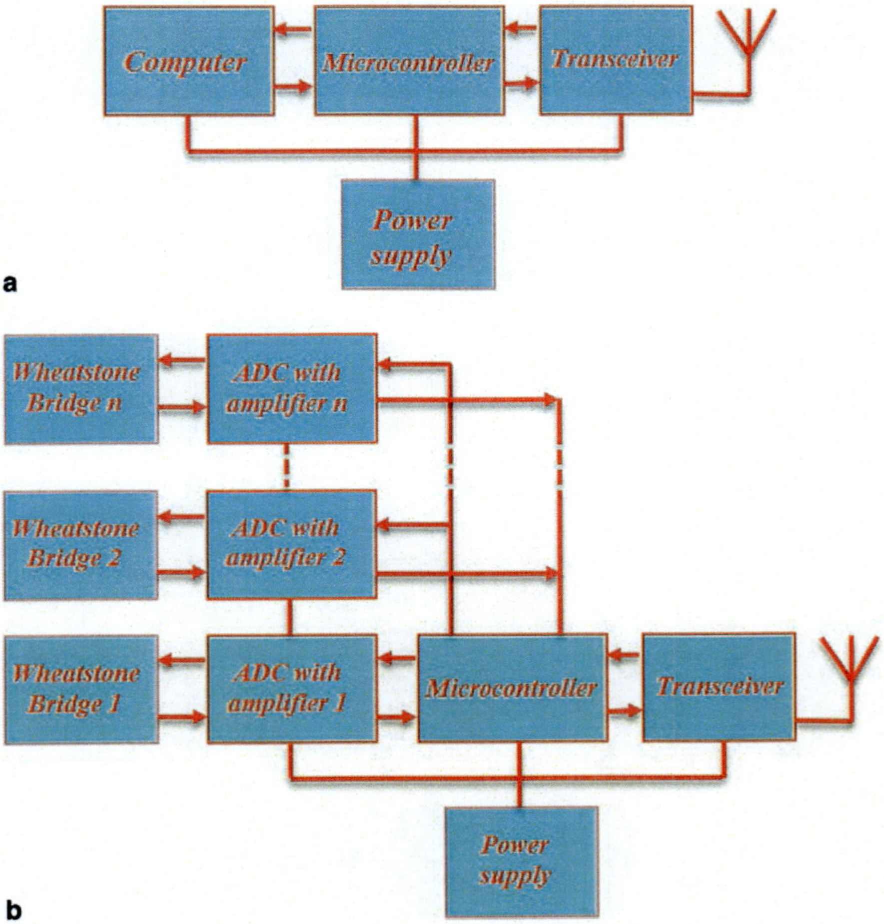

Fig. 4 Block diagrams of the SGC for steel structure monitoring complex: (**a**) central block; (**b**) wireless sensor

structure strain gauge control – Fig. 5b. Its performance is similar to the central unit. In addition, temperature compensation is implemented, which is important for obtaining correct monitoring results [19].

One wireless sensor can have several Wheatstone bridges and an ADC with an amplifier (see Fig. 4b). This parameter is limited only by the functional inputs/outputs of the microcontroller and the required data transfer rate. Also, one central unit can exchange information with six wireless sensor modules. It is possible to use several central modules simultaneously on one computer (see Fig. 4a).

The maximum frequency of updating and sending information from the wireless sensor is limited by the measurement frequency of the ADC with the HX711 amplifier and is a maximum of 80 Hz. In addition, each measurement is filtered by a recursive Kalman filter, which is important for obtaining correct monitoring results [20].

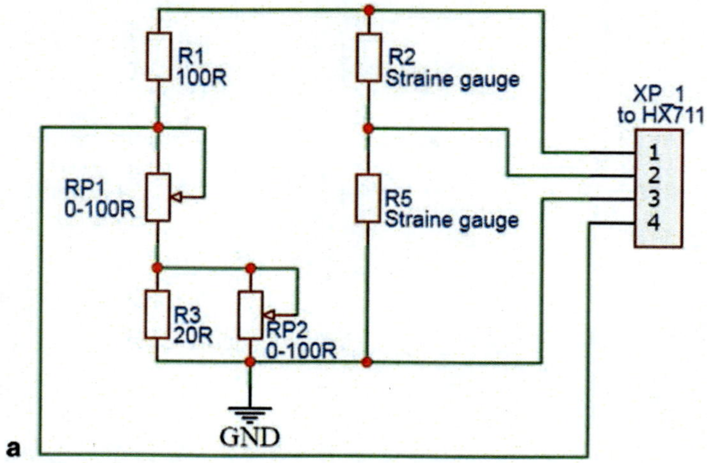

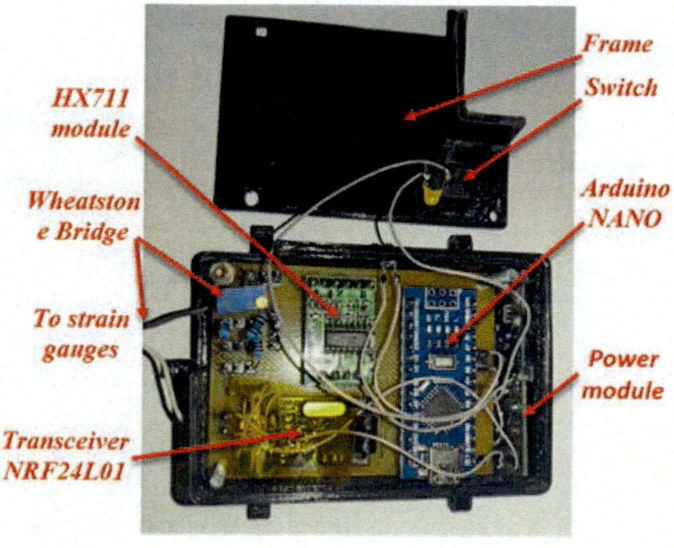

Fig. 5 Wireless sensor of the SGC of steel structure monitoring complex: (**a**) schematic diagram of the Wheatstone measuring bridge; (**b**) appearance of the wireless sensor

Therefore, the proposed obstacle-resistant complex of wireless monitoring of steel structure can simultaneously monitor the parameters of several tens of sensors with an update frequency of tens of Hertz.

One of the main disadvantages of various wireless strain gauge control systems is the insufficient range of operation and poor protection against industrial obstacles, despite digital signal transmission.

In this wireless complex, it is possible to use NRF24L01 modules with normal transmitter power (communication distance up to 50-80 m) and with increased transmitter power (communication distance up to 1 km) without changes in the program code and the schematic diagram.

NRF24L01 modules operate at a frequency of 2.4 Hz and have an allowable operating temperature of -80...+85 °C. Although this frequency is used by Wi-fi and Bluetooth devices, industrial facilities have possible spectrums of obstacles from electrical equipment (e.g., a frequency converter for a motor). All this can significantly affect the quality of information transmission over the radio channel. Therefore, an algorithm was developed, and the analysis of obstacle-free channels (the total number of channels in the NRF24L01 module is 128) and their automatic selection for work was developed. This significantly increases the stability of the proposed interference-resistant wireless SGC monitoring complex at distances of up to 1 km between wireless sensors and the central unit and in industrial production conditions.

Also, thanks to the communication protocol used in the NRF24L01, all data is optionally transmitted in packets with confirmation of their receipt, which completely eliminates the loss of transmitted information.

Wireless sensors use their computing capabilities to decide for monitoring to send data or switch to "sleep" mode to save battery power.

Synchronization by the time of the measuring channels of wireless strain gauges operation when using several central units on one computer is also implemented. For example, this situation can arise when using several dozen wireless sensors.

4 Calculation and Experimental Studies of Cranes Steel Structures

We propose to determine the most dangerous zones in the steel structure, in which it is necessary to monitor its condition using the approach proposed in work [15] and (or) [17].

According to the paper [17], the local stability of the bridge cranes, main girders, vertical walls, etc., is checked based on limit state calculations. The main calculation results are summarized in a table with the limit vs. calculated loads for the conditions of the first, second, and third limit states. All calculations made are saved in the reporting documentation (database) for subsequent printing and storage. This program base can be supplemented and updated with other "Crane" programs installed on other computers. At the moment, more than 110 units of bridge cranes have already been entered into the program database. The created database of bridge cranes allows to select the right (similar) crane already from the database and make the necessary calculations. The "Crane" program also allows to perform calculations not of new construction but of an actual one, with all changed geometric parameters (e.g., taking into account corrosion).

Steel structure strain gauge monitoring wireless complex can also be used to measure shaft torques, and to transmit any data from sensors (not only strain gauges,

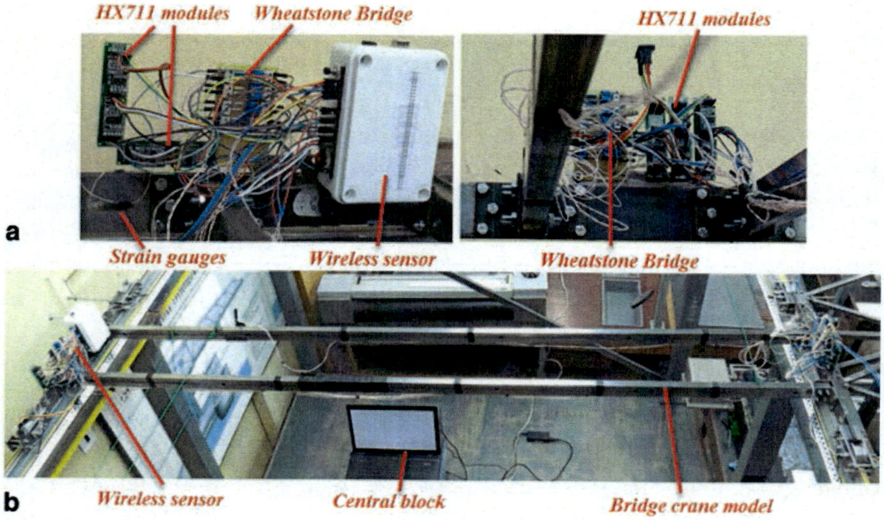

Fig. 6 Use of the SGC for steel structure wireless monitoring complex (**a**) on a bridge crane model (**b**)

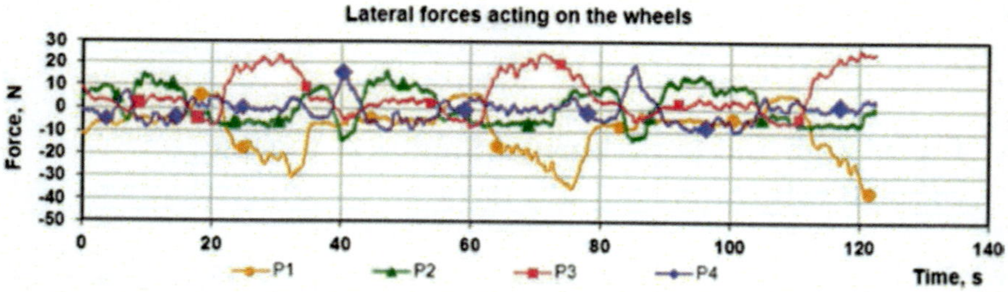

Fig. 7 Oscillogram of registration by the complex of changes in lateral forces on the bridge crane model wheels

e.g., from encoders, accelerometers, etc.). At the same time, it is possible to send commands affecting the operation of wireless sensors from the central unit.

This complex was used to control the skewing loads that occur while moving the bridge crane model – Fig. 6. Wireless data transmission from the encoder was also carried out in this complex to determine the exact placement of the crane bridge.

To fix the parameters that the central unit receives from wireless sensors, a graphic interface has been developed that allows to display data in numerical and graphical form, changes the display scale of the received information, and saves data. In Fig. 7 shows the results of measuring the rail-to-wheel lateral forces P1–P4, that occur during its movement [21].

5 Discussion of the Results Obtained

The proposed approach for creating automated obstacle-resistant wireless SGC of steel structure complexes can be used for various purposes objects. The complex allows the recording of many signals from the strain gauge and other sensors. It is reasonable to use it when creating new equipment and for the equipment in service. To extend the sphere of application of the equipment, time is required to adapt the hardware and software.

To improve the quality of the complex, it is rational to use up-to-date technologies of autonomous decision-making and metrological control.

6 Conclusions

This paper proposes an approach to creating an automated, obstacle-resistant wireless strain gauge complex (SGC) steel structure, which can simultaneously monitor the parameters of several tens of sensors with an update frequency of tens of Hertz.

The complex can be used to diagnose the state of the steel structure based on "risk analysis" factor and increase the safety and quality of the operation of controlled facilities. Its advantages over other similar monitoring systems are resistance to obstacles, large two-way communication radius, and flexibility of composition. These advantages have been proven in practice.

For example, the method under consideration can be used to analyze a certain multi-component process in order to identify the most significant parameters. A typical example is the movement of a bridge crane with a skew. The practical application of this complex to measure the forces acting on the wheels and, as a consequence, on the steel structure of the bridge crane model made it possible to determine the important critical influence of one of the wheels skew (in the horizontal plane) on these forces. For example, a 0.14% increase in one wheel's skew approximately doubled the lateral forces on the wheels.

The use of the developed complex makes it possible to prevent the occurrence of operational defects, to detect them at the nascent stage, that is, the formation of SSS, which can lead to a defect (i.e., to prevent the occurrence of a "bifurcation point"). At the same time, the economic costs for its manufacture and operation are significantly lower compared to similar steel structure monitoring complexes.

References

1. Wu, F.Q., Zhang, J., Yao, W.Q.: Fatigue life analysis of metallurgical bridge crane structure. Appl. Mech. Mater. **437**, 181–185 (2013). https://doi.org/10.4028/www.scientific.net/AMM.437.181
2. Qi, K., Wang, W., Wang, X., Jiang, A., Liu, B., Guo, Z., Liu, J.: Safety assessment and fatigue life analysis of aged crane structures. 13th International Conference on Fracture. 2013 Jun 1–21 (Beijing), pp. 1–4 (2013).

3. Loveikin, V.S., Romasevich, Y.A., Khoroshun, A.S., Shevchuk, A.G.: Time-optimal control of a simple pendulum with a movable pivot. Part 1. Int. Appl. Mech. **54**, 358–365 (2018). https://doi.org/10.1007/s10778-018-0887-x
4. Taher, S.A., Li, J., Jeong, J.-H., Laflamme, S., Jo, H., Bennett, C., Collins, W.N., Downey, A.R.J.: Structural health monitoring of fatigue cracks for steel bridges with wireless large-area strain sensors. Sensors. **5076**, 22 (2022). https://doi.org/10.3390/s22145076
5. Kumar, V., Acot, B., Aygun, L.E., Wagner, S., Verma, N., Sturm, J., Glisic, B.: Detecting, localizing, and quantifying damage using two-dimensional sensing sheet. Lab test and field application. J. Civ. Struct. Heal. Monit. **11**, 1055–1075 (2021)
6. Kotelnikov, V.S.: Metody diagnostirovaniya i risk-analiza metallokonstrukcij gruzopod emnyh mashin v upravlenii ih bezopasnost'yu (Methods of diagnostics and risk analysis of lifting machinery metal structures in their safety management). Doctor of Sciences, Novocherkassk (2006).
7. Dobrotvorskiy, S., Balog, M., Basova, Y., Dobrovolska, L., Zinchenko, A.: Concept of the software for materials selection using .NET technologies. In: Advanced Manufacturing Processes. InterPartner 2019 Lecture Notes in Mechanical Engineering, pp. 32–43. Springer, Cham (2020)
8. Dobrotvorskiy, S., Basova, Y., Dobrovolska, L., Sokol, Y., Kazantsev, N.: Big challenges of small manufacturing enterprises in industry 4.0. In: Ivanov, V., Trojanowska, J., Pavlenko, I., Zajac, J., Peraković, D. (eds.) Advances in Design, Simulation and Manufacturing III. DSMIE 2020 Lecture Notes in Mechanical Engineering, pp. 118–127. Springer, Cham (2020). https://doi.org/10.1007/978-3-030-50794-7_12
9. Chengyin, L., Jun, T., Ning, W.: A wireless strain sensor network for structural health monitoring. Shock. Vib. **2015**(11), 1–13 (2015). https://doi.org/10.1155/2015/740471
10. Bocca, M., Toivola, J., Eriksson, L. M., Hollmén, J., Koivo, H.: Structural health monitoring in wireless sensornetworks by the embedded Goertzel algorithm. In: 2011 IEEE/ACM Second International Conference on Cyber-Physical Systems, pp. 206–214, Chicago (2011). https://doi.org/10.1109/ICCPS.2011.19.
11. Gangone, M.V., Whelan, M.J., Janoyan, K.D.: Wireless monitoring of a multispan bridge superstructure for diagnostic load testing and system identification. Comput. Aided Civ. Inf. Eng. **26**(7), 560–579 (2011). https://doi.org/10.1111/j.1467-8667.2010.00711.x
12. Gasco, F., Feraboli, P., Braun, J., Smith, J., Stickler, P., DeOto, L.: Wireless strain measurement for structural testing and health monitoring of carbon fiber composites. Compos. A Appl. Sci. Manuf. **42**(9), 1263–1274 (2011). https://doi.org/10.1016/j.compositesa.2011.05.008
13. Gregori, A., Giampaolo, E., Carlofelice, A., Castoro, C.: Presenting a new wireless strain method for structural monitoring: experimental validation. J. Sens. **2019**, 1–12 (2019). https://doi.org/10.1155/2019/5370838
14. DiGiampaolo, E., DiCarlofelice, A., Gregori, A.: An RFID-enabled wireless strain gauge sensor for static and dynamic structural monitoring. IEEE Sensors J. **17**(2), 286–294 (2017). https://doi.org/10.1109/JSEN.2016.2631259
15. Gubskyi, S., Yepifanov, V., Chukhlib, V., Basova, Y., Okun, A., Ivanova, M., Panamariova, O.: Integrated approach to determine operational integrity of crane metal structure. Periodica Polytechnica Mech. Eng. **64**(4), 319–325 (2019). https://doi.org/10.3311/PPme.14064
16. He, Y.B., Zhang, Y., Yang, B.K., Liu, S.W., Chen, D.F.: Finite element analysis in dynamic conditions of bridge crane beam. AMM. **331**, 70–73 (2013)
17. Hubskyi, S.: Avtomatizatsiya proverochnykh prochnostnykh raschetov metallokonstruktsiy mostovykh kranov (Automation of verification strength calculations of metal structures of overhead cranes). Collection of scientific papers. Mech. Eng. **14**, 5–9 (2014)
18. Hubskyi S., Okun A.: Komp'yuterna programa «Rozrakhunok metalokonstruktsiї mostovogo krana» («Crane»)» (Computer program "Calculation of the metal structure of the bridge crane" ("Crane")). Ukraine, Copyright registration certificate No. 47890 (2013).

19. Wen, W., Xufeng, X., Shuyao, F., Mengwei, L., Yong, L., Minghui, L.: Development of a wireless and passive temperature-compensated SAW strain sensor. Sensors Actuators A Phys. **308**, 1–10 (2020). https://doi.org/10.1016/j.sna.2021.113183
20. Sancho, J., Almandoz, I., Barandiaran, M., Diaz, J., Mendizabal, J.: Scalable wireless wearing monitoring system for harsh industrial environment. IEEE Trans. Ind. Electron. **69**(1), 1011–1020 (2021). https://doi.org/10.1109/TIE.2021.3053892
21. Hryhorov, O., Hubskyi, S., Turchyn, O., Vyshnevetskyi, H.: Teoretychne i eksperymentalne doslidzhennia protsesu peresuvannia mostovoho krana z perekosom. Collection of scientific papers. Mech. Eng. **19**, 7–17 (2017)

Comparison of Surface Roughness Measured by Contact and Noncontact Methods

Jozef Jurko, Angelina Iakovets, and Khrystyna Berladir

1 Introduction

The surface roughness parameter of mechanical engineering parts is one of the leading indicators of product quality, reliability, and service life. Roughness influences the essential operational characteristics of parts and assemblies: fatigue resistance, wear and corrosion resistance, fit reliability, contact stiffness, and thermal conductivity [1]. Therefore, the surface roughness characteristics must be standardized and analyzed in technological studies and controlled during production.

Among the most common methods of manufacturing parts with specified geometric parameters is metal turning [2], which allows for obtaining a surface with the necessary roughness. The roughness of the surface during the processing of the workpiece depends on the technological process: processing modes (cutting speed and feed) [3]; material and quality of the tool surface [4, 5]; mechanical properties, chemical composition, and structure of the workpiece material [6]; the composition of the lubricating and cooling liquid, etc. [7]. However, after metal processing, any surface is characterized by the acquisition of macro- and micro uniformities, which are not present on a perfectly smooth surface. Thus, determining surface roughness is one of the leading parameters for metal products.

J. Jurko (✉) · A. Iakovets
Faculty of Manufacturing Technologies with the seat in Prešov, Technical University of Košice, Prešov, Slovakia
e-mail: jozef.jurko@uke.sk

K. Berladir
Faculty of Manufacturing Technologies with the seat in Prešov, Technical University of Košice, Prešov, Slovakia

Sumy State University, Sumy, Ukraine

M. Balog et al. (eds.), *EAI International Conference on Automation and Control in Theory and Practice*, EAI/Springer Innovations in Communication and Computing, https://doi.org/10.1007/978-3-031-31967-9_3

An essential stage during the turning operation is product quality control. The physical, mechanical, and geometric properties of parts and products greatly influence the quality and reliability of mechanical engineering parts. First, the accuracy of their manufacture is the geometric dimensions and roughness of the details; the equipment's quality, service life, and functional properties depend on this. Therefore, quality control of turning results must be carried out using devices that measure surface roughness.

Today, contact methods using profilometers and profilographs [8–11] and noncontact methods using optical devices [12–16] are widely used to measure the surface roughness of various materials. Contact methods for measuring surface roughness are among the means of sequential profile transformation [8]. In particular, Vukelic D. et al. [9] used the Talysurf measuring device for high-precision measurement of the surface roughness of AISI 1045 steel. The results of experimental studies on the influence of processing technological parameters on the surface roughness measured using contact methods are also presented for carbon steel AISI 1045 in [10, 11]. The authors noted several advantages regarding using the contact method of measurement. Wang Y. et al. [12] presented a review paper in which noncontact surface roughness measurement methods based on optical interferometry were comprehensively characterized. The authors also described in detail the working principles of each technique and supplemented them with examples in each category. Valíček J. et al. [13] developed a modified optical method for measuring surface roughness and presented its implementation. A high degree of coincidence between optically determined parameters and standard surface values was noted. Aulbach L. et al. [14] proposed a new noncontact method of surface roughness measurement based on online monitoring. The paper presents a detailed theoretical approach to the technique, given simulation and experimental results for a defined set of surface roughness. Patil S.H. et al. [15] demonstrated the effectiveness of using the laser speckle visualization method. They used a single laser source in the experiments, which showed the prospects for further use of the proposed method. A design of a 3D laser profilometer for roughness measurement was described and analyzed in [16]. In the paper [17], the authors compared the quality of measuring the surface roughness of a thin metal coating on a composite substrate using a profilometer and optical coherence tomography. According to the results of the experiments, the advantage of using a noncontact measurement method for metal surface layers was established, which provides more reliable data.

Thus, one of the main disadvantages when measuring roughness using the contact method is the direct contact of the sensors with the surface of the part, which can lead to the appearance of scratches on the part's surface. Also, a relatively long time for the measurement process was noted. But this method provides a clear wave profile and possible measurement at a long distance. Instead, the noncontact method allows to control of the surface in a way that does not damage the part or the product as a whole [18]; the time spent on measurements is much less, but there are limitations in the size of measurement objects.

2 Research Methodology

Technological processes with different outputs can be expressed by Grey Relational Grade (GRG) parameter [19]. The Grey relational analysis (GRA) method can represent the incomprehensibility of system dependence for evaluating multiple output parameters and optimizing using one parameter. The optimization result of multiple complex factors can be transformed into a separate (individual) Single Grey Relational Grade.

To ensure the quality of data processing, the input data should be optimized by GRA and provide these steps [20]:

1. Determination of input factors and output parameters.
2. Defining the number of levels of input factors.
3. Factorial matrix design.
4. Data preprocessing is the process of normalizing data to reduce its variability. After data normalization, it should be transformed, i.e., the reference data should be changed on comparable data. Therefore, it is necessary to design a suitable value from the original value.

For our output parameters intended for normalization to change to an acceptable range, it was proposed the smaller, the better methodology (smaller is better), and we will determine the data according to the formula (1):

$$x_i^*(k) = \frac{\max x_i^0(k) - x_i^0(k)}{\max x_i^0(k) - \min x_i^0(k)} \tag{1}$$

where $i = 1,\ldots,m$ $k = 1,\ldots,n$; m is a number of experimental data; n is a number of parameters; $x_i^0(k)$ indicates the original sequence; $x_i^*(k)$ indicates the sequence after data processing; $\max x_i^0(k)$ is the maximum value of $x_i^0(k)$; $\min x_i^0(k)$ is the minimum value of $x_i^0(k)$; x_i^0 is the required value of $x_i^0(k)$.

The process of data processing of the proposed research was realized according to these steps:

- Calculate Grey Relational Coefficient (GRC) $\xi_i(k)$, according to formula (2),
- Calculate Grey Relational Grade (GRG) based on the results of GRC,
- Analyze the input parameters using GRG,
- Design optimal combinations of input factors of the cutting process:

$$\begin{gathered} \xi_i(k) = \frac{\Delta_{\min} + \xi\Delta_{\max}}{\Delta_{0i}(k) + \xi\Delta_{\max}} \\ \Delta_{oi}(k) = \left| x_o^*(k) - x_i^*(k) \right| \\ \Delta_{\max} = 1.00, \Delta_{\min} = 0.00 \end{gathered} \tag{2}$$

where $\Delta_{oi}(k)$ is the sequence deviation of the reference sequence $x_0^*(k)$ and comparison sequence $x_i^*(k)$; ξ is the identification coefficient that measures according to the range $0 \leq \xi \leq 1$ and depends on the requirements of the system. The coefficient determination is important for the demonstration of the relational level between the reference sequence $x_0^*(k)$ and comparison sequences $x_i^*(k)$; i=1,2...,m a k=1,2,...,n.

The determination of GRG in the GRA procedure should define the degree of influence between sequences of output parameters. $\left[x_0^*(k) \text{and}\, x_i^*(k), i = 1,2,3....16\right]$. Such a process is based on determining the influence degree between sequences of output parameters according to (3):

$$\gamma_i = \frac{1}{n}\sum_{k=1}^{n}\xi_i(k) \tag{3}$$

The γ_i is the required GRG for the ith experiment, and n is the number of output characteristics.

GRG indicates the reference sequence and the comparable sequence correlation level and shoves the overall quality level of the input factors in the defined system. This process converts an optimization problem with multiple output parameters into a single GRG optimization problem through the GRA method and with the Taguchi procedure. According to GRG, the optimal level of input process factors is determined, and its highest values indicate a better machined–functional surface quality. Next, it is necessary to calculate the average values of GRG for each level of input process factors.

Higher values of average GRG correspond to a combination of optimal factors. The design of optimal combination after analyzing will be predicted:

- the variance analysis (ANOVA) for assessing the significance of factors influencing the multiple responses at the 95% confidence level.
- recommended significant information about the experiment data. The P-value (probability of significance) is analyzed to assess the significance of the output parameter.

3 Experimental Design

3.1 Technological System in Laboratory Conditions

The proposed research was organized according to the technological system "machine tool" (Fig. 1). S-machine: Optimum TU 2807 is shown in Fig. 1. This lathe has a certified hardened spindle face (according to the standard DIN 55021) with a runout of less than 0.009 mm.

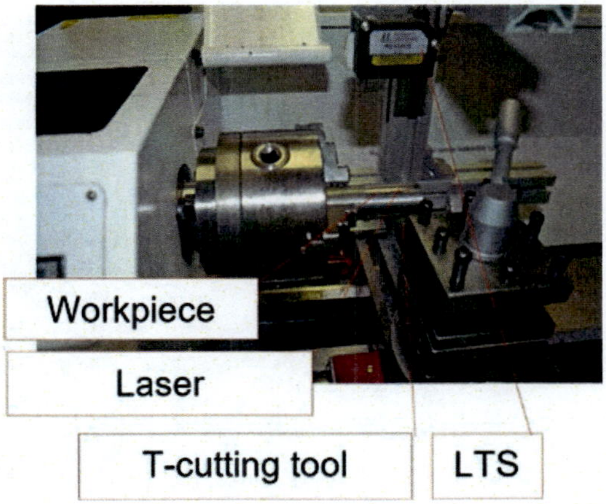

Fig. 1 Overall view and work zone of the Optimum TU 2807 machine

Workpiece – round rod d=30 × 200 mm, rod material – steel C45 (EN 10083-2-91). Chemical composition of steel: 0.42% C, 0.65% Mn, 0.20% Si, max 0.20% Cr, max 0.30% Ni, max 0.035% P, max 0.035% S.

Cutting tool – cutting insert SCMTT09T308 TTR – sintered carbide without coating. Knife holder – SSDCN1212K12-M-A.

3.2 *Measuring System with LTS*

Analysis of the elements affecting cutting process quality is essential [21]. The elements of the cutting zone include the cutting edge, the chip, and the surface after cutting [18, 22, 23]. This study presents the measurement effect of the change on the cutting edge, especially its wear on the front surface, and how these changes affect the quality parameter of the machined surface (surface roughness Ra).

The parameter was measured after the operation's end (turning the cylindrical surface) on the defined sample. The sample and area were chosen based on the requirements of production companies that manufacture shafts from various materials. A mobile measurement system was designed for the experiment (Fig. 2). The basic element of the measuring system is LTS IL100 (by Keyence). Its technical characteristics are a measurement range of 70–130 mm; spot diameter of 400 × 1350 mm (depending on the distance from the workpiece); linearity of ± 20 mm; repeatability of 4 mm; the weight of 75 g.

The elements from Fig. 2:

1. PLC Siemens-1511C,
2. Communication module Keyence CL-3000,
3. Amplifier Keyence CL-P070N,
4. Communication module Keyence DL-PNI, and
5. Amplifier Keyence IL-1000.

Fig. 2 LTS-IL100 and mobile measurement system (MMS) [24]

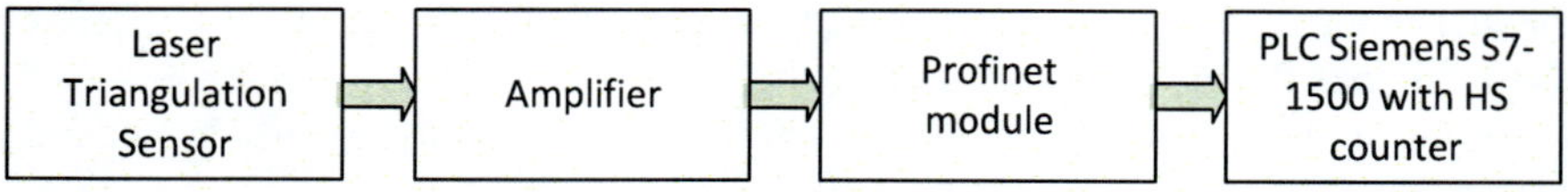

Fig. 3 Measuring chain with data acquisition [25]

The measurement system is further formed by a communication unit (DL-PN1-type DIN rail mount, 150 g) with an amplifier (IL-1000, 170 g, type for communication speed 500 kbps) [24]. Using an extension cable, these elements were connected from the sensor head to the PLC (Siemens 1511C) through the profinet (Fig. 3.) module. The following result of the research is the software design, with which we can process and analyze the data from the measurement and propose a solution for the continuation of production directly on-site or online for further activity in the production process.

The research for verifying the LTS application in parameter measurement was done for the operation of turning a cylindrical surface. The steel C45 for the experiment was chosen (test sample with a diameter of 30 mm and a length of 200 mm). The surface roughness (Ra) and wear of the face of the cutting insert (KB) were measured according to ISO 3685 (Fig. 4).

Each surface roughness and face wear value of the cutting insert was obtained from eight replicates (using two cutting inserts). The surface roughness was measured directly in the working zone of the machine tool using a laser triangulation sensor (LTS) (after machining the surface on the workpiece without unfastening it). In addition, the front surface wear of the cutting disc was investigated with a Carl Zeiss Primotech D/A ESD microscope (Fig. 5).

All experiments showed regular wear on the front surface and therefore analyzed the effect of KB on Ra for steel C45. The research results and methods are applicable for other materials verification.

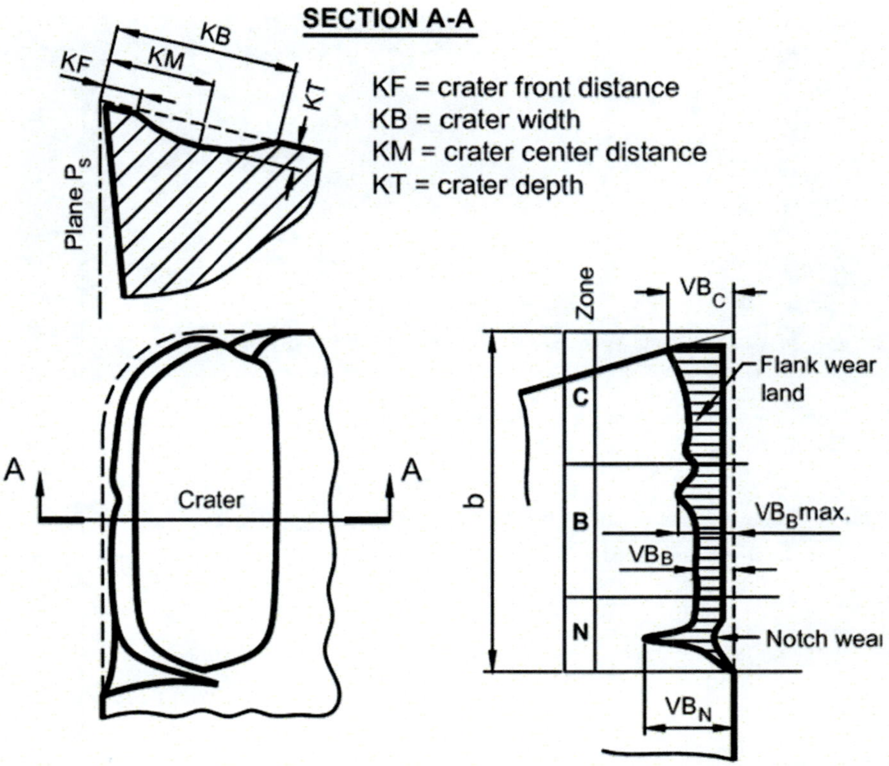

Fig. 4 Cutting tool wear parameters according to ISO 3685 [25]

The experimental research part was planned using Taguchi L16 OA (orthogonal array) for three input factors: cutting speed (A), movement (B), and cutting depth (C) at four levels (Table 1).

Taguchi L_{16} OA (orthogonal array) represents a total of 16 experiments, which were turned with a new cutting edge on an experimental sample with a length of 200 mm (3 cuts with a length of 70 mm). After finishing the turning, the machined surface was measured by LTS and a MarSurf PS10 contact manual roughness meter. Ra and KB were used as average values to process the measured parameters' results, and after filtering (mainly in terms of negative impact on the treated surface – listed at the end of this study). The results of the experimental face wear and surface roughness are shown in Table 2.

Analysis of the obtained data from Fig. 6 for Ra and Fig. 7 for KB showed that the resulting parameter KB wear on the front surface gradually increases with the growth of the parameter values of the cutting speed, movement, and cutting depth.

The most significant influence of input factors was found in the following order: cutting speed, movement, and cutting depth (Fig. 6). The surface roughness significantly decreases when the cutting speed changes from 1 to 4 levels, which is the same for 2–4 feed levels. The lowest value was recorded at a cutting speed of level

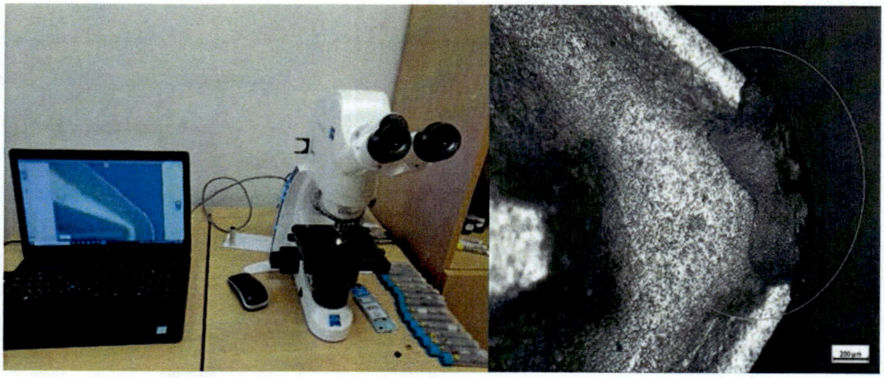

Fig. 5 Investigation process of the face wear of the cutting insert using a Carl Zeiss Primotech D/A ESD microscope

Table 1 Cutting process condition levels for Taguchi L16 Orthogonal Array

Parameter level	Cutting speed *A* (m/min)	Movement *B* (mm per rev.)	Cutting depth *C* (mm)
1	25	0.05	0.02
2	50	0.075	0.05
3	75	0.1	0.1
4	100	0.2	0.5

Table 2 Experimental results for parameters – wear of the front surface and surface roughness according to Taguchi L_{16} Orthogonal Array

Number of experiments	Cutting speed *A* (m/min)	Movement *B* (mm per rev.)	Cutting depth *C* (mm)	*KB* (mm)	*Ra* (μm)
	Process factors			Experimental results	
1	25	0.05	0.02	0.052	5.102
2	25	0.075	0.05	0.048	5.427
3	25	0.1	0.10	0.084	5.364
4	25	0.2	0.20	0.118	5.187
5	50	0.05	0.05	0.076	5.204
6	50	0.075	0.02	0.054	5.021
7	50	0.1	0.20	0.126	5.111
8	50	0.2	0.10	0.087	4.894
9	75	0.05	0.10	0.082	4.987
10	75	0.075	0.20	0.128	5.008
11	75	0.1	0.02	0.053	5.006
12	75	0.2	0.05	0.092	4.872
13	100	0.05	0.20	0.165	4.881
14	100	0.075	0.10	0.087	4.925
15	100	0.1	0.05	0.071	4.648
16	100	0.2	0.02	0.051	4.906

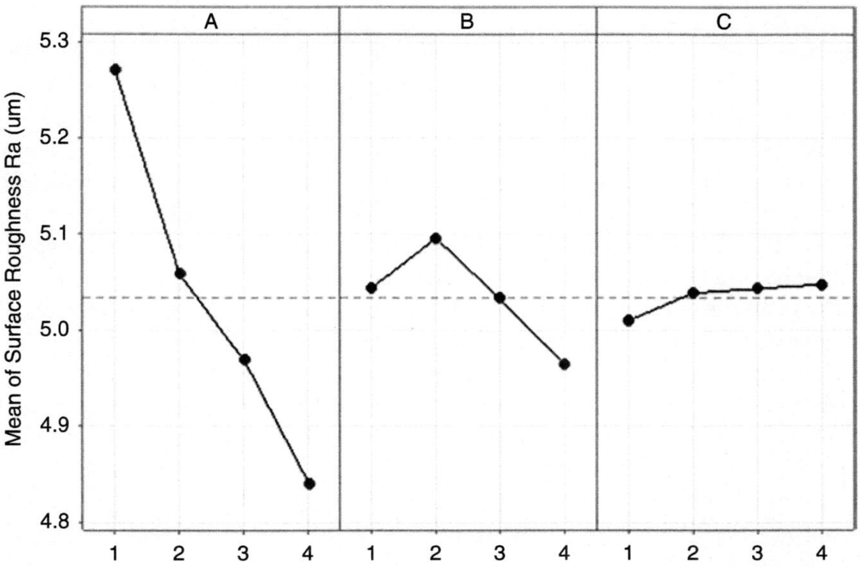

Fig. 6 Main effect plot of surface roughness (Taguchi method, orthogonal array matrix)

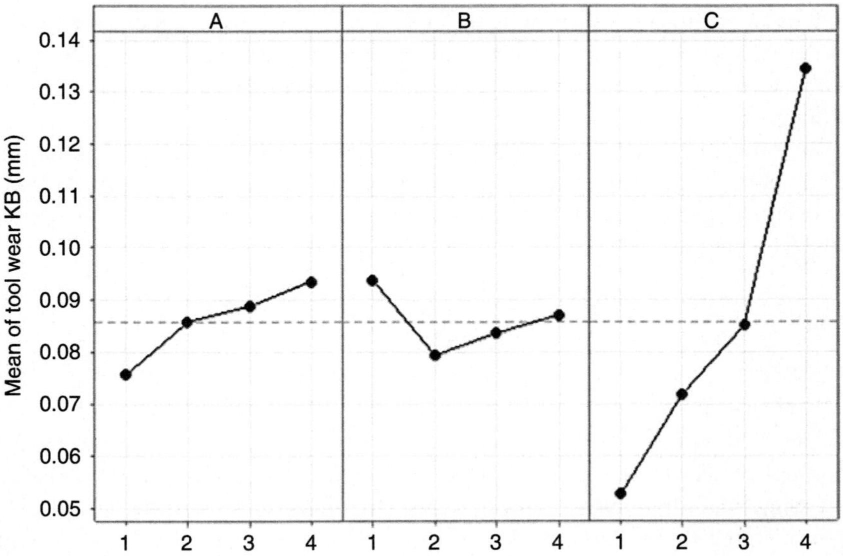

Fig. 7 Main effect plot of tool wear KB

4. During displacement/movement, lower values were measured from 2 to 4 levels. The lowest surface roughness value was recorded at a displacement of 0.2 mm. The effect of the depth of cut on surface roughness is gradual from 1 to 4 levels.

The lowest average surface roughness value was measured at a cutting speed of 100 m/min and was 4.648 μm. The graphs of the main effects show the uncertainty of the levels of the process parameters to be designed for the turning of C45 steel, as the standard of the group of materials – steel, that show that for certain lower levels of the parameters – the wear of the front surface is lower. Still, in this experiment, there is a change in the size of the surface roughness.

The results of measurement show that four levels of cutting speed, four levels of movement, and one level of cutting depth are factors for achieving the lowest surface roughness. For the defined output parameters, the dependence is important because the wear of the front surface of the cutting disc influences the deviation of the diameter as one of the parameters of the quality of the machined surface. Therefore, the investigation of this dependence and the design of optimal input factors are important, also from the point of view of using LTS in measuring quality parameters. In the same way, applying the measurement on the standard test sample provides prerequisites for using the obtained data in processes when the method of comparison with the standard will rarely be used.

The most significant influence of the input factors for KB was found in the order of cutting depth, movement, and cutting speed, as shown in Fig. 6. Wear is reduced when changing the cutting speed from 4 to 1 level and 4 to 2 feed levels. The depth of cut has a significant influence on the size of KB. The lowest value was recorded at a depth of cut of one level. During displacement, lower values were measured from 2 to 4 levels. The lowest wear value was recorded at a displacement of 0.075 mm.

The lowest average value of KB was measured at a cutting speed of 25 m/min, i.e., 0.048 mm. The measurements show that one level of cutting speed, two levels of movement, and one level of depth of cut are factors for achieving the least wear on the front surface.

4 Results

The data were normalized for the output parameters of front surface wear and surface roughness according to Eq. 1 (Table 3). The GRC ratio coefficients were calculated using Eq. 2 from the normalized data set of Table 3. A value of 0.5 was chosen for the distinguishing coefficient because the output parameters have equal weight. The smaller the wear value of the tool's cutting part, the higher the quality of the output parameters of the machined surface (Table 3). The GRG was calculated using Eq. 3 and according to the GRC ratio coefficients.

The results from Table 3 are used for factor optimization when they are converted to one degree. The influence of GRG for the selected experimental conditions is shown in Fig. 8. Experiment number 4 can be determined, for input factors

Table 3 Grey relational values

Experiment	Grey relational generation values		Deviation Sequence		GRC		GRG	Rank
	VB_Bmax	*Ra*	VB_Bmax	*Ra*	VB_Bmax	*Ra*		
Ideal sequence	1	1	1	1				
1	0.9658	0.4172	0.0342	0.5828	0.9360	0.4618	0.6989	5
2	1.0000	0.0000	0.0000	1.0000	1.0000	0.3333	0.6667	6
3	0.6923	0.0809	0.3077	0.9191	0.6190	0.3523	0.4857	12
4	0.4017	0.3081	0.5983	0.6919	0.4553	0.4195	0.4374	16
5	0.7607	0.2863	0.2393	0.7137	0.6763	0.4120	0.5441	11
6	0.9487	0.5212	0.0513	0.4788	0.9070	0.5108	0.7089	4
7	0.3333	0.4056	0.6667	0.5944	0.4286	0.4569	0.4427	15
8	0.6667	0.6842	0.3333	0.3158	0.6000	0.6129	0.6065	7
9	0.7094	0.5648	0.2906	0.4352	0.6324	0.5347	0.5835	10
10	0.3162	0.5379	0.6838	0.4621	0.4224	0.5197	0.4710	14
11	0.9573	0.5404	0.0427	0.4596	0.9213	0.5211	0.7212	3
12	0.6239	0.7125	0.3761	0.2875	0.5707	0.6349	0.6028	8
13	0.0000	0.7009	1.0000	0.2991	0.3333	0.6257	0.4795	13
14	0.6667	0.6444	0.3333	0.3556	0.6000	0.5844	0.5922	9
15	0.8034	1.0000	0.1966	0.0000	0.7178	1.0000	0.8589	1
16	0.9744	0.6688	0.0256	0.3312	0.9512	0.6015	0.7764	2

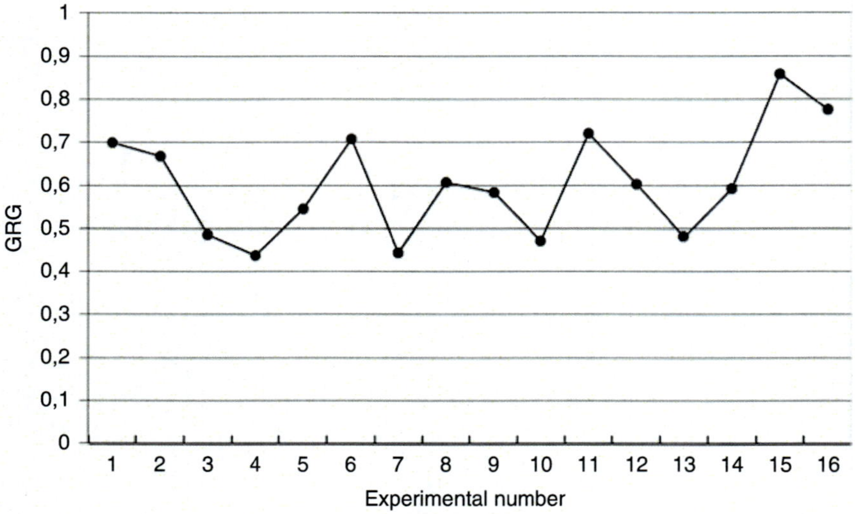

Fig. 8 Grade relational grades for the output parameters wear of the back surface of the cutting disc and surface roughness

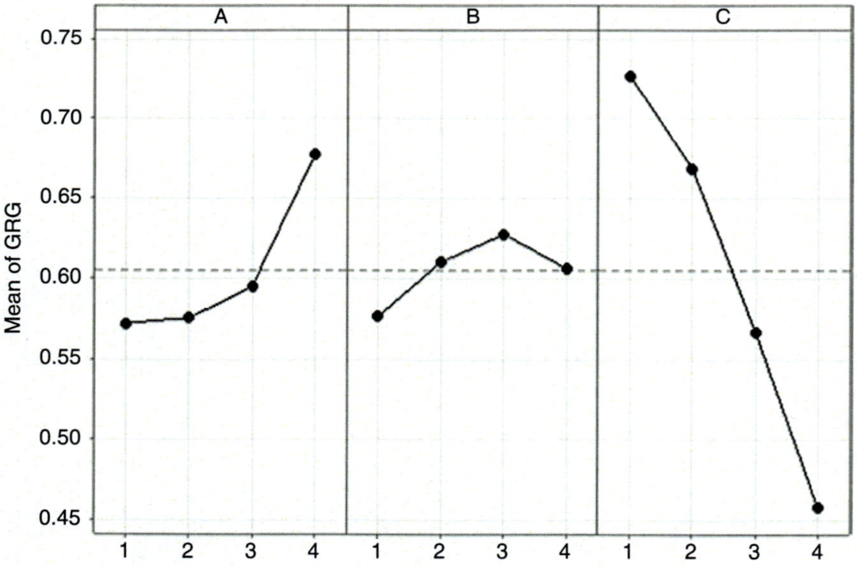

Fig. 9 Main effects plot of mean grey relational grade

A1-B4-C4, as an input experiment. For the GRG values, the effects of each process factor at different levels are shown in Fig. 9.

The optimal combination of the input is selected on higher mean values of GRG basis according to Table 4. A higher value of GRG expresses a stronger correlation

Table 4 Main effects on mean grey relational grade

Factors	Mean Grey relational grade				Max-Min	Rank
	Level 1	Level 2	Level 3	Level 4		
A	0.5722	0.5756	0.5946	**0.6768**	0.1046	2
B	0.5765	0.6097	**0.6271**	0.6058	0.0506	3
C	**0.7263**	0.6681	0.5671	0.4578	0.2687	1

Total mean grey relational grade 0.6048

Table 5 Results of ANOVA on grey relational grade

Source	DF	Adj SS	Adj MS	F-Value	P-Value	Remarks
A	3	0.028805	0.009602	1.60	0.284	Insignificant
B	3	0.005292	0.001764	0.29	0.828	Insignificant
C	3	0.167440	0.055813	9.33	**0.011**	**Significant**
Error	6	0.035898	0.005983			
Total	15	0.237435				

with the reference sequence, so the optimal setting of factors for multiple output parameters is the combination of factors A4-B3-C1, i.e., cutting speed 100 m/min, feed 0.1 mm per rev and depth of cut 0.02 mm.

The higher values of the average GRG (according to Fig. 9) express the minimum values of wear of the front surface of the cutting part KB and surface roughness Ra. Table 4 shows the GRG values as the differences between the maximum and minimum values from the four levels for individual factors. The results can be interpreted so that the depth of cut (0.7263) has the most significant influence on multiple output parameters (responses) compared to the cutting speed (0.6768) and feed (0.6271) when turning C45 steel. The importance of individual process input factors on the output multi-parameters can be listed in the order cutting depth > cutting speed > movement.

In Table 5, the variance analysis regarding GRG is formulated. It indicates the importance of process factors on multiple output parameters (responses). The results showed that the cutting depth is a significant process factor affecting multiple output parameters as its P-value is less than 0.05 at the 95% confidence level. However, for the element's movement and cutting speed, significance was not demonstrated concerning multiple parameters simultaneously.

The next step is to carry out confirmation tests to find the GRG improvement from the initial setting of the input factors to the optimal factors obtained in turning C45 steel with the measurement of the machined area using LTS. For the optimal set of the A4B3C1 input factors, the GRG value is 0.6451, which is very close to 0.6583 (predicted value). The experiment was repeated using the optimal setting of the A4B3C1 input factors to verify the test result. As a result, the validation test results were made with a repeatability of eight for the optimal level of input factors, according to Table 6.

Table 6 Confirmation test

	Initial factor settings	Optimal cutting factors	
		Prediction	Experiment
Level	*A1-B4-C4*	A4-*B3-C1*	A4-*B3-C1*
KB (mm)	0.118		0.096
Ra (mm)	5.187		4.982
GRG	0.4374	0.6583	0.6451
Improvement in GRG=0.2077			

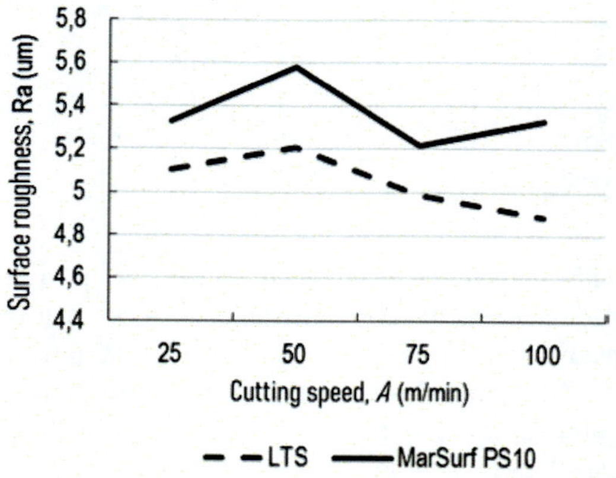

Fig. 10 Dependence of surface roughness on cutting speed during feed 0.05 mm

The validation experiment (Table 6) found that the GRG for multiple output parameters, such as back surface wear and diameter deviation, has been tweaked (by 0.2077) by selecting the optimal parameter combination. The results confirm the resulting output parameters, surface roughness, and face wear performed under operating conditions on a CNC machine. Furthermore, the analysis of the input factors for the minimum values of face wear and surface roughness are optimized and investigated in the turning of C45 steel by GRA based on the Taguchi procedure.

We compared the result of measuring the surface roughness with LTS with the results of the measurement on the MarSurf PS10 device (Table 7). An example is shown in Fig. 10.

Table 7 shows that for the range of input factors, the resulting average values of surface roughness Ra measured on the MarSurf PS10 device are greater than those measured with LTS.

The experiments had shown the opportunity for implementation of the noncontact measurement sensors. Today's trends dictate the use of intelligent sensors and devices that continuously send relevant data immediately to the information system of enterprises, which allows for increasing the quality of products and the technical condition of production machines. Further research will focus on other quality indicators, such as microhardness, directly affecting the production process.

Table 7 Comparison of average results of surface roughness Ra measured with LTS and on MarSurf PS10

Experiment number	Cutting speed *A* (m/min)	Movement *B* (mm per rev.)	Depth of cut *C* (mm)	MarSurf PS10 *Ra* (μm)	*LTS Ra* (μm)
	Process factors			Experimental results	
1	25	0.05	0.02	5.324	5.102
2	25	0.075	0.05	5.684	5.427
3	25	0.1	0.10	5.562	5.364
4	25	0.2	0.20	5.782	5.187
5	50	0.05	0.05	5,687	5.204
6	50	0.075	0.02	5.578	5.021
7	50	0.1	0.20	5.369	5.111
8	50	0.2	0.10	5.210	4.894
9	75	0.05	0.10	5.214	4.987
10	75	0.075	0.20	5.842	5.008
11	75	0.1	0.02	5.629	5.006
12	75	0.2	0.05	5.187	4.872
13	100	0.05	0.20	5.326	4.881
14	100	0.075	0.10	5.628	4.925
15	100	0.1	0.05	5.026	4.648
16	100	0.2	0.02	5.247	4.906

5 Conclusions

The proposed research shows the results of the decision-making process based on using of multi-criteria optimization of process factors in dry turning of C45 steel and measurement of wear parameters of the face of the cutting plate on a microscope, and measurement of surface roughness by applying a new generation laser triangulation sensor through the GRA method and Taguchi's L16 Orthogonal Array. Based on the analysis, the following conclusions are obtained:

- information on the reference material – C45 steel, can be used in further research by the comparison method for verification on other materials;
- surface roughness values were measured and evaluated in the interval (4.2–6.0 mm) when applying LTS;
- design and verification of the program that provides the communication of the PLC and other components in the measuring system by software for recording the data obtained from the laser triangulation sensor;
- the optimal set of input factors for multiple output parameters (responses) is A4-B3-C1, i.e., depth of cut 0.2 mm, feed 0.1 mm per rev., and a cutting speed of 100 m/min. The importance of individual process input factors on output multi-parameters can be listed in the order cutting speed > movement > cutting depth;

- from the ANOVA table, the depth of cut is a significant factor in the process of influencing multiple output parameters at a confidence level of 95% when turning C45 steel as a case study regarding the effect of the wear of the back surface of the cutting part on the deviation of the diameter, which is investigated using LTS for its deployment in CNC machine tools. Thus, the results of the optimization of input factors can be implemented to turn technology in manufacturing companies to reduce production costs in production and product control;
- the proposed GRA methodology with Taguchi's L16 Orthogonal Array effectively solves the relationship of input technological factors of processes with different output parameters as a case study in turning;
- experience from researching the application of a laser triangulation sensor in the measurement of surface roughness after steel turning – standard C45:
- provide compliance with certified deviations of selected dynamic modules of CNC machines (e.g., spindle).
- direct measurement in the working zone of the machine,
- noncontact method of measurement, and the speed of the obtained data: high efficiency
- the need to create software for data processing and evaluation.

Acknowledgments This work was supported by the Slovak Research and Development Agency under contract No. APVV-19-0590, and by the projects VEGA 1/0403/23, KEGA 022TUKE-4/2023 granted by the Ministry of Education, Science, Research and Sport of the Slovak Republic.

References

1. Satbayeva, Z.A., Bayatanova, L.B., Kozhanova, R.S.: Research of electrolyte plasma treatment impact on wear resistance and roughness of 18HN3MA-SH steel. MSF. **989**, 793–798 (2020). https://doi.org/10.4028/www.scientific.net/msf.989.793
2. Ivchenko, O., Ivanov, V., Trojanowska, J., Zhyhylii, D., Ciszak, O., Zaloha, O., Pavlenko, I., Hladyshev, D.: Method for an effective selection of tools and cutting conditions during precise turning of non-alloy quality steel C45. Materials. **15**, 505 (2022). https://doi.org/10.3390/ma15020505
3. Shvets, S.V., Machado, J.: Numerical model of cutting tool blade wear. J. Eng. Sci. **8**(2), A1–A5 (2021). https://doi.org/10.21272/jes.2021.8(2).a1
4. Pavlenko, I., Trojanowska, J., Gusak, O., Ivanov, V., Pitel, J., Pavlenko, V.: Estimation of the reliability of automatic axial-balancing devices for multistage centrifugal pumps. Periodica Polytechnica Mech. Eng. **63**(1), 52–56 (2019). https://doi.org/10.3311/PPme.12801
5. Dobrotvorskiy, S., Basova, Y., Dobrovolska, L., Popov, V., Mounif, A.S.Y.: Creation of a superhydrophilic surface with anti-icing properties for X18H10T stainless steel using a nanosecond laser. In: Cioboată, D.D. (ed.) International Conference on Reliable Systems Engineering (ICoRSE) – 2022. ICoRSE 2022 Lecture Notes in Networks and Systems, vol. 534, pp. 172–184. Springer, Cham (2023). https://doi.org/10.1007/978-3-031-15944-2_17
6. Hovorun, T., Khaniukov, K., Varakin, V., Pererva, V., Vorobiov, S., Burlaka, A., Khvostenko, R.: Improvement of the physical and mechanical properties of the cutting tool by applying wear-resistant coatings based on Ti, Al, Si, and N. J. Eng. Sci. **8**(2), C13–C23 (2021). https://doi.org/10.21272/jes.2021.8(2).c3
7. Wu, X., Li, C., Zhou, Z., et al.: Circulating purification of cutting fluid: an overview. Int. J. Adv. Manuf. Technol. **117**, 2565–2600 (2021). https://doi.org/10.1007/s00170-021-07854-1

8. Butler, D.: Surface roughness measurement. In: Li, D. (ed.) Encyclopedia of Microfluidics and Nanofluidics, pp. 1945–1949. Springer, Boston (2008). https://doi.org/10.1007/978-0-387-48998-8_1506
9. Vukelic, D., Prica, M., Ivanov, V., Jovicic, G., Budak, I., Luzanin, O.: Optimization of surface roughness based on turning parameters and insert geometry. Int. J. Simul. Model. **21**(3), 417–428 (2022). https://doi.org/10.2507/IJSIMM21-3-607
10. Hwang, Y.K., Lee, C.M.: Surface roughness and cutting force prediction in MQL and wet turning process of AISI 1045 using design of experiments. J. Mech. Sci. Technol. **24**, 1669–1677 (2010). https://doi.org/10.1007/s12206-010-0522-1
11. Esteves Correia, A., Paulo Davim, J.: Surface roughness measurement in turning carbon steel AISI 1045 using wiper inserts. Measurement. **44**(5), 1000–1005 (2011). https://doi.org/10.1016/j.measurement.2011.01.018
12. Wang, Y., Xie, F., Ma, S., Dong, L.: Review of surface profile measurement techniques based on optical interferometry. Opt. Lasers Eng. **93**, 164–170 (2017). https://doi.org/10.1016/j.optlaseng.2017.02.004
13. Valíček, J., Držík, M., Hryniewicz, T., Harničárová, M., Rokosz, K., Kušnerová, M., Barčová, K., Bražina, D.: Non-contact method for surface roughness measurement after machining. Meas. Sci. Rev. **12**(5), 184–188 (2012). https://doi.org/10.2478/v10048-012-0028-3
14. Aulbach, L., Salazar Bloise, F., Lu, M., Koch, A.W.: Non-contact surface roughness measurement by implementation of a spatial light modulator. Sensors. **17**, 596 (2017). https://doi.org/10.3390/s17030596P
15. Patil, S.H., Kulkarni, R.: Surface roughness measurement based on singular value decomposition of objective speckle pattern. Opt. Lasers Eng. **150**, 106847 (2022). https://doi.org/10.1016/j.optlaseng.2021.106847
16. Mitaľ, G., Dobránsky, J., Ružbarský, J., Olejárová, Š.: Application of laser profilometry to evaluation of the surface of the workpiece machined by abrasive waterjet technology. Appl. Sci. **9**, 2134 (2019). https://doi.org/10.3390/app9102134
17. Korzeniewska, E., Sekulska-Nalewajko, J., Gocławski, J., Rosik, R., Szczęsny, A., Starowicz, Z.: Surface morphology analysis of metallic structures formed on flexible textile composite substrates. Sensors. **20**, 2128 (2020). https://doi.org/10.3390/s20072128
18. Leššo, I., Flegner, P., Pandula, B., Horovčák, P. New principles of process control in geotechnics by acoustic methods. Metalurgija. **46**(3), 165–168, (2007). ISSN 0543-5846.
19. Deng, J.: Control problems of grey systems. Syst. Control Lett. **1**, 288–294 (1982). https://doi.org/10.1016/S0167-6911(82)80025-X
20. Lin, C.L.: Use of the Taguchi method and grey relational analysis to optimize turning operations with multiple performance characteristics. Mater. Manuf. Process. **19**(2), 209–220 (2004). https://doi.org/10.1081/AMP-120029852
21. Neslušan, M., Mrkvica, I., Čep, R., Kozak, D., Konderla, R. Deformation after heat treatment and their influence on cutting process. Tehnički vjesnik/Technical Gazette, **18**(4), 601–608 (2011). ISSN 1330-3651.
22. Mrkvica, I., Konderla, R., Faktor, M.: Turning of inconel 718 by cemented carbides. In: Morgan, M.M., Shaw, A., Mgaloblishvili, O. (eds.) Precision Machining VI. Key Engineering Materials, 496, 138–147, (2012). ISSN 1013-9826. ISBN 978-3-03785-297-2.
23. Peterka, J., Pokorný, P. Václav, Š.: CAM strategies and surfaces accuracy. In: Annals of DAAAM and Proceedings of DAAAM Symposium, Proceedings of the 19th International DAAAM Symposium “Intelligent Manufacturing & Automation: Focus on Next Generation of Intelligent Systems and Solutions”, 22–25th October 2008, Trnava, Slovakia, vol. 19, issue 1, pp. 1061–1062 (2008). ISSN 1726-9679.
24. Jurko, J., Miškiv-Pavlík, M., Hladký, V., Lazorík, P., Michalík, P., Petruška, I.: Measurement of the Machined surface diameter by a laser triangulation sensor and optimalization of turning conditions based on the diameter deviation and tool wear by GRA and ANOVA. Appl. Sci. **12**, 5266 (2022). https://doi.org/10.3390/app12105266
25. ISO 3685:1993 Tool-life testing with single-point turning tools. 2nd edn (1993).

An Increase in Energy Efficiency and Vibration Reliability of Centrifugal Pumps for Nuclear Power Plants

Ivan Pavlenko, Vladyslav Kondus, Vitalii Ivanov, Anton Verbovyi, Oleksandr Ivchenko, Frantisek Botko, and Jan Pitel

1 Introduction

Due to Russia's invasion of Ukraine, the problem of an increase in the energy efficiency of power plants becomes existential in providing energy independence for European countries. On the other side, given the global security challenges of nuclear power plants (NPPs), the power of nuclear energy around the world is growing permanently [1]. According to the World Nuclear Association, there are about 440 nuclear reactors with a total capacity of 400 GW in more than 30 countries [2]. These reactors provide 10% of the world's electricity needs. The International Energy Agency forecasts an increase in this indicator by up to 15% in the next 20 years.

I. Pavlenko (✉) · V. Ivanov
Sumy State University, Sumy, Ukraine

Faculty of Manufacturing Technologies with the seat in Prešov, Technical University of Košice, Prešov, Slovak Republic
e-mail: i.pavlenko@cm.sumdu.edu.ua

V. Kondus
Sumy State University, Sumy, Ukraine

Sumy Machine-Building Cluster of Energy Equipment, Sumy, Ukraine

A. Verbovyi
Sumy State University, Sumy, Ukraine

JSC "Research and Design Institute for Atomic and Power Pumpbuilding", Sumy, Ukraine

O. Ivchenko
Sumy State University, Sumy, Ukraine

F. Botko · J. Pitel
Faculty of Manufacturing Technologies with the seat in Prešov, Technical University of Košice, Prešov, Slovak Republic

M. Balog et al. (eds.), *EAI International Conference on Automation and Control in Theory and Practice*, EAI/Springer Innovations in Communication and Computing, https://doi.org/10.1007/978-3-031-31967-9_4

According to the International Energy Agency [3], energy consumption in Poland reaches 5 PJ with stable domination of fuels (about 90%). In 2021, the Council of Ministers approved the Energy Policy of Poland until 2040. After that, however, due to Russia's invasion of Ukraine, importing energy resources to Poland will be substantially reduced or prohibited.

Energy consumption in the Slovak Republic has also been steadily operating recently. The average annual power generation in the Slovak Republic is approximately 28.4 TW·h. Moreover, 81% of Slovak energy is obtained from nuclear power plants (54%), hydropower (16%), and natural gas stations (11%). The government of the Slovak Republic approved a long-term energy plan mainly based on energy generation at nuclear power plants Mochovice 1, 2 and Bohunice 3, 4. Moreover, new blocks at power plants Mochovice 3, 4 are realized, and Bohunice are planned to be designed. This fact makes the Slovak Republic an importer of energy carriers and a reliable exporter of the European Union.

Notably, in Ukraine, since the main types of equipment at power stations are rotary machines. The practical significance of the stated problem is undoubtful for NPPs with water-water energetic reactors WWER-1000. The following pumps are still operating: TX 800/70/8-K-2E – for cooling systems of the holding pool and the industrial circuit, X 8/60-K-2G – for systems of hydro testing of the bubbler and purging of sensors, and X 45/90-K-2G – to supply emergency boron solution for cleaning.

Production of the above pumping units is currently doubtful in Ukraine. Therefore, the needs of NPPs in spare parts are met by purchasing them abroad. These facts negatively affect the price indicators of such products and create challenges in energy security. Therefore, the supply of the specified products is a position of critical import to ensure the reliable operation of NPPs.

Moreover, the existing pumping units should be gradually replaced at existing NPPs or supplied at new blocks. Therefore, developing modernized pumping units with increased energy efficiency and reliability for import substitution is expedient.

However, such an intensification should be realized by an increase in operating parameters (e.g., rotation speed, pressure head) of rotary machines. In this regard, problems with process innovation at power stations become inevitable. Moreover, their solutions open various scientific problems in rotor dynamics, strength, and stability.

Moreover, the increased flow parameters (e.g., pressure and flow rate) require improved approaches to ensuring the energy efficiency of the vibration equipment and industrial growth. Therefore, the research aims to ensure the vibration reliability of rotary machines and increase their energy efficiency under intensified conditions.

2 Literature Review

The primary attention in ensuring energy efficiency and vibration reliability of rotary machines should be paid to developing clarified nonlinear mathematical models of forced oscillations of the coupled system "rotor – gap seals – bearing supports" [4]. Its consequent solutions should be realized using a state-of-the-art scientific and methodological approach based on the comprehensive application of intelligent decision-making systems and artificial neural networks. Such an approach should allow the evaluation of nonlinear characteristics of bearing supports and gap seals and predetermine the impact of flow characteristics on the energy efficiency of the related equipment.

Therefore, improving the energy efficiency of centrifugal pumps in NPPs is a crucial problem of the nuclear industry [5]. Moreover, it is closely related to solving significant scientific and applied problems of improving the pressure characteristics of pumping units, ensuring vibration reliability of functional elements for a complex hydromechanical system "rotor – gap seals – bearing supports."

Different research works have previously focused on ensuring energy efficiency and vibration reliability of power systems worldwide. Particularly, Kindl et al. [6] developed a methodology of vibration origin investigation for pumps and compressors. The methodology is based on detecting undesirable vibration using a diagnostic system on-site and uses mathematical modeling of corresponding mechanical parts to reveal the vibration origin. Also, Rai [7] presented a case study of the last stage performance considerations in low-pressure turbines of power plants. As a result, various factors that affect the last stage performance of low-pressure turbines were analyzed.

Song et al. [8] experimentally investigated the effect of nonuniform flow on the vibration characteristics of the reactor coolant pump for an NPP. The frequency domain analysis of the vibration signal revealed that low-frequency vibration occurs under nonuniform inflow conditions.

Jigang et al. [9] carried out a 3D seismic analysis and safety evaluation for an NPP's pump. As a result, the pump's safety under the seismic load was evaluated for the particular case study.

Wang et al. [10] studied the impact of the boundary effect and high temperature and pressure on reactor coolant pump mode. The obtained results showed that eigenfrequency decreases with the increase of dive depth and increases with the increase of radial distance. Shang et al. [11] also studied multi-state reliability for a coolant pump based on a dependent competitive failure model. They provided an effective method for reliability analysis for coolant pumps in NPPs based on the operating environment.

Yin et al. [12] analyzed fluid-structure interaction characteristics for the impeller of the residual heat removal pump. The results provided theoretical fundamentals to improve system performance and further study for more complicated dynamic and fatigue analyses. Also, Jiao et al. [13] also developed a model for calculating the idling of nuclear power coolant pumps considering kinetic energy in a pipeline. The obtained calculation results have a relatively more comprehensive range of engineering applications.

Zhou et al. [14] investigated residual heat removal pumps numerically based on fluid-structure interaction in 1 GW nuclear power plants. They evaluated the design flow rate operating condition to minimize the significant pressure fluctuations peak amplitude. Conversely, Kissick and Wang [15] conducted a techno-economic analysis of alternative power cycles for light-water small modular reactors. As a result, the transcritical ethanol cycle and regenerative reheat Rankine cycle have shown higher thermal efficiency and lower cost of electricity than the baseline regeneration. Also, Temiz and Dincer [16] proposed ways to enhance an NPP with a renewable energy system. The proposed integration demonstrated an essential overall energy efficiency during the operating mode.

Zaitsev et al. [17] proposed methods for gas chromatographic analysis of technological media in the main circulation pumps at a nuclear power plant. As a result, the oil system's basic technological scheme for the main circulation pumps has been developed.

Thus, recent approaches in studying operating processes in rotary systems, ensuring their energy efficiency and vibration reliability, are predominantly based on conventional simulations and the consequent analysis using computer software [18]. However, analytical dependences for such modeling (e.g., bearing stiffness, damping factor, circulating force, etc.) are based on simplified mathematical models of rotor dynamics. As a result, resonant frequencies of forced oscillations are calculated using graphical means like Campbell's diagram, and the dynamic stability of the systems "rotor – gap seals – bearing supports" are based on the Nyquist criterion.

The following objectives have been formulated to improve power equipment's energy efficiency and vibration reliability. Firstly, an analysis of recent approaches in ensuring the vibration reliability of rotary machines and ways to increase the energy efficiency of power machines and the related technological equipment should be made.

Secondly, for the practical case study, ways to modernize the centrifugal pumps for nuclear power plants should be analyzed. The updated designs should correspond to the standards concerning regulations in the nuclear power industry according to national and international requirements.

3 Research Methodology

During the presented research, technical projects of pumping units ACPA 600–35, ACPA 600–35-1, ACPA 8–60, and ACPA 45–90 have been developed for new third and fourth blocks of the Khmelnytskyi NPP (Netishyn, Ukraine). They also can replace other pumping units at Ukrainian NPPs with reactors WWER-1000 within the framework of import substitution according to the requirements of SE "NNEGC 'Energoatom'" [19].

The pumping unit ACPA 600–35 has been designed to replace the unit TX 800/70/8-K-2E of the cooling system at the holding pool, and ACPA 600–35-1 – to replace the unit TX 800/70/8-K-2E for the industrial circuit systems, X 8/60-K-2G – for systems of hydraulic testing of the bubbler and purging of sensors, and X 45/90-K-2G – to supply emergency boron solution for cleaning.

Also, the pumping unit ACPA 600–35-1 (identifiers – TF31D01, TF32D01, TF33D01) is an element of the industrial circuit water system. It is used for forced circulation in cooling heat exchangers [20], bubble tanks, purging aftercooler of the first circuit, cooler of organized leakages [21], and sampling systems.

The pumping units ACPA 600–35-1 are installed in the premises of the technological equipment of the strict regime zone of the reactor compartment at 6600 m above sea level.

The failure criteria under normal operating conditions of the pump unit are as follows: bearing temperature over 90 °C; the vibration on the bracket in the bearing supports over 11.2 mm/s; leakages through to end seals over $0.83 \cdot 10^{-8}$ m^3/s (0.03 l/h); jamming of moving parts; destruction or loss of material density of parts under the pressure of the pumped medium; violation of the tightness of flange joints, which cannot be eliminated by additional tightening with fastening joints.

The limiting criteria at which further operation of the pump is impossible are loss of tightness for the housing or cover and the end of the parts' service life (30 years). Also, pumps should ensure long-term operation at the operating flow rates of 0.108–0.200 m^3/s (390–720 m^3/h). Moreover, the pressure characteristic should be stable (an increase in the flow rate decreases the pressure head).

Vibration technical characteristic (the root mean square value of the vibration velocity measured at the bracket) should be not more than 4.5 mm/s on the nominal mode and 7.1 mm/s – at the operating flow rates.

Noise technical characteristics of the pump should be no more than 90 dB at the nominal mode. The equivalent sound pressure level is the average value at 1 m from the unit's contour.

Numerical research methodology using the Ansys CFX software for the designed pump is presented for the case study of the pumping unit ACPA 600–35.

According to the developed design scheme of the pumping unit ACPA 600–35, 3D impeller models have been built using the Solidworks software (Figs. 1 and 2).

The wall layer is made as follows. The thickness of the near-wall layer is 0.05 mm. The sizes of the next layers are increased according to the exponential law with a power of 1.5. The total number of layers is 7. In the region of the wall layer, the cells have a prismatic shape, and in the region of the rest of the flow – they have a tetrahedral shape.

The calculation model has been created using the software package CFX-Pre 19.0. The working fluid is water under the temperature of 20 °C at the turbulent mode. The standard k-ε turbulence model was used to close the Reynolds equations [22].

As a boundary condition at the inlet to the calculation area, the mass flow rate through a single channel of the impeller is set according to the following dependence:

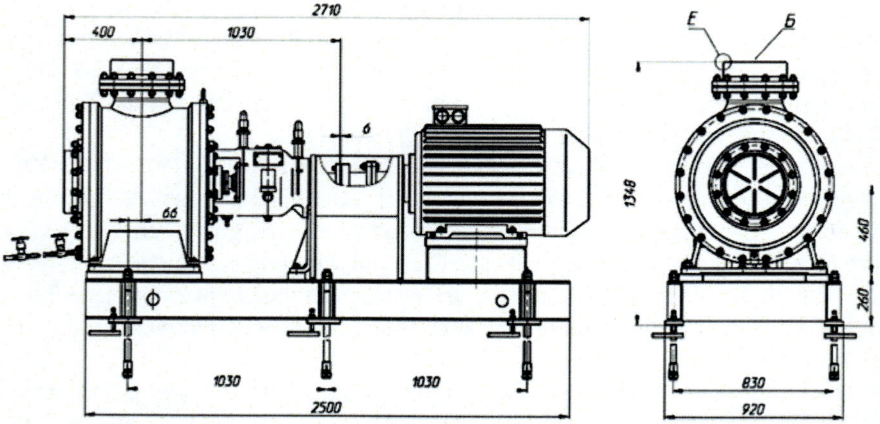

Fig. 1 The dimensional drawing of the pumping units ACPA 600–35 and ACPA 600–35-1

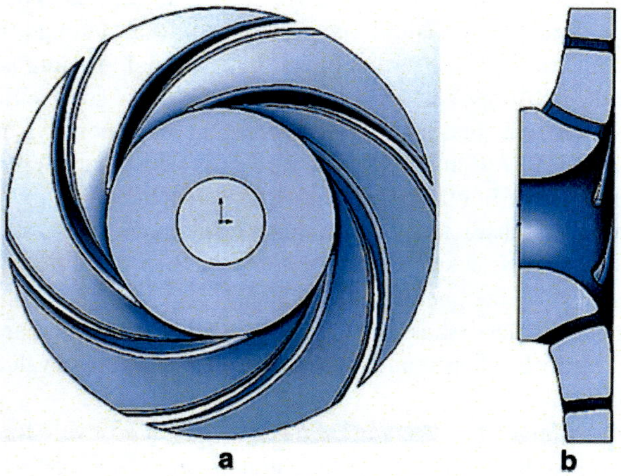

Fig. 2 3D model of the impeller

$$G = \rho \cdot Q / Z, \tag{1}$$

where ρ – density of the working fluid; Q – flow rate at the calculation mode; Z – the total number of the interblade channels.

During the simulations, the standard k-ε turbulence model was chosen as the turbulence model. This particular turbulence model proved itself well in terms of the convergence of the numerical research results with the results of the parametric tests of the pumps in a previous paper [23].

Construction of the finite-volume grid for the impeller was realized using the module ICEM CFD with a total cell number of $2.1 \cdot 10^6$. The following boundary conditions were chosen: at the inlet to the calculation area – fluid flow rate, at the outlet – static pressure of 1 MPa. A rough wall with a roughness of 12.5μm was chosen as the boundary condition on all the walls. The boundary condition at the transition from the rotary element (impeller) to the stator (guide vanes) has been selected, which allows for obtaining the most accurate convergence with the results of parametric tests of the pump.

Finite element analysis has been applied to evaluate eigenfrequencies of rotor's free oscillations for the pumping unit ACPA 600–35 using the software package ANSYS APDL.

The shaft's finite element model has been designed by three-dimensional beam elements BEAM189 based on the Timoshenko model considering shear stresses [24]. The dimensions of the shaft's sections are indicated in Fig. 3. The material of the shaft is steel AISI 431 with the following properties: density – 7750 kg/m^3, Young's modulus – $1.97 \cdot 10^{11}$ Pa, and Poisson's ratio – 0.3.

The local MASS21 element considers the impeller with 6 degrees of freedom (three displacements along axes *X*, *Y*, and *Z*, and three rotations about these axes). As a result, the inertial characteristics of the impeller are as follows: the total mass – 24.37 kg; the coordinate of the gravity center along the *X* axes – 0.036 m; the axial moment of inertia along the *X* axis – 0.411 kg·m^2; the axial moments of inertia along the *X* and *Y* axes – 0.391 kg·m^2.

The shaft is fixed in the radial direction by movable hinged support in the mass center of the spherical roller bearing 22,314-E1-XL (DIN 635–2). Hinged fixed supports have also been considered for a couple of angular contact ball bearings FAG 7315-B-XL-JP.

The Block Lanczos algorithm has been applied to evaluate eigenfrequencies and the corresponding mode shapes. In addition, the sparse matrices method has been used to solve systems of linear equations.

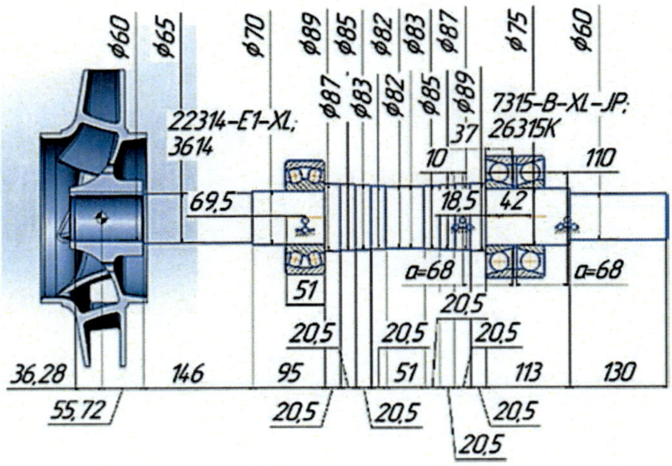

Fig. 3 The design scheme of the rotor

The applied methodology was previously proven for developing centrifugal pumps for other industrial processes [25]. The results of preliminary studies showed high accuracy of the developments compared with the results of parametric tests of the experimental samples of the pumps. In the reviewed studies, the deviation of the results of the numerical research from the results of the parametric tests in the operating range did not exceed 5%, which is acceptable for engineering calculations.

4 Results

As a result of the application of Ansys Workbench software, the first two eigenfrequencies of the rotor have been evaluated: 439 Hz and 1293 Hz, respectively. The corresponding mode shapes are presented in Fig. 4.

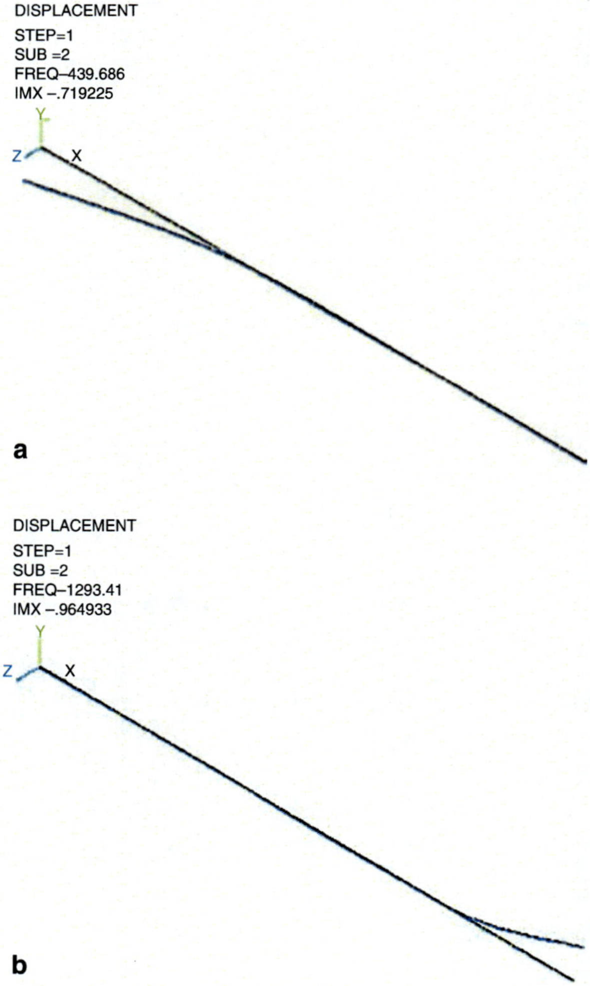

Fig. 4 The first (**a**) and second (**b**) mode shapes

According to ISO 1940-1:2003, the balancing accuracy class G 2.5 has been chosen. The exciting force's frequency resulting from residual imbalances is significantly lower than the first eigenfrequency (25 Hz < < 439 Hz). Therefore, the rotor is rigid.

Figure 5 presents the maximum normal stresses that occurred in hinged support.

The overall normal stress.

$$\sigma_{\Sigma} = \left(\sigma_{My}{}^{2} + \sigma_{Mz}{}^{2}\right)^{1/2} \tag{2}$$

has been calculated considering the most significant values without specifying belonging to a specific section.

The fluid flow has been calculated using Ansys CFX. The obtained results have allowed for obtaining qualitative features of the liquid flow in the pump and making corrections to its structural elements. Relative flow velocities are presented in Fig. 6.

Figure 6 does not contain significant flow separation zones (blue and red colors). This fact indicated the compliance of the flow structure to the design of the interblade channels of the impeller.

The results of the parametric tests are summarized in Table 1. This data has been evaluated for the range of fluid flow from 30 m^3/h to 840 m^3/h (0.05–1.40 of the optimal flow rate).

This data allows concluding about the pressure and energy characteristics of the developed pumping unit ACPA 600–35 (Fig. 7).

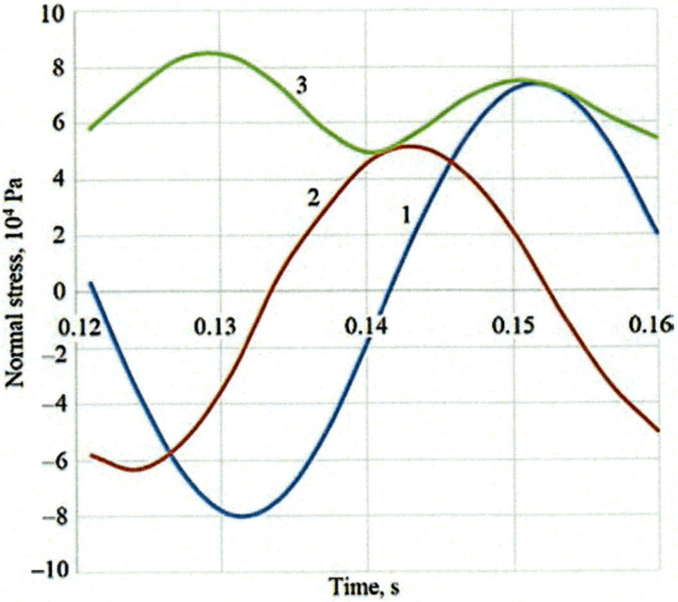

Fig. 5 Time-variable maximum stresses σ_{My} (1), σ_{Mz} (2), and the overall stress σ_{Σ} (3)

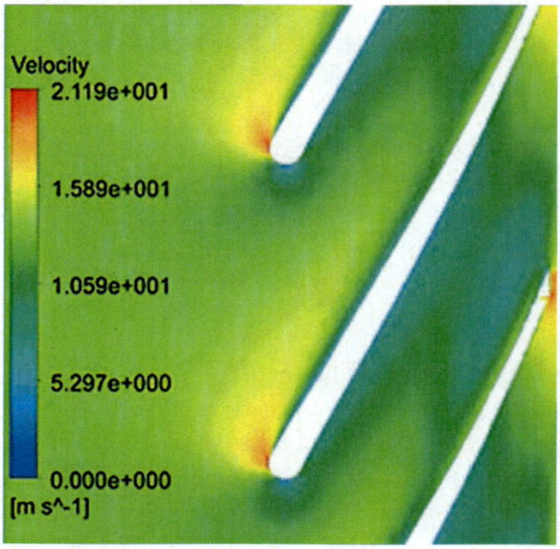

Fig. 6 Velocity distribution contours for the relative flow in the impeller channel

Table 1 Results of parametric tests for pumping unit ACPA 600–35 under the operating speed of 1500 rpm and fluid density of 997 kg/m^3

Flow rate, m^3/h	Pressure head, m	Power consumption, kW	Energy efficiency, %
30	49.91	27.2	13.6
60	47.21	28.4	24.5
120	45.14	33.7	39.6
240	46.85	47.3	58.4
360	47.78	63.8	69.3
480	45.32	75.6	73.8
600	38.44	85.3	69.8
720	27.64	90.8	56.3
840	12.95	93.2	30.0

Such a pressure characteristic is gently sloping, which allows operating the pumping unit even for the flow rates of 5–50% of the nominal mode without the surge phenomenon.

The maximum energy efficiency of 74% is reached at flow rates of about 500 m^3/h. Remarkably, this value is 12% more than the analog TX 600–35.

5 Discussion

The practical significance of the project is in the modernization of the pumping equipment of the operating nuclear power plants (NPPs) of Ukraine (Zaporizhskaya NPP, Rivne NPP, South Ukrainian NPP, and Khmelnytsky NPP) and its use in NPP

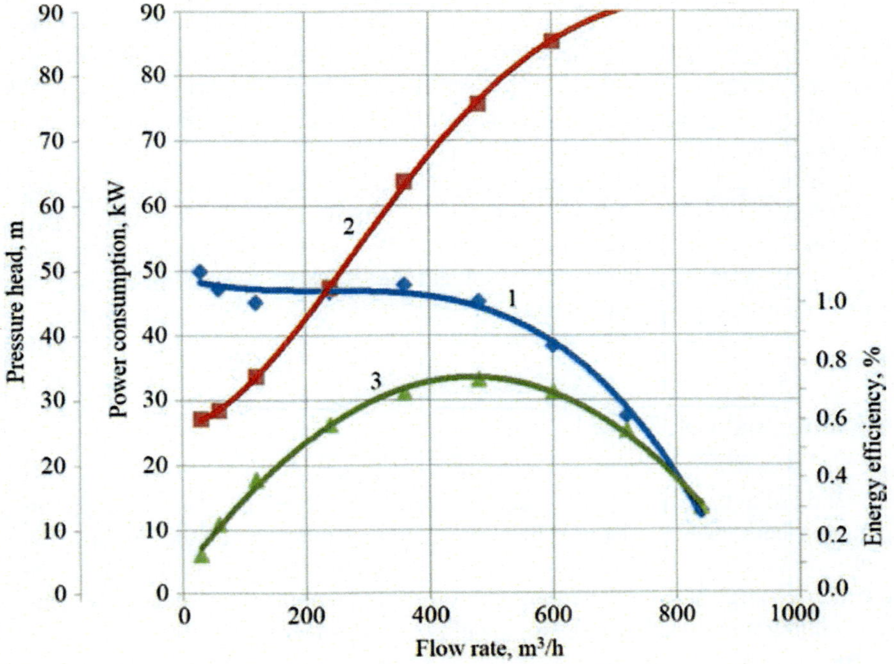

Fig. 7 The main characteristics of the developed pumping unit: (1) pressure head, m; (2) power consumption, kW; (3) energy efficiency, %

units under construction (units 3 and 4 of Khmelnytsky NPP). These NPPs use pumps for cooling systems of the holding pool and the industrial circuit, which were previously imported. Such an import actualizes the risks of pump equipment failure and the emergency calls of Ukrainian NPPs.

In addition, the introduction of developed pumps for the cooling systems of the holding pool and the industrial circuit allows for significantly reduced energy consumption since the efficiency of the designed pumps is up to 12–13% higher than that of existing outdated pumps for this area.

Currently, Sumy State University has established a Scientific Expert Center for Atomic and Power Engineering (SEC APE), which is engaged in introducing developed pumping equipment into the operating processes of Ukrainian NPPs. Currently, SEC APE has received the status of a supplier for the State Enterprise "National Nuclear Energy Generating Company 'Energoatom'" (SE "NNEGC 'Energoatom'"), which is the operator of all those mentioned above Ukrainian nuclear power plants. By the decision of SE "NNEGC 'Energoatom'," the considered pump is included in the list of pumping equipment of Ukrainian NPPs.

Moreover, consideration of the economic feasibility of the work performed is one of the essential stages of pump design. It is planned to reduce the energy consumption of prospective pumps for the cooling systems of the holding pool and the industrial circuit compared with existing analogs.

Simultaneously, the scope of the presented research does not allow for a comprehensively present the project's economic feasibility. It is important to note that the considered paper is only the first in a series of works on increasing the technical level of pumping equipment of Ukrainian NPPs, particularly due to increasing energy efficiency.

Thus, it is planned to comprehensively and fully present the question of the economic feasibility of the development of this pump in one of the following papers. Nevertheless, it can be confirmed that replacing outdated pumping equipment for the cooling systems of the holding pool and the industrial circuit will increase the efficiency of pumping stations by up to 12%.

6 Conclusions

An increase in the capacity of nuclear power plants is directly related to improving pumping equipment's energy efficiency, including main circulating pumps, feed pumps, and pumps for neutral water. Simultaneously, the increased requirements for their ecological safety and the safety of other pumping equipment, emergency feed pumps, and condensate pumps should be completely satisfied.

As a result, the following scientific and practical results have been achieved. Particularly, there have been proposed ways to increase the energy efficiency of rotary machines for nuclear power plants. In addition, detuning from resonance and predictive maintenance have been realized according to international standards.

Increased energy efficiency indicators characterize the developed pumping unit ACPA 600–35 compared with the analog TX 800/70/8-K-2E. The pump's operating flow range is from 360 m^3/h to 660 m^3/h. This fact ensures a wide range of operations with high energy efficiency. At the flow rate of 600 m^3/h, the pressure head is 38.5 m, which is 10% more than the analogs. Also, the power consumption in a range of 63–81 kW allows exploiting the pumping unit with an electric motor by the power of 75–110 kW.

Overall, the improved characteristics of centrifugal pumps for NNPs can be characterized by various indicators. Nevertheless, the primary indicator is the increase in pressure head (both in absolute and percentage terms) for different operating modes. Such production of pumping equipment at industrial enterprises should be the next stage in the energy independence of the European power industry.

Acknowledgments The scientific results were partially obtained within the Ulam NAWA Programme, grant number BPN/ULM/2022/1/00042. The results have also been partially obtained within the research project "Fulfillment of tasks of the perspective plan of development of a scientific direction "Technical Sciences" Sumy State University" ordered by the Ministry of Education and Science of Ukraine (State Reg. No. 0121 U112684). The numerical simulations were partially realized at the Technical University of Kosice within the Joint Ukrainian-Slovak R & D Projects from 2022 to 2023 granted by the Slovak Research and Development Agency (SK-UA-21-0060) and the Ministry of Education and Science of Ukraine. The International Association for Technological Development and Innovations partially supported this research.

Funding This work was funded by the Slovak Research and Development Agency under contract No SK-UA-21-0060 and the project VEGA1/0704/22 granted by the Ministry of Education, Science, Research and Sport of the Slovak Republic. This publication is also a result of the Project implementation: Development of excellent research capacities in the field of additive technologies for the Industry of the twenty-first century, ITMS 313011BWN5, supported by the Operational Program Integrated Infrastructure funded by the ERDF.

References

1. Skalozubov, V.I., Spinov, V.M., Luzhanskaya, A.V., Klimchuk, I.A., Gablaya, T.V.: Emergency management with complete loss of long-term power supply on nuclear power plants with steam drive pumps. J. Autom. Inf. Sci. **52**(9), 51–59 (2021). https://doi.org/10.1615/JAUTOMATINFSCIEN.V52.I9.50
2. World Nuclear Association. https://world-nuclear.org, last accessed 2022/11/14.
3. International Energy Agency. https://www.iea.org, last accessed 2022/11/14.
4. Tkachuk, M.M., Grabovskiy, A., Tkachuk, M.A., Zarubina, A., Lipeyko, A.: Analysis of elastic supports and rotor flexibility for dynamics of a cantilever impeller. J. Phys. Conf. Ser. **1741**(1), 012043 (2021). https://doi.org/10.1088/1742-6596/1741/1/012043
5. Skalozubov, V., Pirkovskiy, D., Spinov, V., Kozlov, I., Komarov, Y.: Reliability of steam-operated emergency feed pumps. In: Tonkonogyi, V., Ivanov, V., Trojanowska, J., Oborskyi, G., Edl, M., Kuric, I., Dasic, P. (eds.) Advanced Manufacturing Processes. InterPartner 2019. Lecture Notes in Mechanical Engineering, pp. 600–607. Springer, Cham (2020). https://doi.org/10.1007/978-3-030-40724-7_61
6. Kindl, V., Byrtus, M., Skala, B., Kus, V.: Key assembling issues relating to mechanical vibration of fabricated rotor of large induction machines. Commun. Sci. Lett. Univer. Žilina. **21**(2), 58–68 (2019)
7. Raj, K.S.S.: Last stage performance considerations in low-pressure turbines of power plants: A case study. J. Eng. Gas Turbines Power. **130**(2), 023004 (2008). https://doi.org/10.1115/1.2795781
8. Song, Y., Xu, R., Yin, J., Wang, D., Xia, S.: Experimental study on the effect of non-uniform flow on the vibration characteristics of the reactor coolant pump. In: The Proceedings of the International Conference on Nuclear Engineering. Book of Abstracts: ICONE, Vol. 27, 1439. The Japan Society of Mechanical Engineers, Tokyo (2019). 10.1299/jsmeicone.2019.27.1439.
9. Jigang, M., Shuhua, Y., Yuefang, W.: Three-dimensional seismic analysis and safety evaluation for nuclear pump of nuclear power plant based on the RCC-M code. Adv. Vibrat. Eng. **10**(4), 343–352 (2011)
10. Wang, M., Wang, X., Zhou, L., He, X., Xin, J.: Influence of boundary effect and high temperature and pressure on reactor coolant pump mode. J. Drain. Irrig. Machin. Eng. **37**(8), 668–672 (2019). https://doi.org/10.3969/j.issn.1674-8530.18.0009
11. Shang, Y.-L., Cai, Q., Zhao, X.-W., Chen, L.: Multi-state reliability for coolant pump based on dependent competitive failure model. Atomic Energy Sci. Technol. **47**(10), 1852–1856 (2013). https://doi.org/10.7538/yzk.2013.47.10.1852
12. Yin, T., Pei, J., Yuan, S., Wang, W.: Analysis of fluid-structure interaction characteristics for impeller of residual heat removal pump. Nongye Gongcheng Xuebao/Transact Chinese Soc. Agricult. Eng. **33**(9), 76–83 (2017). https://doi.org/10.11975/j.issn.1002-6819.2017.09.010
13. Jiao, Z., Cai, L., Zhang, L., Hu, L., Liu, X.: Model for calculating idling of nuclear power coolant pump considering coolant kinetic energy in pipeline. Nuclear Power Eng. **43**(1), 48–51 (2022). https://doi.org/10.13832/j.jnpe.2022.01.0048
14. Zhou, B., Yuan, J., Lu, J., Sun, W., Luo, Y., Pei, J.: Numerical investigation of residual heat removal pumps based on fluid-structure interaction in 1000 MW nuclear power plants. In:

Proceedings of the 6th International Symposium on Fluid Machinery and Fluid Engineering, ISFMFE 2014, Vol. 2014(CP658), 112495. IET Conference Publications, Wuhan (2014). https://doi.org/10.1049/cp.2014.1183
15. Kissick, S.M., Wang, H.: Techno-economic analysis of alternative power cycles for light-water small modular reactors. Int. J. Energy Res. **46**(14), 19467–19479 (2022). https://doi.org/10.1002/er.8516
16. Temiz, M., Dincer, I.: Enhancement of a nuclear power plant with a renewable based multigenerational energy system. Int. J. Energy Res. **45**(8), 12396–12412 (2021). https://doi.org/10.1002/er.6582
17. Zaitsev, S., Kyshnevsky, V., Chichenin, V., Tykhomyrov, A.: Development of methods of gas chromatographic analysis of technological media of the main circulation pumps of a nuclear power plant. East. Eur. J Enterpr. Technol. **6**(6–108), 59–70 (2020). https://doi.org/10.15587/1729-4061.2020.217234
18. Yang, B.: Research on performance optimization of nuclear power system based on improved genetic algorithm. Mechan. Machine Sci. **97**, 245–251 (2021). https://doi.org/10.1007/978-3-030-64690-5_24
19. SE "NNEGC "Energoatom". https://energoatom.com.ua, last accessed 2022/11/14.
20. Nogueira, E.: Applying the concepts of efficiency and effectiveness to analyze the influence of the number of passes in the shell and tubes condenser thermal performance. J. Eng. Sci. **8**(1), F1–F10 (2020). https://doi.org/10.21272/jes.2021.8(1).f1
21. Zhu, W., Wang, F., Wang, Y., Wang, Y.: Technology and management of reactor coolant pumps assembling in pressurized water reactors. Nucl. Power Eng. **40**(6), 144–148 (2019). https://doi.org/10.13832/j.jnpe.2019.06.0144
22. Rode, B.R., Khare, R.: Numerical simulation for design improvement in diffuser of multistage centrifugal pump. In: Saha, S., Sajith, A.S., Sahoo, D.R., Sarkar, P. (eds.) Recent Advances In Materials, Mechanics And Structures. Lecture Notes in Civil Engineering, vol. 269, pp. 153–163. Springer, Singapore (2023). https://doi.org/10.1007/978-981-19-3371-4_14
23. Kondus, V., Gusak, O., Yevtushenko, J.: Improving the efficiency of the operating process of high specific speed torque-flow pumps by upgrading the flowing part design. J. Phys. Conf. Series. **1741**, 012012 (2021). https://doi.org/10.1088/1742-6596/1741/1/012012
24. Li, Z., Wang, Q., Qin, B., Shao, W.: Vibration characteristic analysis of flexible rotor-bearing system subjected to external combined loads. Eur. J. Mech. **96**, 104688 (2022). https://doi.org/10.1016/j.euromechsol.2022.104688
25. Kondus, V., Kalinichenko, P., Gusak, O.: A method of designing of torque-flow pump impeller with curvilinear blade profile. Eur. J. Enterpr. Technol. **3/8**(93), 29–35 (2018). https://doi.org/10.15587/1729-4061.2018.131159

The Relief of the Structured by Nanosecond Laser Stainless Steel Surface Inspection by Sliding Reflection of a Laser Beam

Sergey Dobrotvorskiy, Borys A. Aleksenko, Mikołaj Kościński, Yevheniia Basova, Vadym Prykhodko, Ludmila Dobrovolska, and Jana Mižáková

1 Introduction

When creating multilayer composite materials, the nano and microstructure of the joined surfaces can play a significant role in joining layers. Joining by gluing is a complex physical and chemical process in which both internal factors due chemical composition of the materials and adhesives to be joined and external factors that determine the gluing process in a cold or hot state are involved. All these factors, in general, and the microstructure of surfaces in particular, determine the final strength of the joint. Currently, much attention is paid to the study of surface modification and the creation of LIPSS structures [1]. The reason for this is that LIPSS structures give the surface a unique feature that is radically different from the untreated surface's properties [2], enables a high degree of surface wetting [3], and superior corrosion resistance [4]. However, many studies on the creation of laser-induced periodic surface structures (LIPSS) found that hydrophilic surfaces are formed immediately after irradiation, and only after a long time, from 17 to 50 days can they become hydrophobic [5]. Therefore, the problem arises of controlling changes in the surface structure both during this time and after. At the same time, an effective and inexpensive method for the operational control of the hydrophobicity of structures (LIPSS) has not been developed.

S. Dobrotvorskiy (✉) · B. A. Aleksenko · Y. Basova · V. Prykhodko · L. Dobrovolska
National Technical University "Kharkiv Polytechnic Institute", Kharkiv, Ukraine
e-mail: sergiy.dobrotvorskyy@khpi.edu.ua

M. Kościński
Poznań University of Life Sciences, Faculty of Food Science and Nutrition, Poznań, Poland

J. Mižáková
Faculty of Manufacturing Technologies with the seat in Prešov, Technical University of Košice, Prešov, Slovakia

M. Balog et al. (eds.), *EAI International Conference on Automation and Control in Theory and Practice*, EAI/Springer Innovations in Communication and Computing, https://doi.org/10.1007/978-3-031-31967-9_5

Based on the data on the surface roughness of the material and taking into account the technological properties of the fastened materials, the viscosity (fluid and viscous), its porosity, and tendency to adhesion selected adhesive compositions. It is known that roughness leads to an increase in adhesive strength. Therefore, control over the state of the surface of the material, both before processing and after the process of creating a surface microstructure, plays a huge role. At the same time, the use of specialized equipment and expensive processes, such as electron microscopy, is not a convenient, generally accessible, and technological control method. In this connection, there is a need for a relatively inexpensive, accessible, and simple method for monitoring the surface microrelief, which makes it possible to quickly monitor the surface. Furthermore, the materials to be joined must have a highly developed relief of their surfaces (roughness). Given that the real picture of adhesion phenomena is extremely complex, experimental and analytical research methods are often used to describe them. This makes it necessary to use available technologies to control the surfaces' microrelief to be joined. The presence of a microrelief, increasing adhesion, prevents the displacement of the glued layers and quadratically increases the strength of the joint.

In connection with the development of technologies that allow the creation of a controlled relief of texture [6] and surface structure [7], recently, there has been a need to control the geometry of the microstructure obtained during processing operations. This chapter describes a method for monitoring a hydrophobic surface with a grazing incidence of a helium-neon laser beam.

2 Literature Review

Information about the quality of the surface can be obtained through the use of various optical methods of nondestructive testing [8], such as the light section method [9]; raster method; reflectometric method; cast method; method of comparison with an exemplary part; interference method [10]. Speckle measurement of displacements/strains can also be used to determine the scattering surface [11]. Reference [12] showed that the offset value was 0.6 microns, which satisfied the standard calibration method. The B/D surface ratio results are in good agreement with the stylus-analysis results. To diagnose the roughness parameters of polished optical surfaces [13], a laser infrared (IR) interferometry method was developed, which was successfully tested in production conditions [14]. Modern devices operating on the principle of light section make it possible to measure surface irregularities with a height of 0.8 to 63 microns with an error of readings from 24 to 7.5% [15]. These devices make it possible to determine the parameters Rz, Rmax, and S and photograph microroughness [16]. The work shows that applying the experimental stand with a simple design equipped with a helium-neon laser, where inexpensive components are used, gives good results [17].

Typical methods for measuring surface roughness based on light scattering are total integral scattering [18] and scattering with angular resolution [19]. In general, it is more convenient to use total integral scattering to measure surface roughness than scattering with angular resolution because the evaluation of the results is simple, and the conversion from total integral scattering to root-mean-square roughness height is straightforward. Unlike angle-resolved scatter, data analysis is more complex [20]. Many methods have been developed for measuring the scattering surface roughness with an angular resolution [21] which continues to be improved by scientists [22]. The light scattering method uses a beam of light of known wavelength, which is projected onto a surface with an angle of incidence θ. A beam of electromagnetic radiation can be reflected from the surface in three ways: specular, diffuse, or mixed [23]. Depending on the surface roughness, radiation of a certain wavelength may be reflected specularly, while radiation of a different wavelength may be reflected diffusely. Thus, the sum of specular and diffuse reflection can be used to determine the surface roughness. If the surface is perfectly smooth, the light will be reflected at the same angle as θ, according to the law of reflection. However, if the surface is rough, the reflections will be scattered toward the specular reflection. The scattered light intensity is close to a linear dependence on surface roughness. From a surface with a Ra value of 2 μin, 95% or more, the incident radiation will be reflected specularly. From a surface with a Ra value of 20 μin, 95% or more, the incident radiation will scatter [24].

It is known that optical methods have disadvantages. Some implementations of optical methods require the use of special standards, i.e., they are not fully quantitative. In addition, the fundamental drawback of all optical methods is that the minimum longitudinal (along the surface) size of roughness that can still be detected is limited in order of magnitude by the wavelength of the probing radiation, i.e., it cannot be less than 0.5 ..0.1 μm. However, the light scattering method is relevant and widely used. However, surfaces with the LIPPS structure differ significantly from traditional rough surfaces with a probability distribution of parameters. Therefore, this requires a separate study of the scattering of laser radiation by LIPPS structures.

3 Materials and Methods

3.1 Justification of the Selection of Experimental Equipment

For the experiments, AISI 321 stainless steel plates were cut out. The overall dimensions of the plates were 35x55 mm. The surface, as delivered, had a smooth, mirror-like surface, characteristic of steel. The chemical composition of steel (in weight%): Fe 67.0; Cr 17.0′19.0; Ni 9.0′11.0; Mn <2.0; Si < 0.8; Ti 0.6 = 0.8. To obtain polymodal surface roughness, a laser setup with an IR ytterbium fiber laser (wavelength 1064 nm) and a BM 2500+ biaxial galvanometric scanner was used [25, 26]. This study used laser exposure with a pulse duration of 100 ns, a repetition rate of 20–100 kHz, and a peak power of up to 0.1 MJ in the TEM00 mode. The samples were scanned with a raster at a linear speed of up to 6 mm/s with a parallel line density of up to 1.25 μm. The surface structure after laser exposure was analyzed by atomic force microscopy (Fig. 1). The contact angle of wetting has been determined.

The study of the microrelief of the metal surface was carried out by the proposed method of sliding the reflection of the laser beam. The study was carried out using specially designed laboratory equipment (see Fig. 2a), a system consisting of a helium-neon laser emitter, a bracket with a test plate fixed in a vertical position, and a flat surface with a printed measuring scale.

The measuring plane is perpendicular to the initial laser beam emitted by a radiator with a diameter of 3 mm. The plate under test has an inclination angle of 10° relative to the direction of the laser beam, i.e., the beam falls on the sample at an angle of 10°. The beam irradiates on the surface of the plate, which has a microstructure formed by preliminary laser processing. The reflected beam falls on a plane with a printed measuring scale placed in the path of the beam reflected from the plate.

The experiment is carried out in two stages. During the experiment, the metal studied plate is sequentially installed in two positions relative to the direction of incidence of the laser beam (Fig. 2b, c).

In the first stage of the experiment, the laser beam is directed along the paths of the processing laser. In the second stage of the experiment, the laser beam is directed

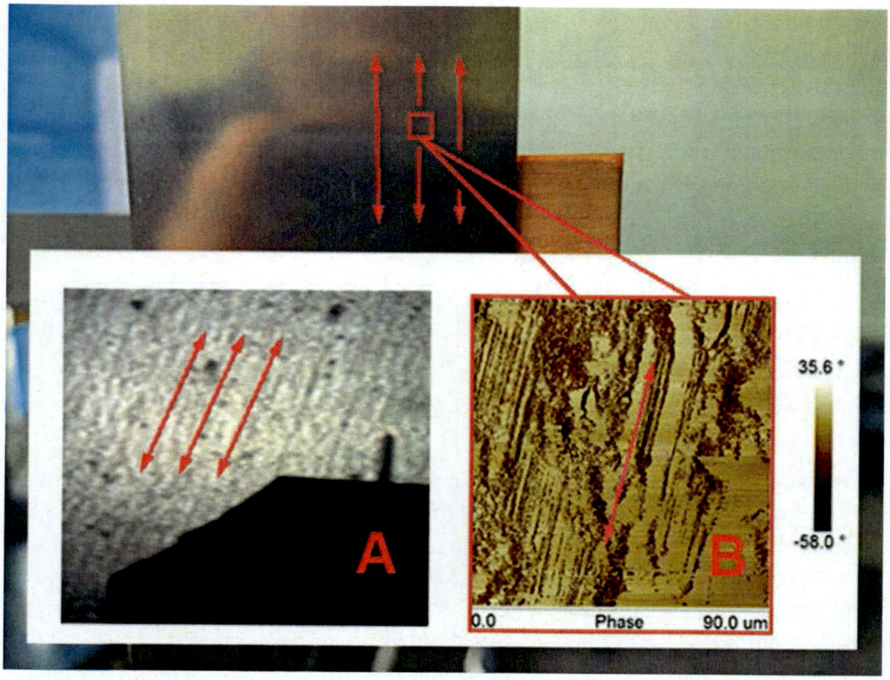

Fig. 1 The studied plate: (**a**) general view of the surface during atomic force microscopy (AFM) scanning; (**b**) distribution of the phase state of the surface; arrows show laser-processing direction

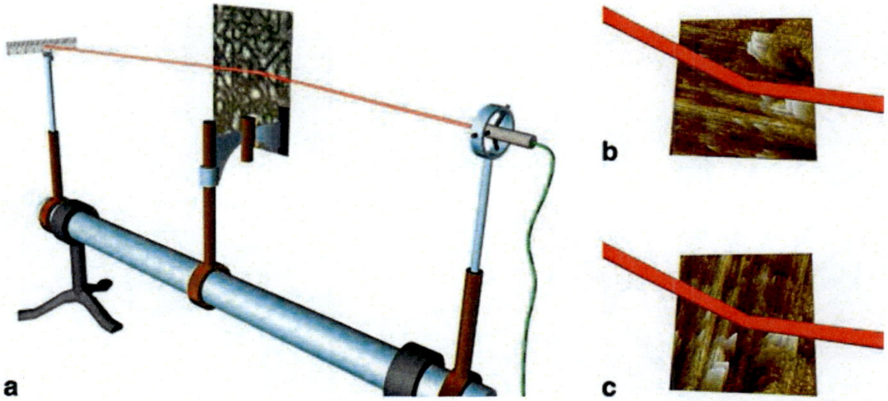

Fig. 2 The practical study: (**a**) is the stand for studying the microrelief by the method of sliding reflection. (**b**) is the incidence and reflection of the laser beam along the microstructure; c – is the incidence and reflection of the laser beam across the microstructure

across the paths of the processing laser (Fig. 1). The change in the direction of incidence of the laser beam on the studied plate was carried out by turning the plate in the holding bracket by 90°. In this case, the deviation of the point of incidence of the laser beam on the studied plate at different stages of the experiment did not exceed $r = \pm 5$ mm.

During the experiment, the geometric characteristics of the reflected laser beam were noted by the light spot formed by the reflected beam on a plane with a printed measuring scale calibrated in millimeters, and the scale value is 1 mm.

A detailed analysis of the surface roughness was performed by atomic force microscopy.

3.2 Beam Model of Light

If we use the ray model of light, then the specular reflection is characterized by the equality of the angles of reflection of all rays at different points on the surface of the object. A beam of parallel rays, after reflection from a flat mirror, changes the direction of its propagation but remains parallel. Specular reflection is obtained if the dimensions of the irregularities on the surface do not exceed the wavelength of light radiation. The light of a point source is reflected from an ideal diffuser according to the Lambert cosine law (Fig. 3a):

$$I = I_i k_d \cos\theta, \tag{1}$$

where $\boldsymbol{I}$– is the reflected light intensity,

$\boldsymbol{I_i}$– is the light source intensity,
$\boldsymbol{k_d}$ – is the diffuse reflectance coefficient,
$\boldsymbol{\theta}$ – is the angle between light direction and surface normal.

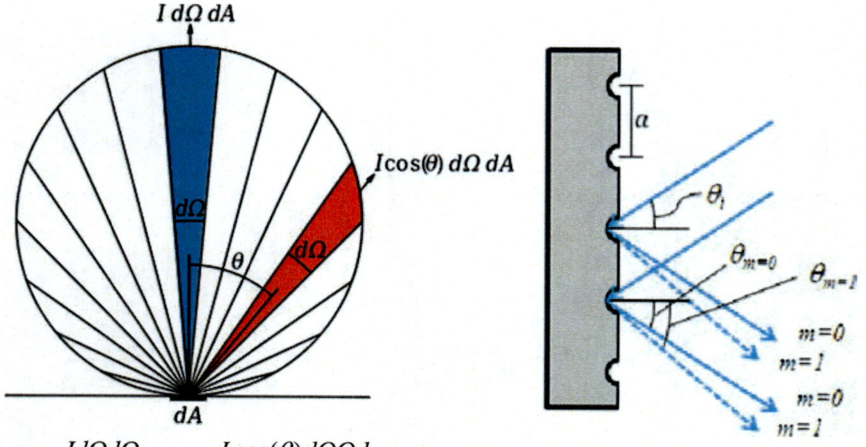

$$I_0 = \frac{Id\Omega d\Omega}{d\Omega_0 dA_0} \quad I_0 = \frac{I\cos(\theta)d\Omega\Omega d}{|d\Omega_0 \cos(\theta)|dA_0}$$

a b

Fig. 3 The *models of reflection:* (**a**) is *the observed intensity of reflection from the Lambertian scattering surface* (dA_0 – is the area of the observation aperture; and dΩ is the *dΩ*– the solid angle forming the aperture from the point of view of the element of the radiating region); (**b**) is the *reflective grating*

If the surface is not perfect, then the amount of light reaching the observer depends on the spatial distribution of the specularly reflected light. Since the distribution is narrow or focused on smooth surfaces and wider on rough ones, the practical result obtained allows us to estimate the roughness parameter of the surface under study.

Diffuse reflection of light occurs from rough surfaces, and the course of the beam reflected from the surface at each point is determined by the angles of incidence and reflection with respect to the plane tangent to the given surface.

The criterion for surface roughness by reflection is the Rayleigh criterion:

$$h < \gamma / 8 \cos\theta, \tag{2}$$

where h is the height of the irregularities;

γ is the wavelength of the incident light;
θ is the angle of incidence on the surface.

The situation for the Lambertian model of diffuse reflection of a scattering surface is illustrated in Fig. 3a: each sector in a circle represents an equal angle dΩ of an arbitrarily chosen size, and for a Lambertian surface, the number of photons per second emitted into each sector is proportional to the area of the sector. The length of each sector is equal to the product of the diameter of the circle and cos θ. The maximum rate of photon emission per unit solid angle occurs along the normal and decreases to zero at $\theta = 90°$. From a mathematical point of view, the brightness along the normal is I photons/(s • m^2 • sr), and the number of photons per second emitted into the vertical sector is dΩdA. The number of photons emitted into the sector at an angle θ per second is:

$$I = \cos\theta \;\; d\Omega dA. \tag{3}$$

In the case of mixed reflection of light, part of the radiation is reflected specularly, and part is diffuse. With such a directionally scattered reflection (Fig. 5b) the axis of the reflected beam of rays is directed in accordance with the law of specular reflection. However, the solid angle of the reflected beam of rays is increased due to the scattering of inhomogeneities of the reflecting surface. Directional-scattered reflection is possessed by matted surfaces of bodies. They scatter light on a surface that has irregularities commensurate in size with the wavelength of the incident radiation.

3.3 The Surface Microrelief as a Reflective Grating

Reflective gratings are obtained by forming parallel strokes on a reflective surface. Thus, light is reflected at different angles, corresponding to different orders and wavelengths. An example of a reflective grating is shown in Fig. 3b. Using a geometric scheme similar to the diffuse and mixed reflection schemes above, and we obtain the reflection grating equation:

$$a\left[\sin\left(\theta_m\right)+\sin\left(\theta_i\right)\right]=m\lambda, \tag{4}$$

where θi is a positive angle, and θm is a negative angle if the incident and diffracted light are on opposite sides of the normal to the grating surface, as shown in Fig. 3b. If these rays are on the same side of the lattice normal, then both angles should be considered positive.

The type of the diffraction pattern is related to the value of the diffraction type criterion, the dimensionless parameter *P*:

$$P = b^2 / l\lambda, \tag{5}$$

where *l* is the distance from the diffraction grating to the projection screen.

Since, in our case, the width of the recess (b) does not exceed 4e-3 [cm], and the wavelength (λ) when using a red laser is about 7e-5 [cm], the diffraction type criterion is <1, which corresponds to the diffraction type by Fraunhofer. In this case, a diffraction pattern of alternating maxima and minima appears from slit b on the observation screen. Therefore, the Fraunhofer diffraction minima condition is a function of the diffraction angle:

$$b\sin\theta = k\lambda. \tag{6}$$

When this condition is met, the difference in the path of the waves emitted by the elementary zones adjacent to the edges of the slot will be equal to an integer number of wavelengths.

The number of main maxima N that can be observed at given values of the wavelength λ and period of the diffraction grating d:

$$N = 2k_{\max} \left(\text{or } N = 2k_{\max} + 1 \text{ counting the central maximum}\right). \tag{7}$$

The intensity distribution in the diffraction spectrum of each slit is determined by the direction of the diffracted rays, and the diffraction patterns created by each longitudinal recess will be the same.

3.4 Visual Evaluation Criterion

The nature of the reflection is estimated by the projection of the reflected beam visually. According to the Rayleigh criterion, images of two nearby identical point sources or two nearby spectral lines with equal intensities and identical symmetrical contours are resolvable (separated for perception) if the central maximum of the diffraction pattern from one source (line) coincides with the first minimum of the diffraction pattern from another. When the Rayleigh criterion is fulfilled, the intensity of the "dip" between the maxima is 80% of the intensity at the maximum, which is sufficient to resolve the maxima of the diffraction pattern. If the Rayleigh criterion is violated, then one line is observed. During the experiment, the light gaps in the intensity of the reflected beam are not visually resolved, which indicates either that the Rayleigh criterion is not met, or that they are completely absent.

4 Results and Discussion

Before carrying out the experiment to control the surface microrelief by the method of sliding reflection of a laser beam, the microrelief of the surface under study was studied using force atomic microscopy. In the topography of the surface, a line structure is clearly distinguished (Figs. 1 and 4a), due to the alternation of lines and the overlap of laser spots. The bottom of the lines is the least susceptible to modification. This is clearly seen in (Fig. 4b) when examining the phases, where the surface of the bottom of the rows looks homogeneous. The protrusions between the rows are clearly pronounced, formed by the displacement of the melt and the shock wave from the expansion of the plasma. The phase composition of the ledges is significantly different from the bottom (Fig. 4b).

In the process of conducting a practical experiment on controlling the surface microrelief by the method of sliding reflection of a laser beam, a difference was

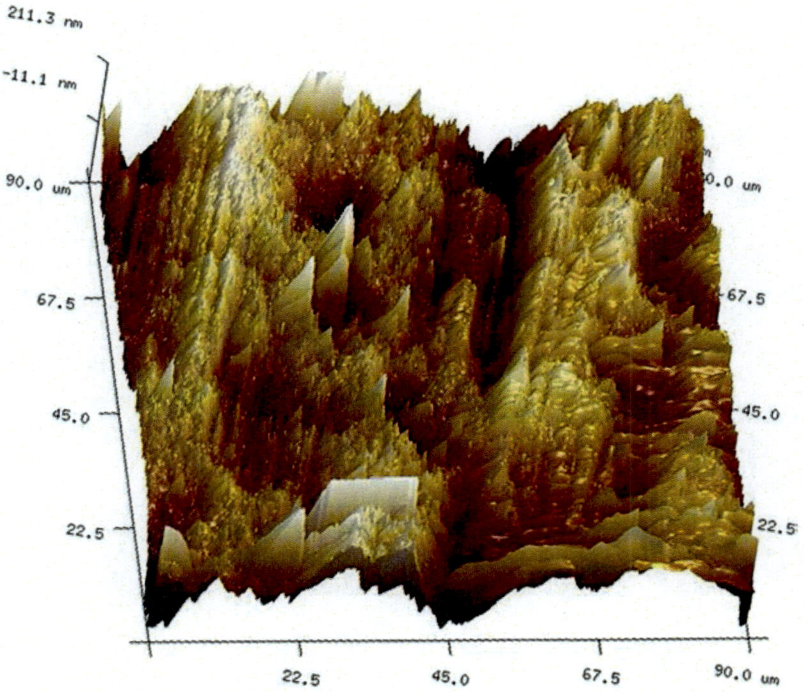

Fig. 4 The height structure of the microrelief of the studied plate according to atomic force microscopy data

found in the shape and size of the light spot formed by the reflected beam on a plane with a printed measuring scale (see Fig. 5).

In the first stage of the experiment, the laser beam is directed along the paths of the processing laser. Diffuse reflection of light is noted (see Fig. 5a), indicating surface roughness without pronounced ordered microstructure elements in the beam incidence plane. The focal spot, characteristic of reflection from a mirror surface, is visible. The general deformation of the projection occurs due to the acute angle of incidence of the beam on the studied plate and errors in the relative positioning of the plate under study and the surface with the measuring scale when installed on the stand.

In the second stage of the experiment, the laser beam is directed across the paths of the processing laser. A change in the scattering angle of the reflected laser beam incident in the direction transverse to the scanning trajectory of the laser beam is noted. In this case, the rays, which before reflection went in a parallel beam, after reflection are scattered in a plane perpendicular to the processing plane. Thus, a mixed directionally scattered reflection is observed. The axis of the reflected beam of rays is visible, directed following the law of specular reflection, while the solid angle of the reflected beam of rays is increased (α, Fig. 5b) due to scattering by a reflecting surface with an ordered microrelief (Fig. 6).

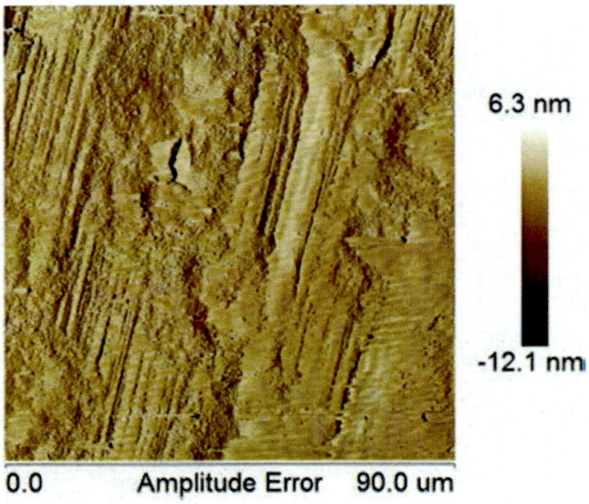

Fig. 5 The amplitude error layer of the atomic microscopy data by the microstructure of the studied plate

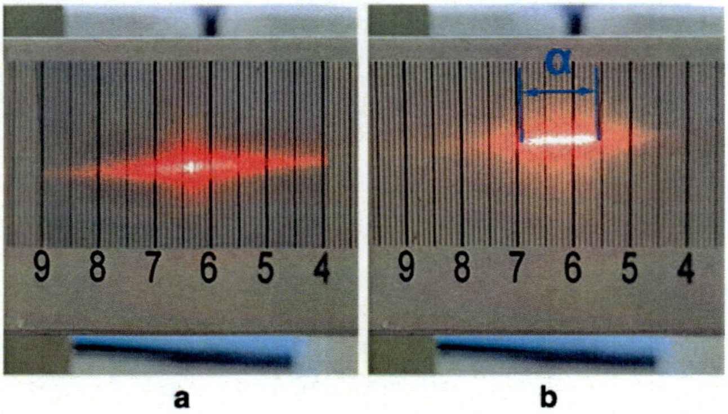

Fig. 6 Projection of the reflected spot during irradiation of the plate under study with a nanosecond laser: (**a**) along the processing trajectory; (**b**) across the processing path

A wide scattering angle in a single plane is possible only if the reflecting surface has such an ordered microstructure in which the normal to each point of the surface lie on the same axis and are rotated relative to each other at different angles, ranging from 0 to 1/2 of the scattering angle. Such a surface is a set of hollows and hills elongated along one axis, positioned parallel to each other, and located perpendicular to the axis of incidence of the beam. In the described case, the irregularities of the microstructure are elongated along the laser processing axis vertically.

Thus, the method of grazing reflection of a laser beam fully correlates with the data of force atomic microscopy (Fig. 4) and confirms that the surface microrelief is a complex structure that includes a set of depressions located rather orderly. This is evidenced by the observed pattern of uniform scattering (Fig. 5a), observed at the first stage of the study. In this case, the relief alternates elongated recesses and hills, which is confirmed by the directionally scattered nature of the reflection observed at the second stage of the study (Fig. 5b). The recesses are parallel and present on the entire machined surface, which causes directional scattering in a plane transverse to the scanning direction. The measurements are made visually up to the limit of the visual resolution of the projection of light spots on a matte measuring scale and ensured by the quality of the scale used, calibrated in millimeters.

The absence of a diffraction pattern is caused by a small number of parallel depressions on the surface under study, their low density, and rather large, on the order of 10 μm or more, their transverse dimensions. The described surface is characterized by a wetting angle of 67.2°, determined experimentally at the minimum laser radiation power. The contact angle of the untreated surface was 52.8°. The absence of pronounced diffraction maxima indicates the absence of a strictly ordered microrelief with a depression width of about 10 μm or less, which also correlates with microscopy data. With a greater order of the microstructure and a decrease in the relief of linear parallel depressions, the reflection pattern of a beam incident across the scanning path (Fig. 5b) will presumably be diffractive, and the wetting angle will increase. Therefore, in order to carry out operational control, it is necessary to once study the change in the shape of light spots on the control ruler and subsequently use them as calibration.

The results of our study are consistent with those obtained by other scientists [27], who have studied the correlation between 3D surface roughness and speckle structure using a helium-neon laser and a spatial filter. In this case, also an experimental setup using inexpensive components was applied. The relevance and correctness of the direction of research we have chosen, the adequacy of the results obtained, and the correctness of the conclusions are also confirmed by other researchers [28], who noted that this method provides the possibility of high-speed noncontact measurement without physical contact with the surface under study. The speed and efficiency of big-area sampling are also emphasized, as well as the convenience of using an optical instrument in a mass production environment.

5 Conclusions

The proposed unique technique is universal. The specified control can be carried out to determine the compliance of the surfaces of the parts to be joined with the technological requirements before their connection, during the honing process, and before applying adhesive and paint coatings. Thus, using the proposed method, it is possible to improve the quality of the connection, determine the suitability of the surface for applying adhesive compositions or protective coatings, and optimally position the parts to be joined during the gluing process.

The nature of the reflected light spot indicates a high uniformity of the finely dispersed structure on the surface of the sample under study. At the same time, the relief has a pronounced pattern, representing a set of similar rectilinear depressions on the surface, relatively equidistant from each other and arranged in parallel. Thus, it is possible to state both sufficient accuracy and repeatability of the process of forming a microrelief using laser processing and high information content of the technology for monitoring the microrelief of a metal surface by the method of sliding reflection of a laser beam, with its simultaneous simplicity, low cost, and availability.

This method can be used to control changes in the structure of the surface of stainless steel during its aging and changes in the degree of its hydrophobicity.

The proposed technology allows you to successfully control such parameters of the microstructure of the material surface as its uniformity, roughness, and directionality of the microrelief pattern.

Acknowledgments This work was supported by the Slovak Research and Development Agency under contract No. APVV-19-0590, and also by the projects VEGA 1/0700/20, KEGA 022TUKE-4/2023 granted by the Ministry of Education, Science, Research and Sport of the Slovak Republic.

References

1. Simões, J.G.A.B., Riva, R., Miyakaw, W.: High-speed laser-induced periodic surface structures (LIPSS) generation on stainless steel surface using a nanosecond pulsed laser. Surf. Coat. Technol. **344**, 423–432 (2018). https://doi.org/10.1016/j.surfcoat.2018.03.052
2. Moldovan, E.R., Doria, C.C., Ocaña, J.L., at all.: Wettability and surface roughness analysis of laser surface texturing of AISI 430 stainless steel. Materials. **15**(8), 2955 (2022). https://doi.org/10.3390/ma15082955
3. Liu, Z., Niu, T., Lei, Y., Luo, Y.: Metal surface wettability modification by nanosecond laser surface texturing: A review. Biosurf. Biotribol. **8**(2), 95–120 (2022). https://doi.org/10.1049/bsb2.12039
4. Kedia, S., Bonagani, S.K., Majumdar, A.G.: at all: Nanosecond laser surface texturing of type 316L stainless steel for contact guidance of bone cells and superior corrosion resistance. Coll. Interf. Sci. Commun. **42**, 100419 (2021). https://doi.org/10.1016/j.colcom.2021.100419
5. Dobrotvorskiy, S., Basova, Y., Dobrovolska, L., Popov, V., Mounif, A.S.Y.: Creation of a Superhydrophilic surface with anti-icing properties for X18H10T stainless steel using a nanosecond laser. In: Cioboată, D.D. (ed.) International conference on reliable systems engineering (ICoRSE) - 2022. ICoRSE 2022. Lecture notes in networks and systems 534, 172–184. Springer, Cham (2023). https://doi.org/10.1007/978-3-031-15944-2_17
6. Gregorčič, P., Šetina-Batič, B., Hočevar, M.: Controlling the stainless steel surface wettability by nanosecond direct laser texturing at high fluences. Appl. Phys. A Mat. Sci. Process. **123**(12), 766 (2017). https://doi.org/10.1007/s00339-017-1392-5
7. Conradi, M., Drnovšek, A., Gregorčič, P.: Wettability and friction control of a stainless steel surface by combining nanosecond laser texturing and adsorption of superhydrophobic nanosilica particles. Scientif. Rep. **8**(1), 7457 (2018). https://doi.org/10.1038/s41598-018-25850-6
8. Nevlyudov, I.S., Zharikova, I.V., Perepelitsa, I.D., at all.: Analysis of methods for controlling the roughness of substrates for electronic products [Analizmetodovkontrolyasherohovatostipodlozhekdlyaizdeliyelektronnoytehniki]. East. Eur. J. Enterpr. Technol. **2**(5), 25–30 (2014) (in Russian)
9. Sprague, R.A.: Surface roughness measurement using white light speckle. Appl. Optics. **11**(12), 2811–2816 (1972). https://doi.org/10.1364/AO.11.002811
10. Zhenrong, Z., Jing, Z., Peifu, G.: Roughness characterization of well-polished surfaces by measurements of light scattering distribution. Optica Applicata. **40**(4), 811–818 (2010)
11. Sirohi, R.S., Metrology, S.: 1st. Academic Press, New York (1993)

12. Balamurugana, R., Muruganand, S.: Displacement measurement and study of surface roughness using laser speckle technique. J. Inst. Industr. Appl. Eng. **3**(2), 96–99 (2015). https://doi.org/10.12792/JIIAE.3.96
13. Denisov, D.G.: Measuring the parameters of roughness of ground and polished optical surfaces using high-precision methods of laser interferometry. Adv. Appl. Phys. **5**(4), 393–411 (2017) (in Russian)
14. Abdulkadyrov, M., Semenov, A.: Modern ways of production of astronomical and space mirrors. Photonics. **3**(51), 62–79 (2015)
15. Makeev, A.: The application of laser methods for monitoring of surface microrelief. Intexpro Geo-Siberia. **5**(1), 43–50 (2015) (in Russian)
16. Anurev, V.I.: Handbook of the designer-machine builder. In 3 volumes. 8th edn. Publishing house "engineering". Moscow. (2001) (in Russian)
17. Whitley, J.Q., Kusy, R.P., Mayhew, M.J., at all.: Surface roughness of stainless steel and electroformed nickel standards using a HeNe laser. Optics Laser Technol. **19**(4), 189–196 (1987). https://doi.org/10.1016/0030-3992(87)90065-X
18. Bennet, J.M., Mattsson, L.: Introduction to surface roughness and scattering. 1st end. DC, Optical Society of America, Washington (1993)
19. Bennett, J.M.: Recent developments in surface roughness characterization. Measurement Sci. Technol. **3(12),001**, 1119–1127 (1992). https://doi.org/10.1088/0957-0233/3/12/001
20. Lu, R.-S., Tian, G.Y.: On-line measurement of surface roughness by laser light scattering. Measurement Sci. Technol. **17**(6), 1496–1502 (2006). https://doi.org/10.1088/0957-0233/17/6/030
21. Ohlídal, M., Unĉovský, M., Ohlídal, I., at all.: Determination of the basic parameters characterizing the roughness of metal surfaces by laser light scattering. J. Modern Optics. **46**(2), 279–293 (1999). https://doi.org/10.1080/09500349908231271
22. Cheng, C., Qi, D., Teng, S., at all.: Extraction of height probability density of random rough surfaces from the central δ-peak of angle-resolved light scattering using the optical inversion algorithm. Appl Phys. Lett. **81**(11), 2124–2126 (2002). https://doi.org/10.1063/1.1506950
23. Suryawanshi, V.B., Inamdar, K.H.: Method of surface roughness evaluation using ruby laser beam. Int. J.Eng. Res.Applic. **2**(3), 1512–1515 (2012)
24. Vincent, D., Lit, J.W.Y., Lavigne, P., at all.: Optical guided-beam splitter: An experimental study. Appl. Optics. **19**(8), 1285–1293 (1980). https://doi.org/10.1364/AO.19.001285
25. Sotnikov, V.T., Dobrotvorsky, S.S., Zapechelnyuk, E.F.: Threshold effects in the emission of charged particles from the surface under the influence of laser radiation. Ukrainian Phys. J. **11**(5), 744–748 (1986)
26. Sotnikov, V.T., Dobrotvorsky, S.S., Zapechelnyuk, E.F., Dobrotvorskaya, M.V.: Self-oscillations of the emission of charged particles from the surface of optical transparent media with nonlinear absorption of laser radiation. Surf. Phys. Chem. Mech. **9**, 103–110 (1986)
27. Jayabarathi, S.B., Ratnam, M.M.: Comparison of correlation between 3D surface roughness and laser speckle pattern for experimental setup using he-ne as laser source and laser pointer as laser source. Sensors. **22**(16), 6003 (2022). https://doi.org/10.3390/s22166003
28. Suryawanshi, B.K., Inamdar, D.K., Suryawanshi, V.B.: Method of surface roughness evaluation using ruby laser beam. Int. J. Eng. Res. Appl. **2**(3), 1512–1515 (2012)

Part II
IoT, IIoT, Optimization, Technologies

Increasing the Speed and Performance of the Drupal CMS Server for Industrial IoT Technologies

Viktor Satsyk, Dagmar Cagáňová, Oleksandr Reshetylo, Oleg Zabolotnyi, and Anatolii Tkachuk

1 Introduction

The international company IDC conducted a global study of the data market for 2020 and made a forecast for the next 5 years. According to these forecasts, global Internet traffic will increase annually by 23%. The fastest-growing data segment is the Internet of Things. IDC specialists also note the growing popularity of clouds for storing and using data. As a result, the average annual growth rate of storage volume worldwide is forecast at 19.2% [1].

A cloud server becomes an effective tool for collecting and storing a large amount of information. Cloud servers allow you to store and process large amounts of data without significant investments in infrastructure, access them from anywhere in the world, and not worry about their storage [2]. This is especially relevant during the implementation of the Industrial IoT (IIoT) paradigm of Industrial 4.0.

WEB sites create a high load on the server. By reducing the load on the server created by the site – costs for hosting, VDS, and renting a dedicated server – are minimized accordingly [3]. If the site's goal is a fast user experience, like Google, then you should focus on simplicity and provide the resources to deliver pages in 1.5 s or less. If the goal is to have a visually appealing site, then a rich theme with lots of interactive JavaScript and longer loading times may be acceptable [4].

For nonstandard, time-consuming, highly loaded services, you should use Drupal. According to the resource [5], 54% of the sites that were created on CMS Drupal got into the "Top Best Sites" rating. CMS Drupal site development is the optimal solution as it is a free open-source system, with high power, security, and

V. Satsyk · O. Reshetylo · O. Zabolotnyi (✉) · A. Tkachuk
Lutsk National Technical University, Lutsk, Ukraine

D. Cagáňová
Faculty of Management, Comenius University Bratislava, Bratislava, Slovakia

M. Balog et al. (eds.), *EAI International Conference on Automation and Control in Theory and Practice*, EAI/Springer Innovations in Communication and Computing, https://doi.org/10.1007/978-3-031-31967-9_6

reliability. The main advantages of Drupal include ease of installation, simple engine management through administration, multi-functionality, and expansion of capabilities due to the installation of additional modules from the official website [6]. In addition, CMS Drupal, widely used in the world, has full Ukrainian localization of the system core of all current versions [7].

2 Literature Review

This paper discloses the practical and effective methods of reducing the load on the server of functioning and newly created sites based on CMS Drupal [8]. For already functioning sites, the following ways of reducing the load on the server are recommended: application of simple methods (neutralization), use of built-in and additionally installed modules, configuration and maintenance of Drupal, and use of various server configuration functions.

In order to increase the speed and performance of the server of newly created sites based on the CMS Drupal, four mandatory steps must be taken:

1. apply the built-in optimization of Drupal based on caching;
2. use the possibility of additionally installed modules (installation of modules: Authcach; Cache Router, CSS Gzip, and JavaScript Aggregator);
3. configure Drupal configurations and maintenance (reducing user session storage time, reducing the number of site logging messages;
4. use server configuration functions (in each case, server optimization settings will be different, as eAccelerator installation in Windows and Linux are significantly different).

The step-by-step theoretical development and practical implementation of the algorithm of the data transformation system from local devices to the WEB interface based on CMS Drupal are disclosed [9], and the authors [10] describe the operation of the system in more detail. The technical capabilities offered by CMS Drupal and its support are described in detail in [11]. Special Drupal modules for artificial intelligence and their capabilities are presented on the site [12]. The analysis of the most common methods of load balancing for Web servers, which increase their speed and performance and ensure the stability of their work in IIoT, is carried out [13]. The existing methods of balancing the load of cloud systems have limited use, and at the moment, there is no universal system for balancing the load of networks in IIoT technologies [14]. Methods for increasing the performance and speed of servers and ensuring data security are considered in [15]. It is recommended to apply quite practical steps to increase the productivity of high-load projects using CMS Drupal [16].

3 Researches Methodology

A web server is a server that accepts HTTP requests from clients, usually web browsers, and issues HTTP responses to them, usually along with an HTML page, image, file, media stream, or other data. The software that performs the functions of the Web server and the hardware on which this software product works are called Web servers [17]. In our research, we will use the most common web server, Apache.

The HTTP protocol in the IIoT concept is implemented using client-server technology and works on the request-response principle without saving the state [18]. The purpose of the request is one or another resource, which is determined by a single resource identifier – URI (Uniform Resource Identifier), HTTP uses one of the varieties of URI – URL (Uniform Resource Locator) – a universal pointer to the resource, which, in addition to information about the resource, determines its physical location, Fig. 1.

Web server load parameters: Processor – processor time (CPU), or the time of information processing by the processor (in our case: 6; 3.0 GHz); RAM (memory) – the level of RAM usage by the server (in our case: 8GB); Hard disk (disk) – the level of disk memory usage (in our case: 500Gb; 7200rpm; 64MB); Network interface (in our case: 10/100/1000).

The disk subsystem can be loaded when the application is working with files. In addition, the disks can be loaded by database work. Disk load is measured in Mb/s. However, for this work, we will translate these parameters into percentages. There are two types of load on the disk system: Actual DISK READ (the load that occurs when reading) and Actual DISK WRITE (the bag that occurs when data is written). Despite the fact that Actual DISK READ and Actual DISK WRITE create different functional loads in different time periods, and as practice shows, their value is small, in the study, we will use the Actual DISK READ parameter. Each time a user accesses the site, the server receives information that it needs to process – perform queries to the database and provide the user with one of the site's pages as a response. The amount of such information that the server is able to process depends on the RAM on the hosting server. This parameter is measured in Mb. For convenience in

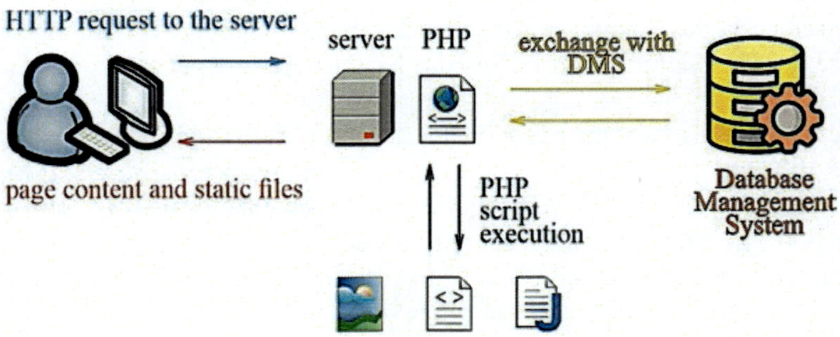

Fig. 1 Data exchange between the client and the server using the HTTP protocol in the IIoT concept

our work, we will translate the RAM load into %. Knowing that the server has 8GB of RAM, let's take this parameter as 100%.

The network interface is the amount of information the server can process in one second, which is measured in Mb/s. Therefore, high network traffic is not a problem in itself. But close-to-peak indicators indicate the need to scale the server in the future. For example, an average traffic of 95 Mb/s on a 100 Mb/s interface will soon mean that the current server will be insufficient. However, in this work, we will not consider this parameter because this site does not access third-party resources on the back-end side, and therefore the load on the network interface does not exceed 3%.

Methods of server load analysis:

- generalized – using the online tool PageSpeed Insights,
- detailed – using access to the server via SSH (which provides the prediction of the occurrence of possible problems in certain functions). SSH is a protocol that uses a client-server model to authenticate remote systems and encrypt data over remote access.

Three pages are selected for server load analysis:

- main page,
- blog list page,
- one blog page.

We will use a scale from 0 to 100 points to evaluate the speed of the site. Where zero means high loads, long page loading, and 100 means the opposite. This evaluation option uses PageSpeed Insights statistics. A single-threaded ApacheBench command-line computer program executed 1000HTTPGET requests with 100 concurrent requests completed in 204.6 s; the server responds to 4.89 requests in one second (Requests per second, which means that the server processed 0.489% of requests in one second Fig. 2)

Analyzing this graph, we can say that the distribution of the load on the main page server without optimization according to three parameters is not carried out evenly. The smallest (6…18%) load on the server by the main page is caused by the disk system of the server (disk). The maximum (13…56%) load on the server by the main page is observed in the used memory resource. The average (10…37%) load on the server by the main page is displayed during CPU usage. It should be noted that the speed of information processing by the server depends on its percentage load.

The blog list page was tested for the load. For this, the command was used: ab-n 1000-c 100http://.

On this page, 1000 HTTPGET requests will be executed in 181.3 s. The server responds to 5.51 requests in one second. This means that the server processed 0.51% of requests in one second. The data collected after monitoring the results of the "htop" and "iotop" commands are shown in Fig. 3.

In order to check the load of one blog page in the console, execute the following command: ab-n 1000-c 100http://. After running this command, the load on CPU

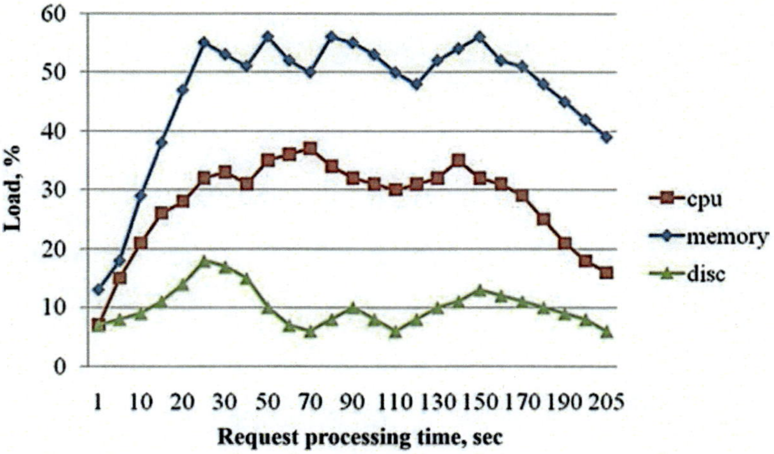

Fig. 2 Graphic representation of the percentage load of the main page of the site without optimization on the server depending on the processing time of 1000 HTTPGETs for 100 parallel requests

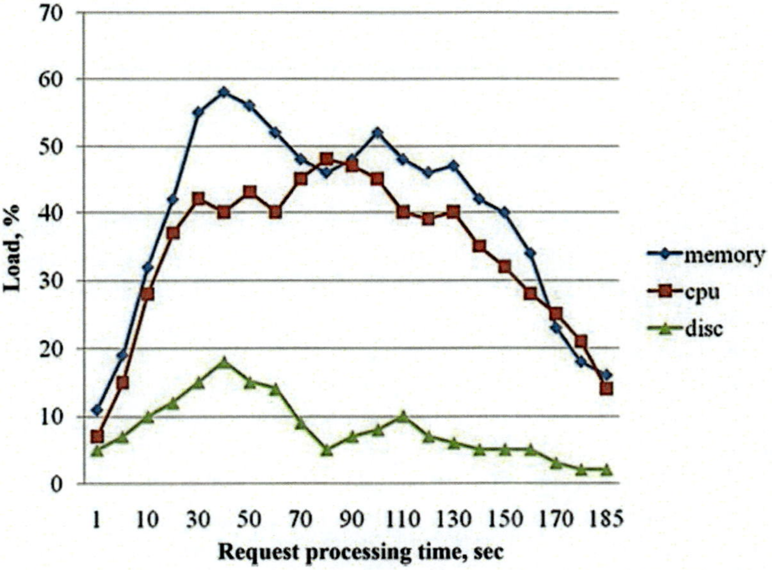

Fig. 3 Graphic representation of the percentage load of the blog list page of the site without optimization per server depending on the processing time of 1000 HTTPGET requests for 100 parallel requests

will be from 88% to 92%, and on mysql from 32% to 78%. However, after some time, the error “Benchmarking (be patient) apr_poll: The timeout specified has expired (70007)” will be displayed in the console. The process ended without processing even 100 HTTPGET requests. This means that the page is too heavy and needs immediate optimization.

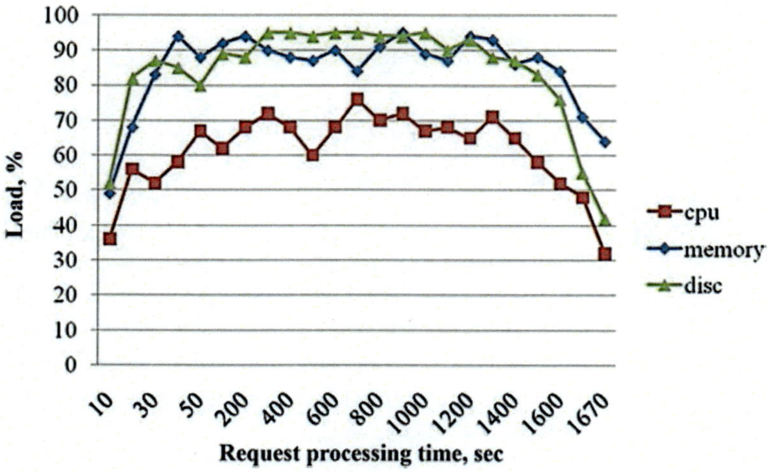

Fig. 4 Graphic representation of the percentage load of one blog site page without optimization per server depending on the processing time of 100 HTTPGET requests for 1 parallel request

Therefore, at the moment, we will reduce the number of requests to 100 and parallel requests to 1. The command in the terminal will look like this: ab-n 100-c 1http://. The server processed these requests in 1662.7 s, and the server answered 0.06 requests in one second. This means that the server processed 0.06% of requests in one second. The data collected after monitoring the results of the "htop" and "iotop" commands are shown in Fig. 4.

Analyzing the graph (Fig. 4), we can say that the distribution of the load on the server page of one blog without optimization according to three parameters is not carried out evenly. The least (10...75%) load on the server of one blog page is caused by the processor (CPU). The maximum (50...97%) load on the server of one blog page is observed when using the RAM resource (49...94%) (memory) and when using the disk system (52...95%) (disk). A generalized test of the load on the server by the statistical tool PageSpeed Insights (simplified) showed that the minimum load on the server without optimization is provided by the blog list page – 53 points. At that time, referring to the page of one blog carries the maximum load on the server – 28 points). The average between them is the main page – 47 points. This load on the server can be explained based on the structure of each element of the site.

Checking the load on the server using access to the server via the SSH protocol (detailed) showed the regularity of the load distribution. The minimum load on the server without optimization creates the blog list page (2...18%) and the main page (6...18%) using minimal server disk system resources. List of blogs: The maximum load on the server is created by using each of the pages of the site (one blog – 10...75%; list of blogs – 11...58%; the main one – 13...56%) using maximum memory resources; The page of one blog creates the maximum load both in the use of the RAM resource (memory) (49...94%) and when using the disk system (disk) (52...95%).

Research Options

Option 1. Built-in (Basic) Drupal Optimization

(Enabling caching, minifying, and combining CSS and JS files, changing the quality of loaded images).

Option 2. Optimization Using Additionally Installed Drupal Modules

(Installing: Advanced CSS/JS Aggregation module to collect and optimize CSS and JS site files; page caching module – Boost; Clean block markup and Clean panels markup to disable some HTML5 elements; ImageMagick module to reduce image quality; execution of the "drush" command "image-flush – all," which will delete old images and generate new ones).

Option 3. Server Optimization.

(Installing PHP eAccelerator, enabling MySQL cache and size 64 megabytes).

Option 4. Optimization of Configurations and Drupal Hooks

(Database normalization; optimization of Drupal configurations; optimization of Drupal hooks).

4 Results

4.1 The Generalized Method of Analyzing the Load on the Server

Option 1. Built-in (Basic) Drupal Optimization

According to the PageSpeed Insights tool, after performing basic Drupal optimization, the main page, blog list page, and single blog scores increase by 16, 16, and 23 points, respectively. However, after this optimization, the site has not become more user-friendly. Pages still take a long time to load.

Option 2. Optimization with the Help of Additionally Installed Drupal Modules

However, after optimizing Drupal with the help of modules, these pages received an additional 20, 14, and 35 points, respectively. This optimization method can be classified as one of the most effective. After all, all pages received the "Good" category according to the scale of the PageSpeed Insights tool. All pages of the site load quickly except the pages with one blog. It should be noted that the results vary depending on the installed modules.

Option 3. Server Optimization

The next step was to configure the server. As mentioned above, this method may not be available to most Drupal site owners. After all, hosting does not provide access to change server configurations. In this work, this service was provided by the

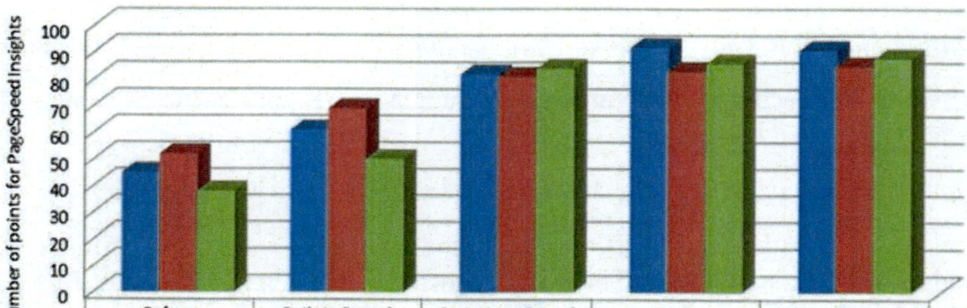

	Before optimization	Built-in Drupal optimization	Optimizing Drupal with modules	Server optimization	Optimization of configurations and Drupal Hooks
Main page	45	61	82	92	91
List of blogs	52	69	81	83	85
One blog	38	50	84	86	88

Fig. 5 Dynamics of changes in server load from step-by-step optimization (data from the PageSpeed Insights tool)

owner of the server on which our site is located. After that, PageSpeed Insights showed the largest increase in scores for the main page at 10. However, the blog list page and the single blog page gained 2 and 5 points, respectively. The data from these studies are shown in Fig. 5.

Option 4. Optimization of Configurations and Drupal Hooks
After optimizing configurations and Drupal hooks, PageSpeed Insights showed no changes on any page. However, all pages have reduced load times in the browser.

4.2 *A Detailed Method of Analyzing the Load on the Server*

Using access to the server via SSH, namely the load that the site creates on the processor (CPU), RAM (memory), and disk system (disk), using the "ab," "htop," and "iotop" utilities.

The results we got are much more accurate than the data from the PageSpeed Insights tool. After all, we can visually see all the problems that arise on the server. Let us depict the load on the server in the form of a correlation graph of the ratio of the number of requests that the server can process in one second to the optimization steps, Fig. 6.

Option 1. Built-in (Basic) Drupal Optimization
Built-in optimization is extremely simple to implement and inefficient. After all, the maximum number of requests the server can process on the main and blog list pages increased by 0.52. And a page with one blog added 0.019 requests in one second.

Option 2. Optimization Using Additionally Installed Drupal Modules
Optimizing Drupal using additional customizable modules is one of the most effective ways to reduce server load. After carrying out this optimization, the server

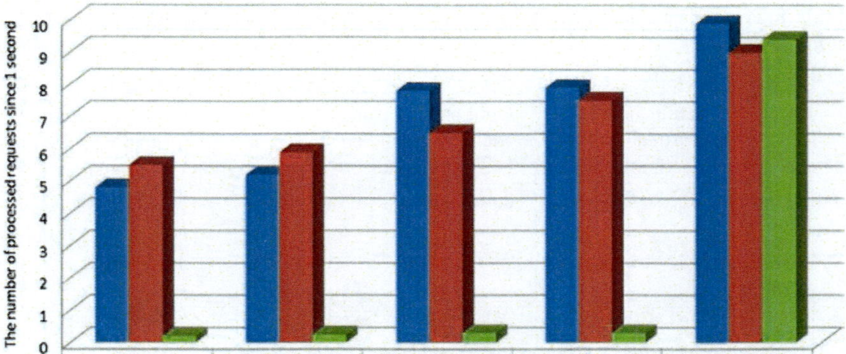

	Before optimization	Built-in Drupal optimization	Optimizing Drupal with modules	Server optimization	Optimization of configurations and Drupal Hooks
Main page	4,8	5,2	7,8	7,9	9,9
List of blogs	5,5	5,9	6,5	7,5	9
One blog	0,2	0,25	0,3	0,3	9,4

Fig. 6 The ratio of the number of requests that the server can process in one second to four steps of optimization (1 – built-in Drupal optimization, 2 – optimization of Drupal using modules, 3 – optimization of the server, 4 – optimization of configurations and Drupal hooks)

increased the number of requests it can process in one second on the main page by 2.46 requests, on the blog list page by 0.77 requests, and on the page of one blog by 0.082. However, this method requires more time to implement, and it cannot be classified as a simple way to reduce the load on the server. This method will be more effective for pages with many nonoptimized images, CSS, and JS files. However, it is mandatory for all Internet sites.

Option 3. Server Optimization
Optimizing the server is the most difficult way to reduce the load because most Drupal sites are hosted on hosts where this setting cannot be performed. You can solve this problem by renting a server. However, it is too expensive. Therefore, before launching the site, you should carefully choose the hosting and the space on the server that they provide. The server was optimized for us by the owner himself, after contacting him. This method made the server response faster for the main page by 0.16 requests per second, the blog list page by 0.91 requests per second, and the single blog page by 0.01 requests per second.

Option 4. Optimization of Configurations and Drupal Hooks
Optimizing configurations and Drupal hooks is also not an easy and slow way. After all, before that, you need to analyze the database architecture and site logic. After the actions were taken, the server increased the response to the main page of the site by 1.77 requests per second. Likewise, the response to the blog list page increased by 1.4 queries per second. But the page with one blog received the biggest increase, with 9.32 requests per second. Table 1 presents numerical indicators of the effectiveness of the applied methods in increasing the speed and performance of the server.

Table 1 Performance indicators of the applied methods of increasing the speed and performance of the server

№	Site pages	Optimization methods (number of processed requests in 1 second)			
		Built-in (basic) Drupal optimization	Optimization using additional installed Drupal modules	Server optimization	Optimization of configurations and Drupal hooks
1	Main page	0.52	2.46	0.16	1.77
2	Blog List Page	0.52	0.77	0.91	1.4
3	One blog page	0.19	0.082	0.01	9.32

This method has shown that it is one of the most effective and necessary. Also, this method highlights problems with certain modules; in our case, the Advanced CSS/JS Aggregation module reduces the number of requests to the server by combining, minifying, and caching CSS and JS files. This module stores the cache in the database. However, the tables of this module are not normalized. This increases the response of the server and increases the load on MySql.

Optimizing configurations and Drupal hooks is a must for every Drupal site. In our case, the normalization of databases was carried out, which reduced the load on the database management system; the configuration settings were correctly performed, which protects the site from unwanted bots and users, and also reduces the load on the database, reduces the storage of user sessions, and logging messages of site operation.

5 Conclusions

The analysis of publications on this issue shows that CMS Drupal is a sufficiently developed content management system. The effort of developers to provide their system with universality and increased functionality compared to competitors leads to an increase in the load on the server and a decrease in its speed. Thus, in this CMS there is a demand for server resources. In most cases, this problem is solved by choosing specialized expensive hosting services by allocating servers or several servers. Our work will be especially interesting and useful for developer practitioners who receive graphic and digital indicators of research on increasing the speed and performance of the server using effective, relatively simple, and free methods for existing sites and sites under development based on CMS Drupal.

Research results:

- built-in optimization is extremely simple to implement and inefficient;
- optimizing Drupal using add-on modules is one of the most effective ways to reduce server load. This method will be the most effective for the main page with many nonoptimized images, CSS, and JS files. However, it is mandatory for all Internet sites;

- server optimization is one of the most difficult, inefficient, and expensive methods;
- optimizing configurations and Drupal hooks is complex and long-term but simultaneously the most effective. Also, this method highlights problems with certain modules; in our case, the Advanced CSS/JS Aggregation module reduces the number of requests to the server by combining, minifying, and caching CSS and JS files, and;
- optimizing configurations and Drupal hooks are mandatory for every Drupal site.

In the future, we can use methods to increase the speed and performance of the Drupal CMS-based server to balance the load on IIoT servers. Load balancing is required to evenly distribute the processing work of two or more devices, network links, storage devices, or other devices, ultimately providing faster service with greater efficiency. Analysis and distribution of load in cloud systems used in IIoT is an urgent task since most cloud systems with open access use simple load schedulers of their physical servers.

References

1. Metinvest Digital.: https://metinvest.digital/ua/page/globalne-dosl-dzhennya-idc-shcho-cheka-rinok-danih-u-nastupn-5-rok-v. Last accessed 2022/12/04.
2. Drivas, I., Kouis, D., Kyriaki-Manessi, D., Giannakopoulos, G.: Content management systems performance and compliance assessment based on a data-driven search engine optimization methodology. Information. **12**(259) (2021). https://doi.org/10.3390/info12070259
3. Satsyk, V., Smolyankin, O., Grudetskym, R.: Decrease of load on server by means of CMS Drupal. Sci. Europe. **32**, 74–81 (2018)
4. Jagamogan, R., Ismail, S., Hafizah, N., Abas, H.: A review: penetration testing approaches on content management system (CMS). In: 7th International Conference on Research and Innovation in Information Systems (ICRIIS), 1–6 (2021). https://doi.org/10.1109/ICRIIS53035.2021.9617087.
5. McDonald, C., Burkhardt, H.: Library-authored web content and the need for content strategy. Inf. Technol. Libr. **38**(3), 8–21 (2019). https://doi.org/10.6017/ital.v38i3.11015
6. Krouska A., Troussas C., Virvou M.: Comparing LMS and CMS platforms supporting social e-learning in higher education. In: 8th International Conference on Information, Intelligence, Systems & Applications (IISA), 1–6 (2017). https://doi.org/10.1109/IISA.2017.8316408.
7. Lapteva, U., Kuzyakov, O.: Rationale for principles of developing control and protection of web content using CMS Drupal. In: International Multi-Conference on Industrial Engineering and Modern Technologies (FarEastCon-2018), 1–6 (2018). https://doi.org/10.1109/FarEastCon.2018.8602487.
8. Satsyk, V., Grudetsky, R., Kuzmych, O., Bahniuk, N., Hlynchuk L., Melnychuk, Y.: Reduction of server load by means of CMS Drupal. In: 2020 10th International Conference on Advanced Computer Information Technologies (ACIT), pp. 523–528 (2020). https://doi.org/10.1109/ACIT49673.2020.9208874.
9. Ferrnandes, J., Lourenço, A.: Bringing named entity recognition on drupal content management system. In: Saez-Rodriguez, J., Rocha, M., Fdez-Riverola, F., De Paz Santana, J. (eds.) 8th International Conference on Practical Applications of Computational Biology & Bioinformatics (PACBB 2014) Advances in Intelligent Systems and Computing, vol. 294. Springer, Cham (2014). https://doi.org/10.1007/978-3-319-07581-5_31

10. Skulimowski, A., Badecka, I.: Software innovation dynamics in CMSs and its impact on enterprise information systems development. In: Tjoa, A., Xu, L., Raffai, M., Novak, N. (eds.) Research and Practical Issues of Enterprise Information Systems. CONFENIS 2016 Lecture Notes in Business Information Processing, vol. 268. Springer, Cham (2016). https://doi.org/10.1007/978-3-319-49944-4_23
11. Li, C., Zhang, Q., Huang, C., et al.: Optimal service selection and placement based on popularity and server load in multi-access edge computing. J. Netw. Syst. Manag. **31**, 15 (2023). https://doi.org/10.1007/s10922-022-09703-2
12. Bouflous, Z., Ouzzif, M., Bouragba, K.: Analysis of load balancing algorithms used in the cloud computing environment: advantages and limitations. In: Arai, K. (ed.) Proceedings of the Future Technologies Conference (FTC) Volume 3. FTC 2022 2022. Lecture Notes in Networks and Systems, p. 561. Springer, Cham. (2023). https://doi.org/10.1007/978-3-031-18344-7_13
13. Hessen, S., Abdul-kader, H., Khedr, A., Salem, R.: Load balancing based on multi-agent framework to enhance cloud environment. Comput. Mater. Continua. **74**(2), 3015–3028 (2023)
14. Xiong, Z., Zhao, M., Yuan, Z., Jianlong, C.: Energy-saving optimization of application server clusters based on mixed integer linear programming. J. Parallel Distrib. Comput. **171**, 111–129 (2023). https://doi.org/10.1016/j.jpdc.2022.09.009
15. Moroz, S., Tkachuk, A., Khvyshchun, M., Prystupa, S., Yevsiuk, M.: Methods for ensuring data security in mobile standards. Informatyka Automatyka Pomiary W Gospodarce I Ochronie Środowiska. **12**(1), 4–9 (2022). https://doi.org/10.35784/iapgos.2877
16. Kumar, N., Sharma, B., Narang, S.: Emerging communication technologies for industrial internet of things: industry 5.0 perspective. In: Singh, P.K., Wierzchoń, S.T., Tanwar, S., Rodrigues, J.J.P.C., Ganzha, M. (eds.) Proceedings of Third International Conference on Computing, Communications, and Cyber-Security Lecture Notes in Networks and Systems, vol. 421. Springer, Singapore (2023). https://doi.org/10.1007/978-981-19-1142-2_9
17. Bakhovskyy, P., Yevsiuk, M., Zabolotnyi, O., Cagáňová, D., Tkachuk, A.: Stages of the virtual technical functions concept networks development. In: Cagánová, D., et al. (eds.) Advances in Industrial Internet of Things, Engineering and Management EAI/ Springer Innovations in Communication and Computing, pp. 119–135 (2021). https://doi.org/10.1007/978-3-030-69705-1_7
18. Rafique, F., Obaidat, M., Mahmood, K., Ayub, J., Ferzund, M., Chaudhry, S.: An efficient and provably secure certificateless protocol for industrial internet of things. IEEE Trans. Industr. Inform. **18**(11), 8039–8046 (2022). https://doi.org/10.1109/TII.2022.3156629

Proposal for Linking MOXA ioLogik and its Application in IIoT

Simona Konečná, Marek Kočiško, Petr Baron, and Martin Pollak

1 Introduction

The Internet of Things refers to a network that allows digital or mechanical devices to connect without human intervention. The Industrial Internet of Things is a superstructure of the Internet of Things in an industrial environment and includes an intelligent network of industrial components. Components are used to increase the performance of industrial and manufacturing processes. We can use components to monitor devices, collect data and analyze this data in real time. Although the Internet of Things and the Industrial Internet of Things has many common technologies, including cloud platforms, sensors, communications, and data analysis, they are used for different purposes [8]. When switching to the Industrial Internet of Things, it is necessary to properly connect sensors and devices to the Internet based on the needs and device types. Wired and wireless mechanisms can be used to connect to the network. This paper focuses on designing and implementing the ioLogik W5340 circuit with a sensor for data acquisition. For illustration, an inductive sensor is used for noncontact proximity detection of metallic objects.

S. Konečná (✉) · M. Kočiško · P. Baron · M. Pollak
Faculty of Manufacturing Technologies with a seat in Prešov, Technical University of Košice, Prešov, Slovakia
e-mail: simona.konecna@tuke.sk

M. Balog et al. (eds.), *EAI International Conference on Automation and Control in Theory and Practice*, EAI/Springer Innovations in Communication and Computing, https://doi.org/10.1007/978-3-031-31967-9_7

2 Literature Review

The network of computers and systems that are connected to the Internet and used to collect data sent to the central cloud service, where it is processed with additional data and shared in a useful way with end-users, is called the Industrial Internet of Things (IIoT) in the manufacturing industry. Industrial Internet of Things uses intelligent sensors and actuators to improve production and industrial processes. Devices and intelligent means transmit information directly to the data communication infrastructure, where it is converted into usable information about how a particular machine works [1]. One of the main advantages of IIoT equipment used in the manufacturing industry is that it enables predictive maintenance. Organizations can use data generated from IIoT systems in real time to predict when a machine must be repaired. In this way, the necessary maintenance can be carried out before malfunction occurs. This can be particularly beneficial on the production line, where machine failure can result in machine coming to a halt at huge costs [2, 6, 7].

Many industries use IIoT. One example is the automotive industry, which makes extensive use of industrial robots and IIoT. This makes it possible to detect potential problems before they disrupt production [4, 8]. Fig. 1 shows the Industrial Internet of Things architecture.

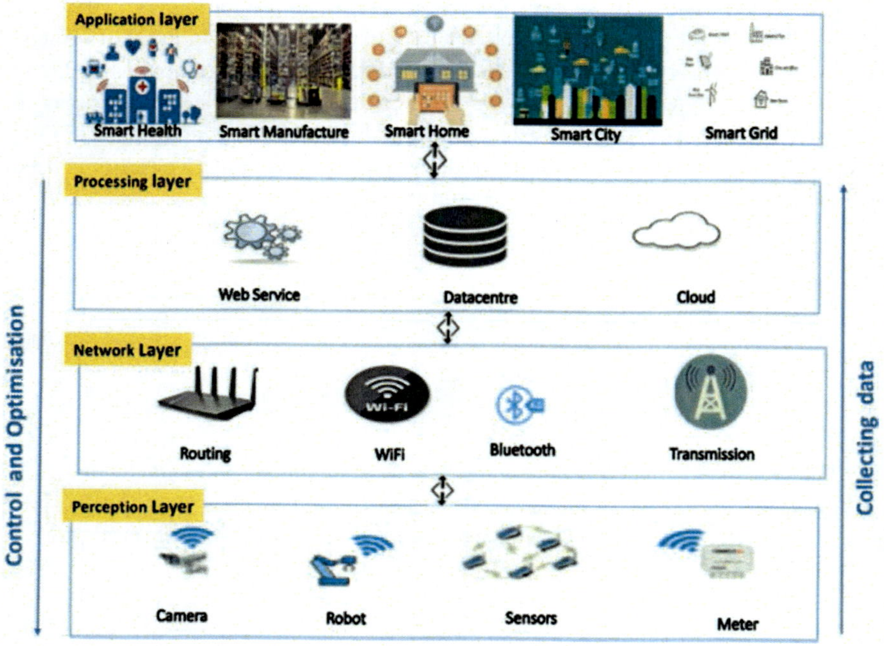

Fig. 1 IIoT Architecture [3]

The Industrial Internet of Things is a network of smart equipment connected to devices that monitor, collect, exchange, and analyze data. Each Industrial Internet of Things consists of:

- connected devices capable of reading information;
- public or private data communication infrastructure;
- data analysis;
- the storage of data generated by IIoT equipment;
- people [1, 5].

2.1 MOXA ioLogik W5340

ioLogik W5340 is a robust and compact remote monitoring device that combines GPRS communication, font-end intelligence, and data logging for information analysis and prediction. Thanks to GPRS technologies, ioLogik offers a wide range of remote monitoring. It is suitable for remote monitoring and securing systems, the connection of which by wire is either complicated or impossible. The ioLogik W5340 combines a mobile modem, a data logger, and a remote I/O device into a single compact box, reducing the amount of effort required to integrate devices. ioLogik W5340 includes an RS-232/422/485 serial port for connecting serial devices (analyzers, sensors, etc.). MOXA ioLogik uses MOXA's proprietary active OPC server with push communication technology. [9]

The mobile interface supports three-bandwidth HSPA/UMTS and four-bandwidth GSM/GPRS/EDGE frequencies, which offer a full range of mobile communication services. Multiple options are available for obtaining remote data protocols such as FTP, email, and Moxa DA-Center™. In addition, this mobile remote I/O unit is a programmable device that supports Click & Go™ control logic to create customized system control. ioLogik W5340 is a durable device that tolerates a wide temperature range and is suitable for hard-to-connect remote monitoring and alarm applications [8, 9]. Fig. 2 shows the ioLogik W5340, and Fig. 3 describes the essential parts of this device.

3 Research Methodology

Before installation, the ioLogik device and WiFi router must be configured separately. The ioLogik device is installed first. Then, it must be connected to the mains using 12 to 36 V DC voltage. Once connected, the READY check light on the front side lights up, and the system is ready for use. Next, we use the Ethernet interface to connect ioLogik to a computer and proceed with its configuration. For configuration and for simple and easy setup, the manufacturer offers ioAdmin software. The device enables selecting an IP address of choice. We set a static IP address of 192.168.127.2. See Fig. 4 for more information.

Fig. 2 MOXA ioLogik W5340

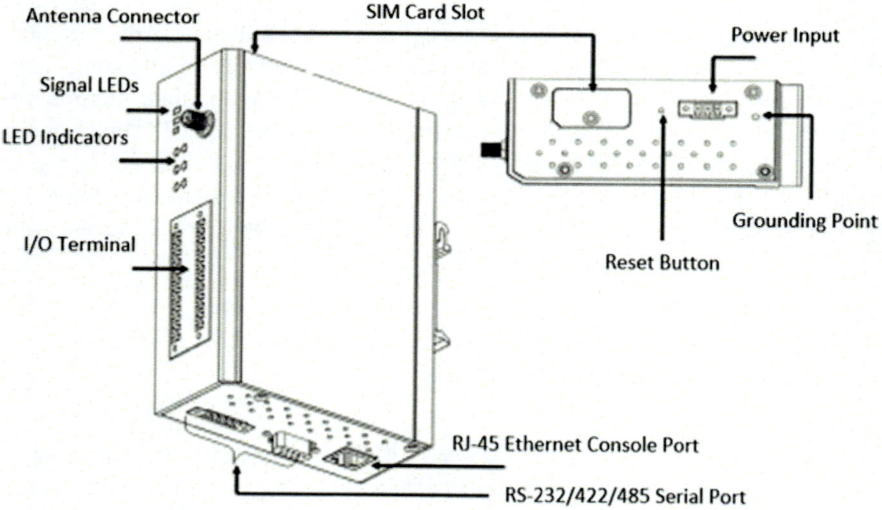

Fig. 3 Basic elements of ioLogik W5340 [9]

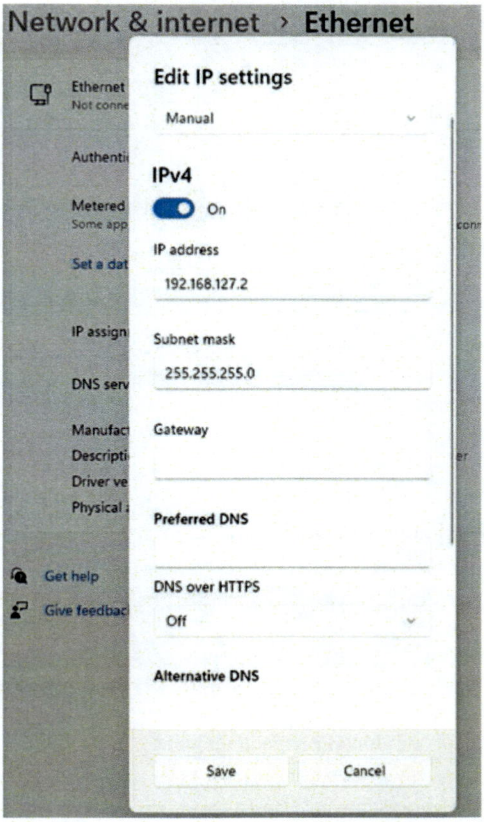

Fig. 4 Setting the IP address on the computer

IP Address	192.168.127.2
Subnet mask	255.255.255.0

Open the ioAdmin software installed in advance. Initially, it shows the option of searching for connected devices. Available in the Ethernet connection RJ-45, RS-232/485 series, and search in a specific range of IP addresses and selection of the supported models. To start the search, press Start Search. If ioAdmin finds more devices, it is necessary to select the correct one.

After selecting the device, the ioAdmin application window opens, divided into three parts – Fig. 5. A program bar appears at the top. A device tree on the left can contain multiple devices, and it is easy to switch between them in this section. The right part shows the complete configuration menu of the selected device. The upper part of this module shows the full name of the device that is currently active.

To fully utilize working with the configuration, it is necessary to log in, which is done on the server setting tab. You only have to click on login, as no password is set by default. If an IP address is assigned and the ioAdmin configured, we can connect sensors, readers, etc., to the device.

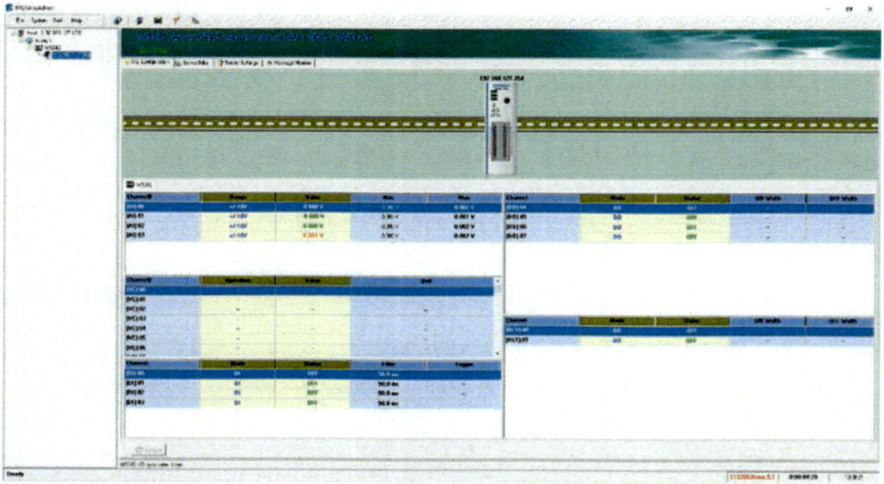

Fig. 5 ioAdmin Software

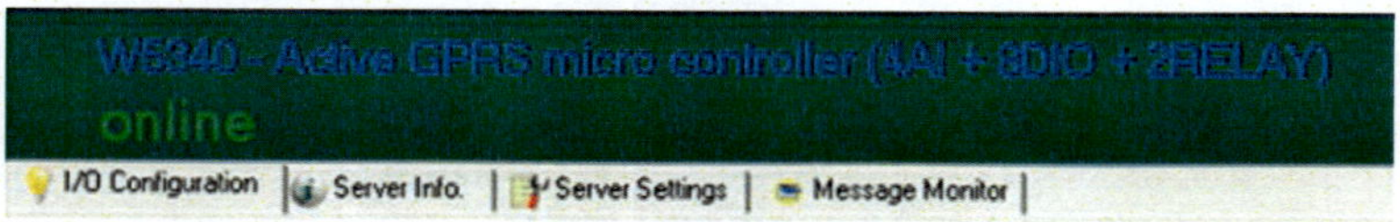

Fig. 6 ioAdmin Configuration Menu

If you want to configure a device that has already been set up before, it is better to first reset the device using the button on the back of the device (Reset Button) before configuring it.

The basic configuration panels are shown in Fig. 6.

I/0 Configuration displays the status of all I/O channels. The input channels are shown on the left-hand side, and the output channels are on the right-hand side of the device. After double-clicking on the relevant channel, a window appears with the configuration option.

The info server contains information about the server, such as the firmware version or the IP addresses that have been set.

The settings server includes the option of logging in as an administrator, either to change the administrator password, or change the server name or its location.

The Message Monitor displays all TCP/UDP active device messages. When run for the first time, the message window is empty because it does not contain any predefined rule messages. Messages may be displayed in ASCII, HEX.

After completing the settings, the ioLogik device can be disconnected and the WiFi router configured. The WiFi router is connected to the computer's LAN port using the RJ-45 cable and the IP address is set to configure the router. Enter the IP address of the WiFi router in the web browser. A login window will appear for

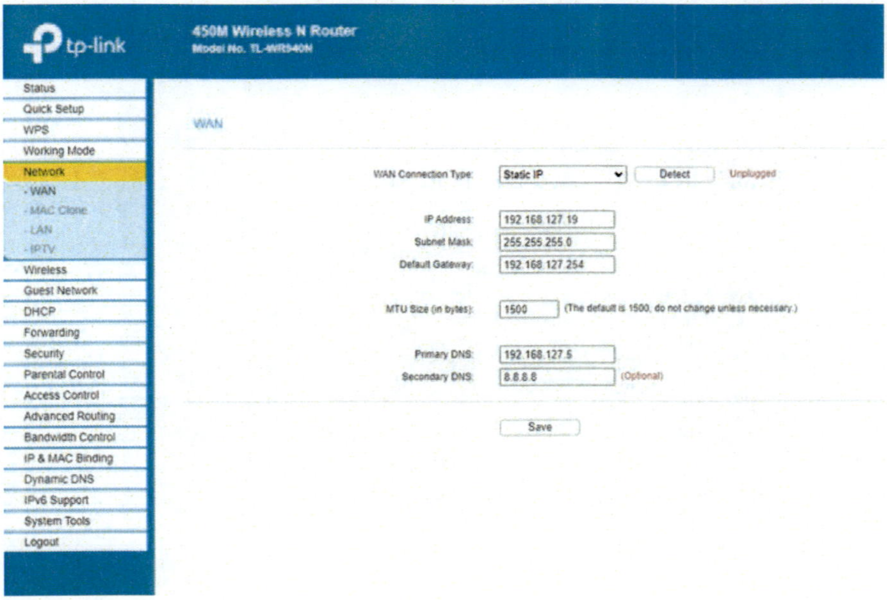

Fig. 7 WAN interface setup window

entering information from the WiFi router. The configuration window opens after the login is successful.

On the wireless tab, select wireless settings and set the SSID. This is the name of the wireless network – e.g., ioLogik W5340. Region – Slovak Republic.

In the next section, it is necessary to enter the MAC address of the device from which the connection will be established. In our case, it is a desktop computer. This is done in the wireless MAC Filtering tab.

The network card must also be configured when setting up the WiFi router. Set the static IP address in the WAN tab. Since we want to connect to the ioLogik W5340 device using the port and its IP address is 192.168.127.2, the WAN IP address of the WiFi router port needs to be set to 192.168.127.19, the mask to 255.255.255.0 and the gateway to 192.168.127.254 (ioLogik IP address)—Fig. 7.

Once this setting is done, we must test whether ioAdmin can find our device. The LAN cable from the computer can be linked with the ioLogik W5340. Once connected to the ioLogik W5340 network, running the ioAdmin, and finding the device, the main application window should open, showing our device. Upon installing the Active OPC Server, a network is in place for data collection using a WiFi network. Fig. 8 shows the ioLogik W5340 connected to a computer.

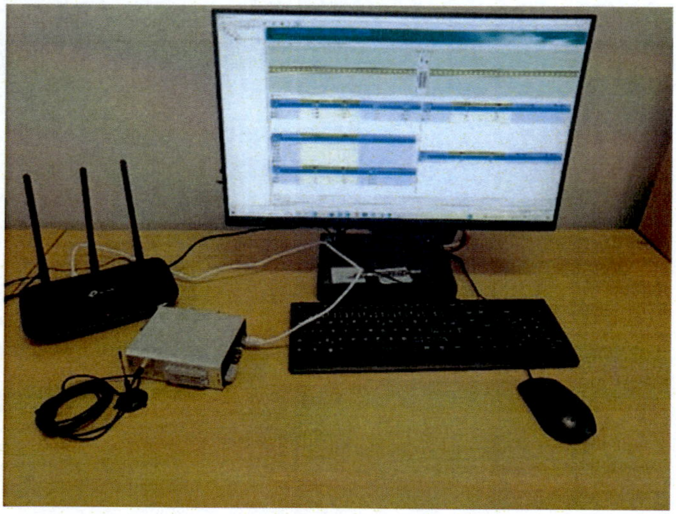

Fig. 8 ioLogik W5340 linked to a computer

3.1 Application

There are many advantages to using the Industrial Internet of Things. We can prevent failure from occurring thanks to the measurement and evaluation of critical parameters. The ioLogik W5340 module can be used to read various variables. The sensors must have a predefined analog output, which can be either a current or a voltage one. The current output should be 4–20 mA, and the voltage should be between 0 and 10 V.

For this paper, we used a contactless inductive sensor. Contactless inductive sensors are integral to almost every automated machine or production line. The sensors respond by activating the output when a metal object draws near.

By default, these sensors are built for use in industrial production environments. The advantage of these sensors is that they enable long-term operation in dusty or otherwise polluted industrial environments. In addition, they come with a wide range of working temperatures.

Tested sensor properties:

- cylindrical design with M12 thread;
- a switching distance of 4 mm;
- PNP output configuration, analog output;
- rated operating current 0–150 mA;
- no load current 15 mA;
- supply voltage of 5–30 V;
- operating temperature – 25 to 70 °C.

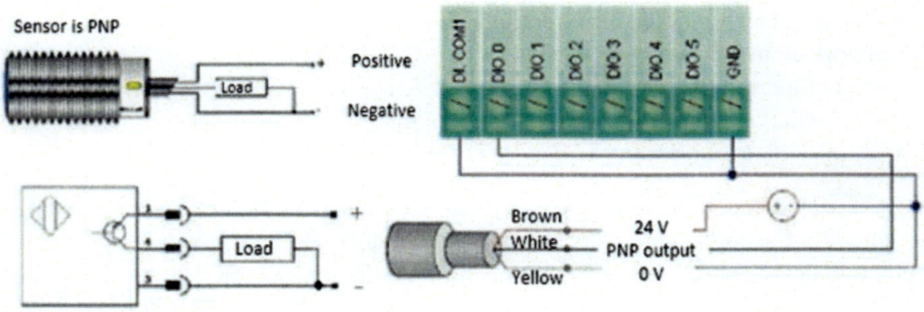

Fig. 9 Connecting the inductive sensor to an active device

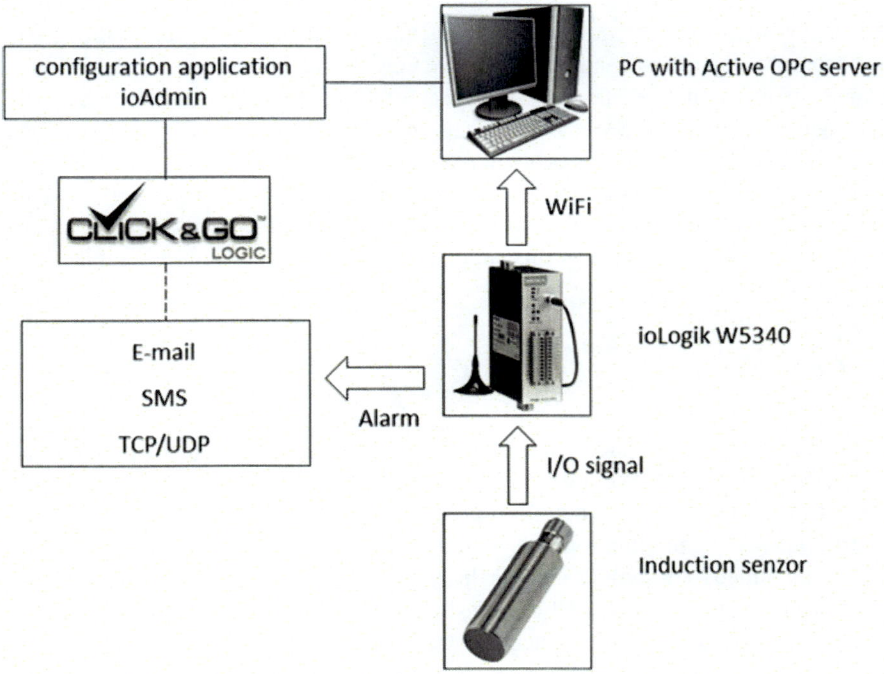

Fig. 10 Network infrastructure of measurement performed

Available to the inductive sensor with a PNP output are three wires, which are connected according to Fig. 9.

- the brown wire representing the positive pole is connected to the DI.COM1 port,
- the white wire is connected to the DIO 0 port,
- the yellow wire represents the negative pole and is connected to the GND port,

The design of the network infrastructure for data collection shown in Fig. 10 includes all components proposed to be used for the input, processing, collection, and transmission of data, i.e.:

- measuring elements (inductive sensor);
- network elements (ioLogik W5340);
- servers (operating system and Active OPC Server);
- data exchanges (database and monitoring center);
- transmission network.

We connected the inductive sensor to the ioLogik W5340, with a PNP output designed for the contactless detection of metal objects. The sensor was connected to the active device and then set in the ioAdmin configuration application via digital DI input to the OFF status. This represents a state in which the sensor does not have a metal object in its range. If a metal object approaches the device, the device status automatically changes to ON.

The data sent to the Active OPC Server was obtained from the server, which indicates contactless detection of the approaching metal objects. Based on the active tags from the connected channel, the Active OPC Server is always updated when a change in the signal occurs. The ioLogik W5340 can send these changes as a message in the form of SMS, email, TCP/UDP packets, and SMNP in real time. These event and alarm messages are defined using Click & Go rules.

Vibration sensors, tilt sensors, laser sensors, and temperature transducers are also available for measuring critical parameters. However, measuring and evaluating the data obtained is not the topic of this paper, so we will not deal with this subject.

4 Conclusion

The main task of IIoT equipment is to digitize and monitor equipment and production lines in an enterprise. Industrial organizations use IIoT sensors to measure various technical values. For example, pressure detectors are used to regulate and control pipelines, and vibration sensors monitor the vibrations on the lathe spindle. This paper aimed to describe a module designed for communication between two or more machines using the available forms of data transmission. For this purpose, a module from MOXA was selected, as the company is dedicated to producing devices both for data collection and communication between machines. The ioLogik W5340 was selected from devices on offer and used for linking measuring devices (sensors, readers, controllers, etc.). The device processes information from these measuring devices and communicates the data to the remote OPC server via a network interface. The device can send messages to the user in predefined or alarm modes (SMS and email). The user can predefine what the device should do if the measured values exceed the specified limits. The tested active device of ioLogik W5340 is a good option for controlling and collecting measured data, especially when we need to be informed about the status of inputs and outputs. The active ioLogik device sends the data obtained from the sensors only upon the occurrence of a change event. For a simple demonstration of sensor connection and settings, we used an inductive sensor designed for contactless detection of approaching metal objects. The paper also describes linking the ioLogik W5340 for data collection with a computer using wireless communication. The network communication infrastructure for the conducted analysis was carried out at the Faculty of Manufacturing Technologies of the Technical University of Kosice.

Acknowledgments The authors thank the Ministry of Education of Slovak Republic for supporting this research with the grant VEGA no. 1/0026/22, KEGA 038TUKE-4/2021 and project APVV-18-0316.

References

1. Anandan, R.: Industrial Internet of Things (IIoT) - Intelligent Analytics for Predictive Maintenance: Intelligent Analytics for Predictive Maintenance. John Wiley & Sons, Nashville, TN (2022)
2. Misra, S., Mukherjee, A., Roy, A.: Introduction to IoT. Cambridge University Press, Cambridge, England (2021)
3. Abosata, N., Al-Rubaye, S., Inalhan, G., Emmanouilidis, C.: Internet of things for system integrity: a comprehensive survey on security, attacks, and countermeasures for industrial applications. Sensors (Basel). **21**, 3654 (2021). https://doi.org/10.3390/s21113654
4. Salazar, C.: Internet of Things-IOT: Definition, characteristics, architecture, enabling technologies, application & future challenges. In: International Journal of Engineering Science and Computing, pp. 6122–6131. IJESC (2016)
5. Javaid, M., Haleem, A., Pratap Singh, R., Rab, S., Suman, R.: Upgrading the manufacturing sector via applications of industrial internet of things (IIoT). Sens. Int. **2**, 100129 (2021). https://doi.org/10.1016/j.sintl.2021.100129
6. Kanagachidambaresan, G.R., Anand, R., Balasubramanian, E., Mahima, V.: Internet of Things for Industry 4.0: Design, Challenges, and Solutions. Springer Nature, Cham, Switzerland (2021)
7. Rholam, O., Tabaa, M., Abbas Dandache, F.M.: Smart device for multi-band industrial IoT communications. Procedia Comput. Sci. **155**, 660–665 (2019). https://doi.org/10.1016/j.procs.2019.08.094
8. Hlavatá, S., Vodilka, A., Kočiško, M., Pollák, M.: Implementácia prediktívnej údržby vretena sústruhu na základe nedeštruktívnej diagnostiky. In: Průmyslové inženýrství 2021: Sborník příspěvků. Západočeská univerzita v Plzni (2021)
9. MOXA, ioLogik W5340. https://www.moxa.com/en/products/phased-out-products/iologik-w5340, last accessed 2022/11/3

The Potential of Using the Internet of Things Instead of Manufacturing Intelligence Systems in Modern Production

Miriam Pekarčíková, Michal Dic, Richard Duda, and Marek Mizerák

1 Introduction

The manufacturing industry can be considered volatile and highly competitive, so most production companies must be agile and continuously improving to survive. However, by adopting the latest technology alone, manufacturers may not always increase production rates or improve quality or profitability. To achieve these changes, operations managers and production managers need to create an environment that allows them to precisely control production while expanding visibility into they are the most critical processes. Extending that visibility and control outside their four walls to sister plants running common best practices, corporate management, customers, and the extended supply chain completes the figure. In the past, many companies relied on simple, manual systems and spreadsheets to manage their plants (Note: MES ≠ Microsoft Excel Spreadsheets). Then, standalone MES solutions and Manufacturing Intelligence systems were introduced. Now, with declining technology costs and software as a service (SaaS), manufacturers of all sizes are implementing specialized software solutions and platforms for Industrial Internet of Things to orchestrate the systems and processes in their facilities.

M. Pekarčíková (✉) · M. Dic · R. Duda · M. Mizerák
Department of Industrial and Digital Engineering of the Faculty of Mechanical Engineering, Technical University of Košice, Košice, Slovakia
e-mail: miriam.pekarcikova@tuke.sk

M. Balog et al. (eds.), *EAI International Conference on Automation and Control in Theory and Practice*, EAI/Springer Innovations in Communication and Computing, https://doi.org/10.1007/978-3-031-31967-9_8

2 Manufacturing Intelligence and Big Data as Obsolete Tools

While raw data has the potential to be explored for insights, several software tools provide real-time statistical analysis of the data itself. In the past, business intelligence was a tool to provide strategic insights from Big Data Analysis but did not present critical real-time manufacturing processes and predictions. One of the most important gains mentioned about business Intelligence implementations is often related to the ease of report generation, not necessarily data analysis. *Enterprise Manufacturing Intelligence* tools (sometimes with the name Enterprise Manufacturing Intelligence) are focused more specifically on data analysis and clear presentation of the analysis. The amount of data that even small to medium companies generate is daunting. Big Data was a buzzword for such enormous data stores. This data is so complex that traditional tools like relation databases and visualization applications are not able to deliver useful insight. Consider this issue with companies operating globally, and the issue is no longer just site-related Big Data in information silos. Those "big" companies employ even more obsolete manually complemented discrete software or other custom legacy systems. In these cases, the results decision-makers need on time have such a big latency that insights are already quite stale. The open question might be that short-term decisions are strictly tactical and not as important to key decision-makers, while long-term decisions can be more important strategic. Unfortunately, business intelligence is also quickly perishable in today's ever-accelerating pace of commerce [1–3]. A "snapshot" taken only days before today may not suffice to help make tactical decisions within a narrower time window. Manufacturing intelligence software has some common features [4–6]:

1. *Database Independence* with the ability to access and analyze from disparate databases across an organization's locations, preferably without creating yet another unique database;
2. *Real-time, manufacturing* specific statistical analysis and reporting; Accumulated Causal Data stored in a searchable knowledgebase with statistical outlier data (CAPA or AC/CA);
3. *Real-time Dashboard* reporting for operators, managers, and executives to quickly identify key outlier events; and present selected Key performance indicator (KPI) which are important for the manufacturing organization.
4. *Sharing and Collaboration* tools across the organization to quickly react while documenting and retaining institutional memory for Best Practices.

Enterprise Manufacturing Intelligence (EMI) software functionality enables real-time visibility into plant-level data across disparate sources and provides context on the interrelationships of data that may be less than obvious due to being stored in different equipment, systems, applications, or databases. The basic framework of EMI software is described in Fig. 1 – Enterprise Manufacturing Intelligence [7]. EMI software's ability to integrate and contextualize data from different systems and data sources and provide real-time actionable data has also

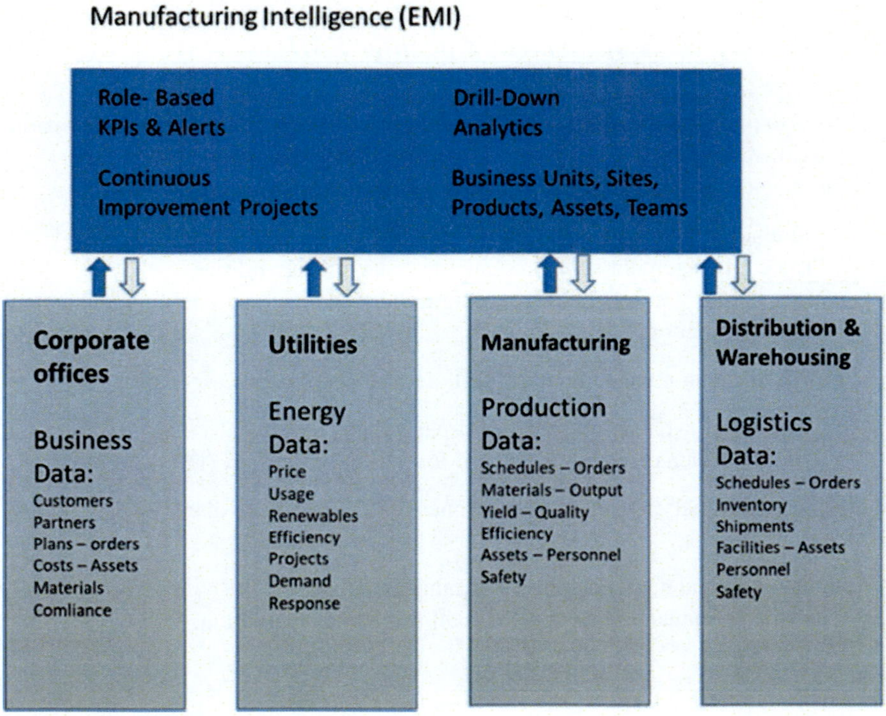

Fig. 1 Enterprise Manufacturing Intelligence

been a quantum leap in manufacturing intelligence capabilities across the enterprise compared to the days of manual collection and analysis. As Big Data capabilities advance, EMI solutions can mine increasing amounts of data at lightning speed, uncovering previously unknown correlations between systems. As a result, the pulse of enterprise operations will be measurable at scope expansive in reach and precision, including unprecedented predictive maintenance capabilities for shopfloor assets, creating the ultimate "smart" shopfloor for manufacturing operations.

Enterprise Manufacturing Intelligence integrates, connects, and utilize data sources such as Manufacturing Execution System (MES), Quality Management System (QMS), Advanced Planning and Scheduling (APS), Laboratory Information Management System (LIMS), Enterprise Resource Management (ERP) and others production systems – into one accessible analytical data model providing capabilities to explore and drill down into contextualized data [8–10].

EMI software tool is used at the plant (shopfloor) level to improve collaboration and data exchange between the factory shop floor and enterprise managing systems and/or at the enterprise level to benchmark and compare production runs or predict various plant scenarios. As data from different sources are merged together, they can be put into a new context and provide operational users with a different and more complete visibility of manufacturing operations regardless of where the data were gathered.

2.1 Key Features of Manufacturing Intelligence (Based on EMI Tools of Siemens Opcenter EMI)

- Uniform platform for operational and enterprise data provides a common vision across the company.
- Governance and integration of data for MES/MOM ecosystem.
  - Manufacturing analytical data model based on standards (ISA-95).
  - Project management and automatic deployment capabilities.
  - Many times Out-of-the-box configurations which are enabling dedicated industry functionalities.
- Self-service data management to fulfill new business needs.
  - Data model extension for any third-party data.
  - Automatic data management – no information technology (IT) skills required.
- Self-service analytics enable everyone to analyze and discover new business values.
  - Historical and near real-time data analysis interface.
  - Key performance indicator (KPI) definition and calculation.

3 Adopting Industrial Internet of Things

Internet of Things (IoT) is described by the Oxford English Dictionary as "a proposed development of the Internet in which everyday objects have network connectivity, allowing them to send and receive data." Simply explained, IoT is the network of internet-connected physical objects that can communicate with each other and other systems over the internet network. For example, in production environment IoT devices capture volumes of process and environmental data about manufacturing from sensors attached to production machines, PLCs, and machine drive's controllers, and communicate that data to a gateway that can be accessed by monitoring and analytic software or applications under the IoT production network. For example, vibration and temperature sensors can detect changes in machinery performance and send alerts if preventive maintenance is required.

3.1 IIoT Performance Insight

Performance insight applications work in the Industrial Internet of things network and provide additional transparency on manufacturing machines across production lines or entire sites. These tools provide production optimization, which can be achieved through deeper insights and analytics based on individual predefined Key Performance Indicator (KPI) calculations and visualization, for example, Overall Equipment Effectiveness (OEE), Quality, Performance, Availability, and many more. One example of IIoT architecture is described in Fig. 2, which also describes the connection and workflows between IIoT Cloud and factory levels and the communication between them. End devices are connected via IoT into the network and collect raw data. Data are then processed and presented by one of the applications working in IoT network, as the example in Fig. 3 shows [11, 13].

Asset performance management (APM) is a term in manufacturing to describe methods of capturing and integrating data, visualizing it, and analyzing it to improve the reliability and availability of physical plant–floor assets. IoT for asset/plant performance optimization is the modern version of APM. It integrates state-of-the-art data-capture and integration tools (like IoT gateways) and software tools (like IoT platforms) to analyze how assets can be run and maintained at optimum levels (e.g., optimized asset speed settings, optimized material input settings, or optimized maintenance intervals) [14].

One of the most important reasons that drive production companies to invest in IIoT APM technologies rely on main benefits that include the improvement of the operational assets' availability, reliability, and safety, the decreased percentage of unplanned re-work, the minimization of the costs associated with machine and equipment maintenance, and the decrease percentage of the failure risks of critical machines. In order to achieve these main goals, some processes must be performed, including the collection of equipment data through sensors or external databases, the elaboration of predictive and intervention models, and the covering of both engineering and financial aspects. In addition, to effectively optimize maintenance-related activities such as determining which operations should be performed on industrial resources and their scheduling, APM technologies are required to be able

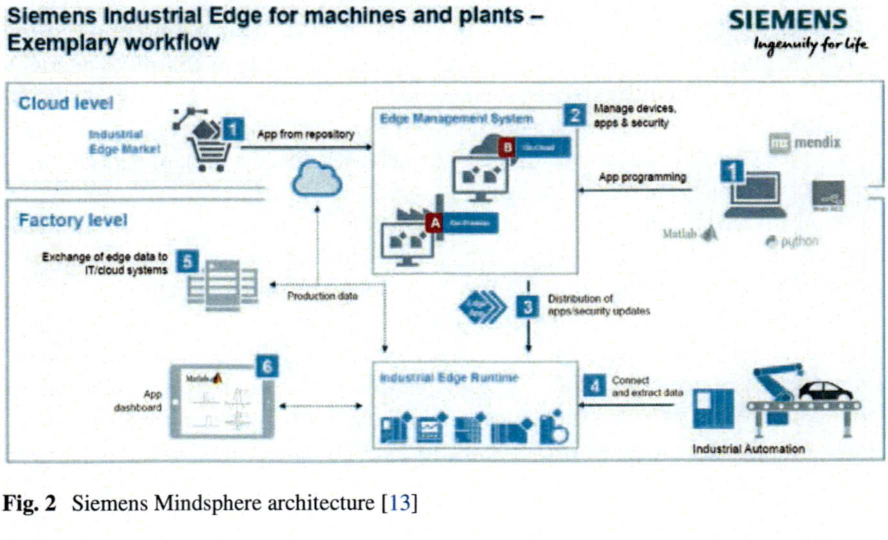

Fig. 2 Siemens Mindsphere architecture [13]

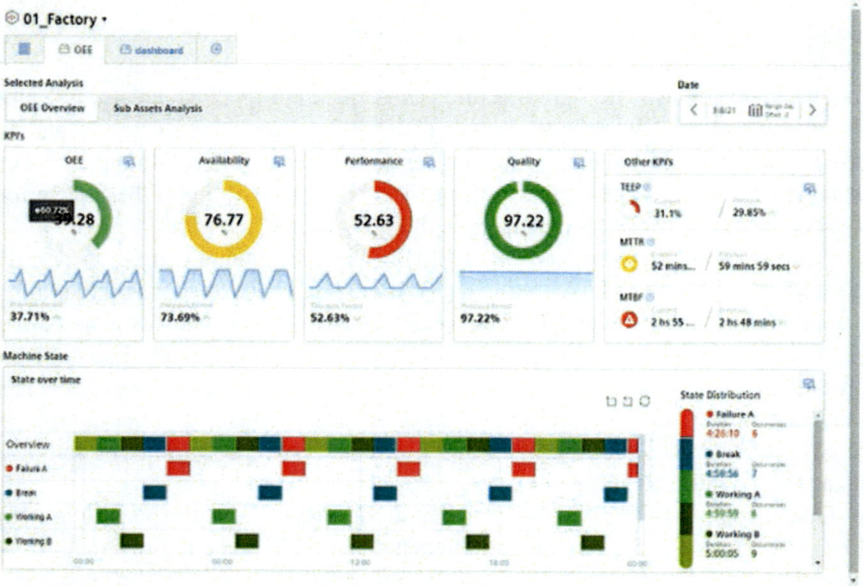

Fig. 3 Siemens Mindsphere dashboard [12]

to capitalize on reliable data management, visualization techniques, risk mitigation tools, advanced analytics, and many more. Data management and its utilization is constantly challenged and changed by the continuous technological progress with the appearance of new technologies of central importance in Asset Performance Management, such as Industrial Internet of Things, machine learning, clouds, web-based software, and image recognition. These technologies allow the connection of computing instruments implanted in industrial machines, store the huge quantity of data created, run analytical models more quickly and streamline decision-making operations, being it all reinforced by an environment of cooperation allowed by web-based software. As most modern machines are IoT ready or retrofit to add IoT modules are not difficult to add, implementation of IIoT networks can be considered as relatively easy and cheap. Furthermore, IIoT networks usually work as Software as a Service (SaaS) in the cloud, which also decreases the cost of hardware needed. The other advantage of having software in the cloud is dashboard accessibility, as the dashboard work on HTML5 web interface, which means – it can be accessed from any device at any time. All of this, allied with domain experience and deep asset performance comprehension, brings great industrial improvements.

The benefits of IoT are so high that most manufacturing companies are thinking in this direction. Company – IoT – Analytics researched, while more than 30% of companies responded that they had adopted IoT for asset/plant performance optimization. With 42% of companies planning to invest significantly in IoT, it is among the use cases with the highest expected growth rate (Fig. 4) [11].

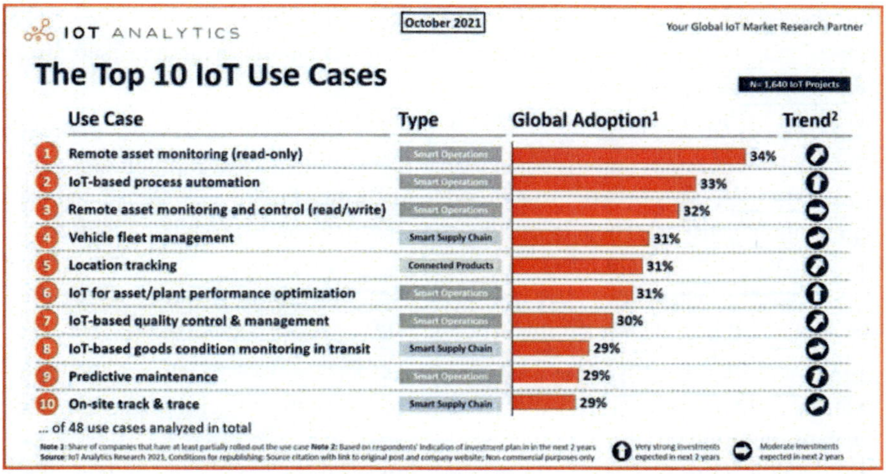

Fig. 4 Results of survey – IoT – analytical about top IoT use cases [11]

4 Testing the IoT in the Laboratory Conditions in TestBed 4.0 Workplaces

The essence of TestBed 4.0 is unique cyber-physics workplaces consisting of several workstations, and each made up of a computer with software dedicated to a specific area of interest. Also, each workstation contains sensors and a control unit. The workstations are interconnected, allowing work in developing and verifying individual machines, equipment, processes, production lines or assembly or robotic workers, or work on multiple subsystems in a single production system. TestBed 4.0 must work with the goal expected by the customer when solving the question of PLM system implementation. However, this idea often hides possible risks associated with the implementation of this system. As the system is, as already mentioned, unique and fully configured for the customer's needs, it is possible that the implementation is not possible or appropriate [15–17].

The workplace provides methods of verification in production and subsequent debugging in structural, electrical, and programming modeling or production of equipment such as conveyor lines or other machines. A PLC program in need of debugging bears the risks associated with downtime of the entire production line. This causes overpricing and extended time before the equipment is even operational. It is these risks that this workplace seeks to prevent, and where a 3D model can be created based on customer requirements, such as functional and design requirements, under the guidance of SOVA Digital's in-house experts. The 3D model provides a digital twin of a production line or single-purpose device, which can also be tested using the technologies applied in TestBed 4.0. Thus, the benefits of this workplace within the process described above are:

- Increasing the share of time for PLC programmers in preparation and debugging,
- Debugging takes place in the safety of the virtual environment, and therefore, there is no risk of damage to the device,
- Debugging the PLC program before the device is finalized.

TestBed 4.0 provides options when preparing to implement new production processes and for changes to established production. Thus, both the personnel and the equipment of TestBed 4.0 help in making changes to the basic production indicators in accordance with the requirements, for example, to increase product volume, production capacity and reduce costs per unit. The following phenomena are distinguished in the implementation of new production:

- Production ramp-up: It is the process when volume growth reaches the highest values quantitatively due to the increase of production capacities within the expansion of the production process.
- Production start-up: Thanks to the improvement of the quality of technological and control processes, the basic indicators of production are improved.
- Production ramp down: The process of reducing the production quantity at the expense of implementing a new type of product; hence, the production volume has a decreasing character over time.

TestBed 4.0 offers solutions related to operations in warehouses and similar areas by implementing a WMS, i.e., a warehouse management system. The Testbed software as well as the technologies used to control and collect information in the warehouse areas are applied in the development and modification of the WMS system for particular companies. The main mission of the WMS system is to manage and visualize data flowing from the warehouse in real time [18–20].

The system works as a standalone application, i.e., software with complete or partial configuration options. The information received on operations or the flow of material, equipment or people is provided by technologies such as AIDS, RFID, and RTLS (discussed in the next chapter). Such software and hardware system implementations offer:

- Increased control over warehouse operations.
- Control over the flow of materials, equipment, and manpower.
- Increased work efficiency.
- Increased storage capacity.
- Reduction in operating costs.
- Reduction of failure rates in material operations such as dispensing and receiving.
- Increased organizational efficiency in storage.

The implementation of WMS is initially costly in both time and money. Metrics have been introduced to make decisions about investing in a WMS system that are directly tied to the business strategies and financial capabilities of the enterprise. TestBed 4.0 provides assistance in evaluating the investment by comparing the perfect metric to achievable performance. The metrics include three main categories: order fulfillment, stock management, and productivity.

The synergy between manufacturing practice and simulation technologies under the principles of the digital twin enables complex analyses of the cyber-physical system, i.e. of a physical device linked to its digital twin. The scope of these analyses is not only the design of possible production situations and conditions but also the reverse applicability of the design output to the physical model. It is the area of analyses or causality of change that brings about the development of areas of digital twin use in the field of sophisticated algorithms for solving situations in production processes. A great contribution of the digital twin lies in the solution of complex logistics tasks and in the construction of electronic kanban [21–23]. In laboratory environments such as TestBed 4.0, the principles of digital twin and Industry 4.0 demonstrate the ability to streamline and optimize critical manufacturing processes while offering a unique visual and economic view of individual work operations and equipment. The implementation of a digital twin under the TestBed 4.0 umbrella leads to

- Efficient utilization of materials, human resources, time, and equipment.
- Impetus for introducing innovations in the company.
- Centralizing the application of information and introducing a collection and evaluation center.
- Significant increase in productivity, flexibility, and reduced error correction time.

TestBed 4.0 uses several technology partners, whether in the field of industrial or academic practice. One of the key partners is SIDAT Digital, member of the SOVA Digital group. SIDAT's role in monitoring, evaluation, and data collection. They emphasize data collection and monitoring of production associated with CNC equipment run on any control system. By connecting directly to the control system of the CNC machine, data is collected and monitored in real-time and stored in more advanced software and hardware storage databases. The data in the repository can be accessed at any time for further use and evaluation.

5 Conclusion

EMI software and IIoT APM are tools that manufacturing companies are leveraging to gain visibility and manage enterprise performance. Effectively managing performance requires optimizing an organization's key resources—people, processes, and technology—to achieve new levels of operational success. We might consider some of their function as similar, and the main difference is their application and the cost of implementation. IIoT software applications are opening a new world of different applications that production companies, after implementation, can use, from asset performance management to predictive maintenance [24–27].

The multipurpose of IIoT is the fact it also decreases the costs of implementation of IIoT. For example, suppose the manufacturing company already has implemented MES/MOM solution. In that case, EMI tools are still a good choice for intelligence systems, but if the production company needs to build a network to have intelligent dashboarding, then IIoT is opening a new world of functionalities and scalability. Important note: One of the differentiators between IIoT and EMI – having cloud software is dismissed as MOM/MES vendors are already providing EMI Software tools in cloud too.

Acknowledgments This article was created by the implementation of the grant projects: APVV-17-0258 Digital engineering elements application in innovation and optimization of production flows. APVV-19-0418 Intelligent solutions to enhance business innovation capability in the process of transforming them into smart businesses. VEGA 1/0438/20 Interaction of digital technologies to support software and hardware communication of the advanced production system platform. VEGA 1/0508/22 "Innovative and digital technologies in manufacturing and logistics processes and system" KEGA 020TUKE-4/2023 "Systematic development of the competence profile of students of industrial and digital engineering in the process of higher education."

References

1. Cooley, J., Petrusich, J.: Optimizing real time manufacturing intelligence (2021)
2. Shukla, S., Agarwal, S., Shukla, A.: Trends in cloud-ERP for SMB's : A review; Int. J. New Innovat. Eng. Technol. (IJNIET) (2012)
3. Tao F., Yhang M., A.Y.C. NEE, Digital Twin driven Smart Manufacturing ISBN 978–0–12-817630-6
4. Goodwin, G.: How EMI Software Has Changed Manufacturing Operations. LNS Research (2013)
5. Beric, D., Stefanovič, D., Lalič, B., Čosič, I.: The implementation of ERP and MES systems as a support to industrial management systems, IJIEM, ISSN 2217-2661 (2018)
6. What is ITOT Convergence? [online] [cit. 10.11.2022]. Web: https://www.arcweb.com/blog/what-itot-convergence
7. How EMI Software has changed manufacturing operations [online] [cit. 10.11.2022]. Web: https://blog.lnsresearch.com/blog/bid/191239/How-EMI-Software-Has-Changed-Manufacturing-Operations
8. IoT Signals Manufacturing Spotlight [online] [cit. 12.11.2022]. Web: https://clouddamcdnprodep.azureedge.net/gdc/gdcmAYP6m/original
9. Annath, V.K., Abinash, M., Rao, L.B.: Intelligent manufacturing in the context of industry 4.0: A case study of siemens industry. J. Phys. Conf. Series. (2021)
10. Lappi, J.: Asset Performance Management Application For Power System Condition Monitoring In An Internet Of Things Platform. Alto University School of Electrical Engineering (2019)
11. The top 10 IoT Use Cases [online] [cit. 10.11.2022]. Web: https://iot-analytics.com/top-10-iot-use-cases/
12. Analyze MyPerformance /OEE-Monitor and Analyze MyPerformance /OEE-Tuning Documentation [online] [cit. 10.11.2022]. Web: https://siemens.mindsphere.io/en/docs/Analyze-MyPerformance
13. Rui Pedro Teixeira da Silveira Soares Morais: Conceptualization and Design of an Asset Performance Management Solution Within a Manufacturing Execution System. Faculdade de engenharia Universidade do Porto (2021)
14. Web: https://www.plm.automation.siemens.com/global/en/products/manufacturing-operations/enterprise-manufacturing-intelligence.html
15. Trebuna, P., Pekarcikova, M., Kliment, M., Trojan J., Mizerak, M.: Modelling and simulation of production systems in industrial and digital engineering, 1. vyd. – Lüdenscheid : RAM-Verlag – 2019. – 441 s. [print]. – ISBN 9783942303927.

16. Trebuna, P., Pekarcikova, M., Kliment, M., Trojan J., Mizerak, M.: Production processes and systems digitalization, 1. vyd. – Lüdenscheid : RAM-Verlag - 2022. - 400 s. [print]. - ISBN 9783942303927.
17. Ojstersek, R., Acko, B., Buchmeister, B.: Simulation study of a flexible manufacturing system regarding sustainability. International Journal of Simulation Modelling. **19**(1), 65–76 (2020). https://doi.org/10.2507/IJSIMM19-1-502
18. Vegsoova, O., Khouri, S., Straka, M., Rosova, A., Kacmary, P., Betus, M.: Using technical means and logistics principal applications to solve ecological water course accidents. Polish J. Environ. Stud. **28**(5), 3875–3883 (2019). https://doi.org/10.15244/pjoes/95183
19. Saniuk, S., Grabowska, S., Straka, M.: Identification of social and economic expectations: Contextual reasons for the transformation process of industry 4.0 into the industry 5.0 concept. Sustainability. **14**(3), 1391 (2022). https://doi.org/10.3390/su14031391
20. Bucková, M., Skokan, R., Fusko, M., Hodon, R.: Designing of logistics systems with using of computer simulation and emulation. Transport. Res. Procedia. **40**, 978–985 (2019). https://doi.org/10.1016/j.trpro.2019.07.137
21. Dulina, L., Edl, M., Fusko, M., Rakyta, M., Sulirova, I.: Digitization in the technical service management system. MM Sci. J. **2018**(1), 2260–2266 (2018). https://doi.org/10.17973/MMSJ.2018_03_2017105
22. Kabele, P., Edl, M.: Increasing the Efficiency of the Production Process due to Using Methods of Industrial Engineering, 2nd International Conference on Design, Simulation, Manufacturing – The Innovation Exchange (DSMIE), Lutsk, Ukraine, 126–137 (2020). https://doi.org/10.1007/978-3-030-22365-6_13
23. Grznar, P., Gregor, M., Gaso, M., Gabajova, G., Schickerle, M., Burganova, N.: Dynamic simulation tool for planning and optimisaton of supply process. International Journal of Simulation Modelling. **20**(3), 441–452 (2021). https://doi.org/10.2507/IJSIMM20-3-552
24. Straka, M., Sofranko, M., Glova Vegsoova, O., Kovalcik, J.: Simulation of homogeneous production processes. Int. J. Simul. Model. **21**(2), 214–225 (2022). https://doi.org/10.2507/IJSIMM21-2-597
25. Knapcikova, L., Behunova, A., Behun, M.: Using a discrete event simulation as an effective method applied in the production of recycled material. Adv. Product. Eng. Manag. **15**(4), 431–440 (2020). https://doi.org/10.14743/apem2020.4.376
26. Fedorko, G., Vasil, M., Bartosova, M.: Use of simulation model for measurement of MilkRun system performance. Open Eng. **9**(1), 600–605 (2019). https://doi.org/10.1515/eng-2019-0067
27. Straka, M., Hurna, S., Bozogan, M., Spirkova, D.: Using continuous simulation for identifying bottlenecks in specific operation. Int. J. Simulat. Model. **18**(3), 408–419 (2019). https://doi.org/10.2507/IJSIMM18(3)477

Optimization of the Production Process of the Statory EM Production Line

Peter Trebuňa, Matúš Matiscsák, Marek Kliment, and Jozef Trojan

1 Introduction

Making changes in the production process necessary for innovation is largely responsible for problems such as production stoppages. Changes are often time-consuming, costly, and the outcome can be uncertain. Therefore, these changes need to be verified before they are implemented, as the application of modern technology is no guarantee of success [1]. Computer simulations play an important role in production planning issues because they help us to determine the behavior of a system before it is put into operation. Simulation programs allow not only a 3D representation and animation of machines, workers, and products but also provide analytical tools to validate proposed solutions. Cost saving is another of the main reasons why we create digital models based on which we optimize the process. For example, a digital model of a line can also serve as a subject for early diagnosis of problems or, if necessary, line analysis.

2 Theoretical Knowledge

This section of the document will explain the basic concepts that are necessary for a deeper understanding of the problem addressed in the paper.

P. Trebuňa · M. Matiscsák (✉) · M. Kliment · J. Trojan
Department of Industrial and Digital Engineering of the Faculty of Mechanical Engineering, Technical University of Košice, Košice, Slovakia
e-mail: matus.matiscsak@tuke.sk

M. Balog et al. (eds.), *EAI International Conference on Automation and Control in Theory and Practice*, EAI/Springer Innovations in Communication and Computing, https://doi.org/10.1007/978-3-031-31967-9_9

2.1 Industry 4.0

A characteristic feature of Industry 4.0 is the significant progress in technology, the trend toward digitization, and the gradual automation of workplaces. Artificial intelligence in manufacturing will take over jobs with simple and repetitive activities, but the consequence will be that there will be higher unemployment of lower-skilled employees and, conversely, there will be greater demand for higher-skilled labor [2].

The modern Industry 4.0 factory will monitor its status, and the systems implemented will have the capability of self-awareness [3]. The operation of the smart factory will be built on pillars such as the Internet of Things or Industrial Internet, Cloud, etc. The biggest weakness of the whole Industry 4.0 idea is the vulnerability of data [4].

The trend in the Industry 4.0 era is the creation of a kind of digital twin. The biggest advantage of such a digital twin is that it can be intertwined with the real operation and thus contains all the information about the operation in real time. Such a functionally coupled digital twin with traffic (see Fig. 1) also allows online data collection [5]. Data and information such as workstation utilization, product movement, downtime occurring, and the completion time of production process operations can be obtained [6].

The information obtained through the digital doppelganger concept can be used to optimize the production process [7]. The digital twin is thus a digital copy of all production processes, the flows affecting the process, and the possible states of the individual processes [8].

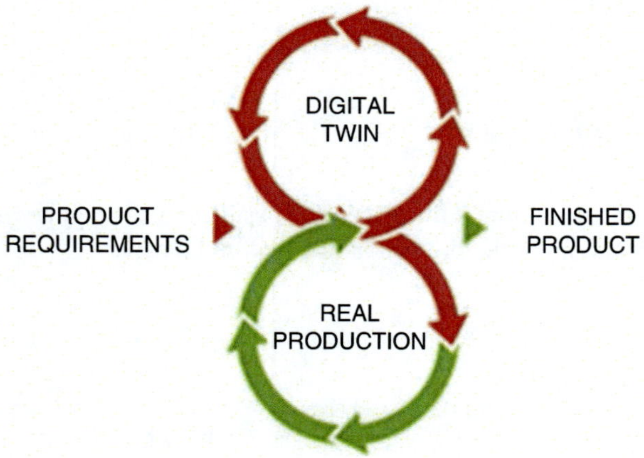

Fig. 1 Continuity of digital twin with real production

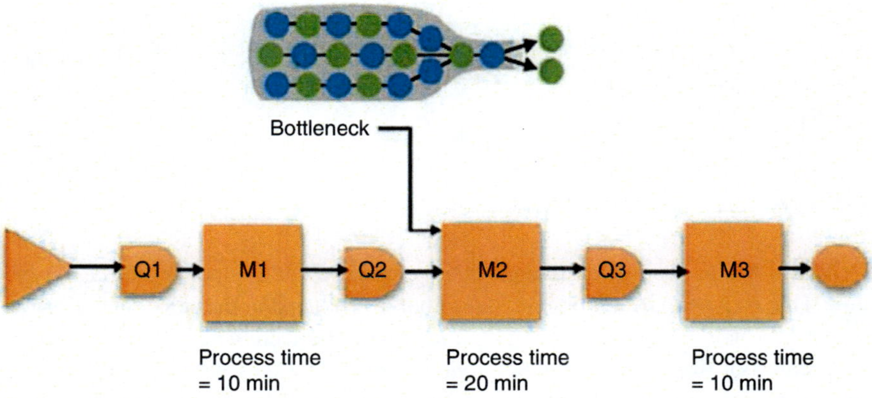

Fig. 2 Bottleneck

2.2 Bottleneck

A bottleneck (see Fig. 2) is a phenomenon in which the performance or capacity of an entire system is limited by one or a limited number of resources or components. A related term in English is "bottleneck," which translates to bottle neck. It means that if liquid spills out of a bottle, the rate of liquid outflow is limited by the width of the bottle neck. Increasing the width of the neck of the bottle makes it possible to increase the rate at which the liquid or product flow exits the neck of the bottle at different frequencies. Such limiting components of the system are often referred to as bottleneck points [9].

A bottleneck is an element in the ongoing production process that limits the speed at which the entire process can be completed. Typically, a bottleneck is the element in the production process with the smallest capacity, and the highest utilization, i.e., the capacity of the unit is less than the workload. Once the manufacturing process is mapped and the bottleneck is identified, the output of the system can be predicted and controlled [10].

Bottleneck analysis is becoming an increasingly important means of optimizing highly complex manufacturing processes and systems [11]. The goal of bottleneck analysis is to detect and eliminate them. Bottlenecks are a perfect example of small changes that can have a huge impact on the productivity of an entire system [12].

3 Statory EM Production Line

A production line is a configuration that contains a series of production steps. It is a flow of items from step to step that move progressively closer to becoming a finished product, part, or component.

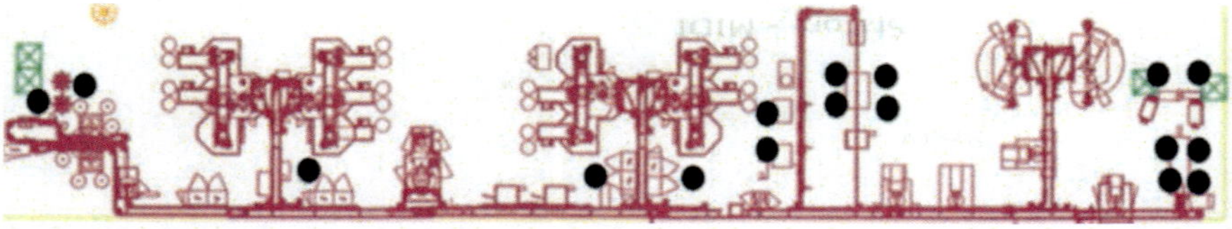

Fig. 3 Layout of the Statory EM production line

The EM Stator production line produces a compressor component, the EM type stator. This is the smallest model of stator that is produced in the selected company. There are 17 workers at the site, and their layout can be seen in Fig. 3. The operators who are part of the production line are deployed so that they can do their work. A work shift on the Statory EM line lasts 12 hours which is 43,200 seconds; however, there are two scheduled breaks during the work shift. In addition, each employee is entitled to one 30-minute, 15-minute, and 10-minute personal break. Of these, one 10-minute break is paid.

The production line consists of manual positions, semi-automatic, and automatic machines. The material flow on the Statory EM production line is linked by conveyors. Belt conveyors are used. Their speed is approximately 0,2 m/s. Figure 3 shows the layout of the production line, where the start of the production process is on the left side. The figure also shows the layout of the production line workers.

Indirect workers such as setters, tool technicians, maintenance technicians, or quality technicians are also part of the production line. These workers may be responsible for multiple production lines.

3.1 Description of Production Line Bottleneck

Pressing of the working winding – the press extrudes the winding into the stator grooves. This operation is carried out on an automatic device, which is the Elmotec VA706 extrusion press. The aim is to mold and extrude the working winding into the stator grooves on both sides. The cycle time of the machine is 8.16 seconds. This workstation represents the bottleneck of the production line. The conveyor capacity before the next operation is 40 pieces. In Fig. 4, we can see the indication of the bottleneck in production.

In this position, it must not stop, and there must be enough material at the entrance.

4 Configuration of the Simulation Model of the Production Process

The model was created in FlexSim software. The first step is to set the simulation time or length. The Statory EM production line currently operates in two-shift operation. Each shift lasts 12 hours. In the simulation, it was necessary to subtract

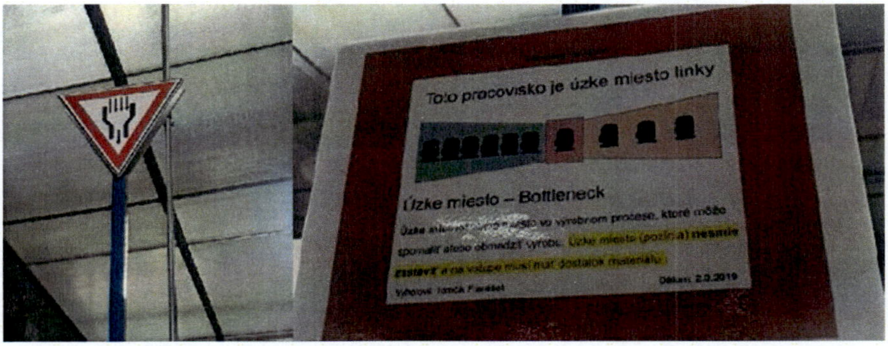

Fig. 4 Production line bottleneck marking

the break time, which is 55 minutes in total. The simulation runs for 11 h 5 min. The simulation software FlexSim allows the use of the Warmup time function. Warmup time translates to warmup time, which in the simulation represents the time when the production line is full, since, even in practice, production lines do not start production in an empty state. By using the Warmup time function [14], the simulation takes one hour longer, just to fill the production line [15]. After this time has elapsed, all counters are reset, and the simulation continues under normal conditions. Warmup time is set to 3600 seconds, which is 1 hour, and the duration of the simulation is from 5:00 to 17:05.

It was essential to insert the production line layout in the area reserved for modeling. The modeling area is also important for inserting individual objects that correspond to the real production line. The insertion of the layout was chosen because of the distances between the individual devices, the lengths of the individual conveyors and the distances traveled by the workers. The layout also ensures the correct arrangement of the production equipment and its placement. First, the backgrounds function was selected in the top bar, which translates to the background. Next, in the File name field, selecting the desired file with the contents of the selected layout was necessary. Then in the dialog box, the position, rotation, and size of the layout was set. The layout (Fig. 5) was provided by the company in .dwg format and created in a CAD program. A more detailed view of it can be seen in the appendixes of this thesis.

After setting the simulation length and inserting the layout, it was then possible to insert the production line objects. The objects are connected to each other by connectors that ensure their interconnection. In the simulation, a simple shape of the production line objects was chosen for better handling and to adapt the size of the objects to the real dimensions of the equipment, also because of the computer's performance capabilities. Blocks named Source, Queue, Processor, Sink, Combiner, Separator, Operator, Conveyor, Decision Point, Straight Conveyor, Curved Conveyor, visual elements, and others were inserted in the modeling area. Operator logic control was implemented by creating algorithms from the ProcessFlow library. The shape, color, size, and rotation of the objects can be changed by clicking on the object itself and then clicking on the General tab. Selecting a suitable object shape

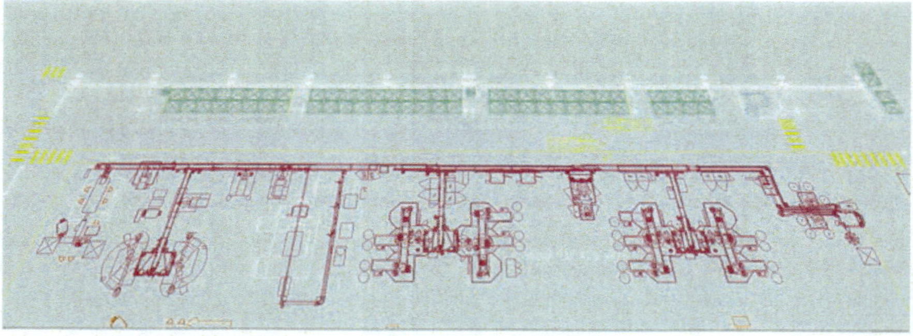

Fig. 5 Inserted layout on the modeling surface

from the software menu or inserting a custom file in the 3D Shapes section is possible. It is important to name each object according to the real names of the production equipment to configure the model accurately and avoid errors and shortcomings. In the same way, the basic shape of the conveyors has been changed. The first source of the model is the input that provides the arrival of the individual stator slats. The creation of the entity consists of an action where in the Toolbox section, it is necessary to right-click on the FlowItem Bin and then click on the green cross to insert a new object that will represent the stators. In the right section, we will adjust the dimensions on the X, Y, and Z axis [13]. In this case, the name chosen for the input was Statory_EM. It is also possible to change the color of the created objects or select a custom object. When creating the simulation model, the cycle times of the individual operations of the production line were entered into the Global Table. The first step was to open the Global Table from the Toolbox, and then the second step was to rename and populate the table according to the length of each operation of the production process. Cycle times are defined for each stator model and vary for some manufacturing operations. Then in the third step, it was possible to assign the cycle times to the individual Processors.

4.1 Results of the Simulation

We can compare the simulation results with the real results, based on the analysis of the production line results of 12-hour shifts, the average production quantity from March 6 to March 28, 2020, was 3912 pieces of stators (Fig. 6).

The simulation resulted in 3998 pieces per 12-hour shift. The result is close to the real results of the Statory EM production line. The difference is 86 pieces which is 2.19%.

Production line - Statory EM							
Year	Month	Week	Day	Shift	R/P/N	Model	Quantity
2021	3	9	6	C	Nocna	1355871	4355
2021	3	9	7	A	Ranna	1355808	4370
2021	3	9	7	B	Nocna	1355823	4508
2021	3	10	8	A	Nocna	1355836	2970
2021	3	10	9	C	Ranna	1355879	4050
2021	3	10	9	D	Nocna	1355870	4200
2021	3	10	10	B	Ranna	1355868	3720
2021	3	10	10	C	Nocna	1355802	3453
2021	3	10	11	A	Ranna	1355882	2500
2021	3	10	11	B	Nocna	1355882	4000
2021	3	10	12	D	Ranna	1355808	3921
2021	3	10	12	A	Nocna	1355864	4170
2021	3	10	13	C	Ranna	1355864	3894
2021	3	10	13	D	Nocna	1355876	4000
2021	3	10	14	B	Ranna	1355876	4070
2021	3	10	14	C	Nocna	1355869	4260
2021	3	11	15	A	Ranna	1355850	3950
2021	3	11	15	B	Nocna	1355835	4330
2021	3	11	16	D	Ranna	1355836	3874
2021	3	11	16	A	Nocna	1355829	4786
2021	3	11	17	C	Ranna	1355876	4015
2021	3	11	17	D	Nocna	1355880	3855
2021	3	11	18	B	Ranna	1355836	4153
2021	3	11	18	C	Nocna	1355876	4360
2021	3	11	19	A	Ranna	1355802	3600
2021	3	11	19	B	Nocna	1355836	4200
2021	3	11	20	D	Ranna	1355838	3840
2021	3	11	20	A	Nocna	1355871	4100
2021	3	11	21	C	Ranna	1355868	4328
2021	3	11	21	D	Nocna	1355882	3370
2021	3	12	22	C	Nocna	1355882	3800
2021	3	12	23	A	Ranna	1355882	3260
2021	3	12	23	B	Nocna	1355811	4200
2021	3	12	24	D	Ranna	1355838	3430
2021	3	12	24	A	Nocna	1355882	4000
2021	3	12	25	C	Ranna	1355836	3865
2021	3	12	25	D	Nocna	1355802	3530
2021	3	12	26	B	Ranna	1355876	3633
2021	3	12	26	C	Nocna	1355882	4206
2021	3	12	27	A	Ranna	1355876	4020
2021	3	12	27	B	Nocna	1355876	3650
2021	3	12	28	D	Ranna	1355868	3600
2021	3	12	28	A	Nocna	1355829	3800
							3912

Fig. 6 Real production results of Statory EM production line

5 Design to Improve the Production Process by Using Existing Operators

Based on the report, the operator in front of the bottleneck labeled OP_Stators_EM_03 is 94.28% utilized, and the operators labeled OP_Stators_EM_01 and OP_Stators_EM_02 are 25.70% and 46.43% utilized, respectively. It is important to note that operator 3 has almost no weight load and does not need to lift anything, but operators 1 and 2 have high weight loads. When solving the designs, it is necessary to take into account the [10] "Government Regulation No. 281/2006 Coll. - Regulation of the Government of the Slovak Republic on minimum safety and health requirements for manual handling of loads" [11]. According to the results of the activities of operator, 3, 22.56% of the 12-hour work shift is Setup status, which includes the activities of the operator, which are the replenishment of material (Fig. 7). During this time, it occurs that bottleneck VA706 is waiting for material as it does not have enough pieces to enter and hence is not efficiently utilized as can be seen in the report, 18.28% of the time this workstation is idle.

The weight load of operators 1, 2 is above the limit set by Government Regulation No. 281/2006 Coll, the operators are males aged between 18 and 29 years (Fig. 8).

The proposal is to assign the activities of Operator 3, which constrain the bottleneck and ensure material shortage at the bottleneck entrance, to Operators 1 and 2, as their utilization is only at 25.70% and 46.43%, respectively. They are also the closest to the site. However, it is not possible for any operator to perform the given activities. Therefore, when assigning operators 1 and 2 to certain activities of operator 3, it is necessary to increase the qualification of operators 1 and 2. Part of the proposal was that operator 3, once the conveyor was filled prior to the bottleneck, would be released to perform other activities and be further utilized. It was necessary to add logic to that operator to ensure that the operator was released after the conveyor had been filled before the bottleneck. A Queue block was inserted at the beginning of the conveyor with a maximum capacity of 2. If this value was reached, the operator would be released. If the conveyor is full, the operator can then be released for material replenishment.

It should be taken into account that operator 3 is assigned the number 1 in Labels, i.e., the responsibility for a given workstation, in case of multiple workstations, but will only carry out the activities on the workstations when the conveyor is full and only then can it leave M_501755.

After assigning operators 1 and 2 to operator 3's bottleneck-constrained activities, bottleneck utilization increased by 6.05%. Also, the utilization of operators is more even than in the previous report (Fig. 9).

The output increased from 3998 products to 4294 units.

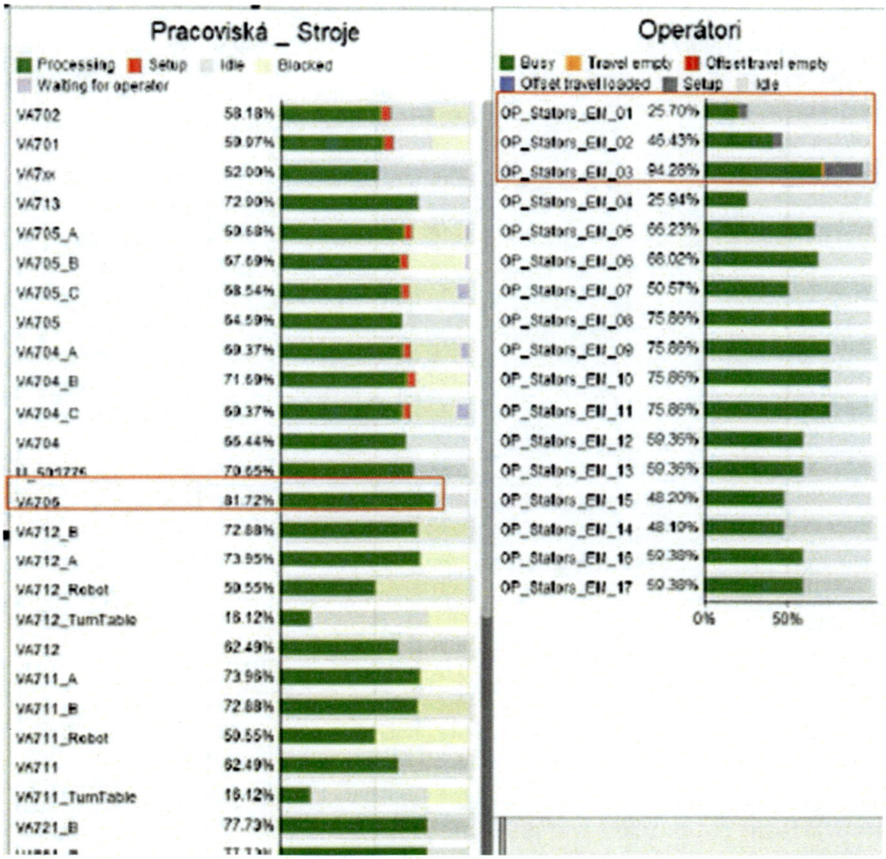

Fig. 7 Report results in the standard model

	weight
OP_Stators_EM_01	14004.09
OP_Stators_EM_02	9425.33
OP_Stators_EM_03	0

Vek	Podmienky	Maximálna hmotnosť bremena		Maximálna celozmenová hmotnosť (kg)	
		Muži	Ženy	Muži	Ženy
18 - 29 r.	priaznivé nepriaznivé	50 kg 40 kg	15 kg 10 kg	10 000 8 000	6 500 5 500

Fig. 8 Weight load of workers

6 Conclusion

The work aimed to create a digital model of the Statory EM production line and then design a solution to increase the bottleneck efficiency. The digital enterprise was produced in FlexSim.

Based on the reports of the model, which show the percentage of utilization of individual operators, equipment, and the overall output of the production line, it was then possible to verify the given results with the results of the real production line.

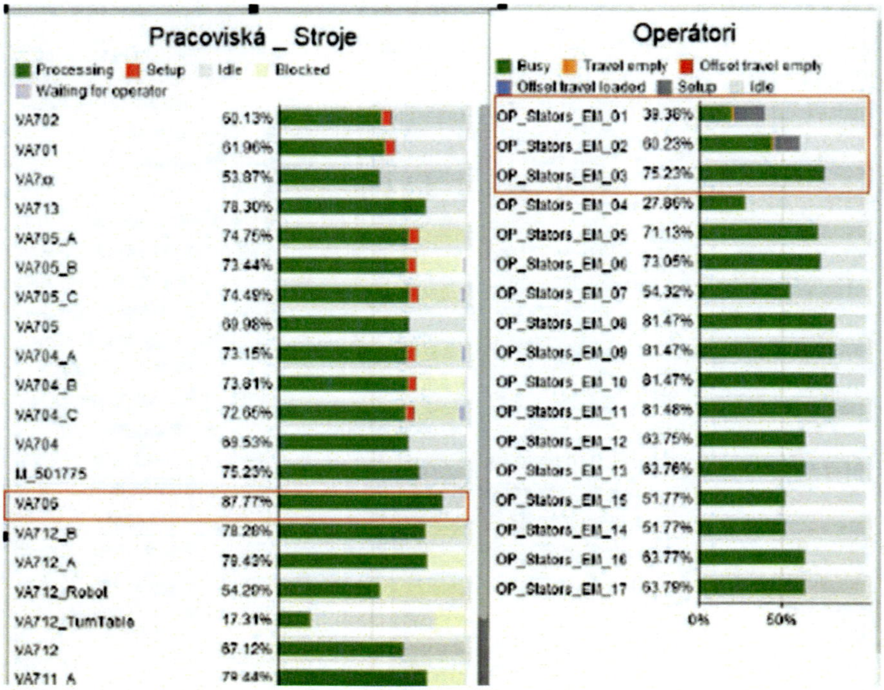

Fig. 9 Report after assigning activities to operators 1 and 2

Therefore, considering the results of the model, we can say that they are almost identical to the results of the real production line.

When designing new solutions for the production line, it is possible to consider the weight loads when handling loads of individual operators. The bottleneck streamlining solution proposed was in the form of reorganizing the responsibilities of individual workers in front of the bottleneck. After implementing the individual suggestions, the objective was achieved as the bottleneck utilization increased from 81.72% to 87.77% in the design.

Acknowledgments This article was created by the implementation of the grant projects: APVV-17-0258 Digital engineering elements application in innovation and optimization of production flows. APVV-19-0418 Intelligent solutions to enhance business innovation capability in the process of transforming them into smart businesses. VEGA 1/0438/20 Interaction of digital technologies to support software and hardware communication of the advanced production system platform. VEGA 1/0508/22 Innovative and digital technologies in manufacturing and logistics processes and system, KEGA 020TUKE-4/2023 "Systematic development of the competence profile of students of industrial and digital engineering in the process of higher education."

References

1. Fedorko, G., Vasil, M., Bartosova, M.: Use of simulation model for measurement of MilkRun system performance. Open Eng. **9**(1), 600–605 (2019). https://doi.org/10.1515/eng-2019-0067
2. Balamaurugan, E., Flaih, L.R., Yuvaraj, D., Sangeetha, K., Jayanthiladevi, A., Kumar, T.S.: Use case of artificial intelligence in machine learning manufacturing 4.0., pp. 656–659. International Conference on Computational Intelligence and Knowledge Economy (ICCIKE), Dubai, United Arab Emirates (2019). https://doi.org/10.1109/ICCIKE47802.2019.9004327
3. Bonilla, S.H., Silva, H.R.O., Silva, M.T., Gonçalves, R.F., Sacomano, J.B.: Industry 4.0 and sustainability implications: a scenario-based analysis of the impacts and challenges. Sustainability. **10**(10), 1–24 (2018). https://doi.org/10.3390/su10103740
4. Straka, M., Taušová, M., Rosová, A., Cehlár, M., Kačmáry, P., Sisol, M., Ignácz, P., Farkas, C.: Big data analytics of a waste recycling simulation logistics system. Pol. J. Environ. Stud. **29**, 2355–2364 (2020). https://doi.org/10.15244/pjoes/108684
5. Brozzi, R., Forti, D., Rauch, E., Matt, D.T.: The advantages of industry 4.0 applications for sustainability: Results from a sample of manufacturing companies. Sustainability. **12**(9), 1–19 (2020). https://doi.org/10.3390/su12093647
6. Raj, A., Dwivedi, G., Sharma, A., Jabbour, A.B.L.S., Rajak, S.: Barriers to the adoption ofindustry 4.0 technologies in the manufacturing sector: an inter-country comparativeper-spective. Int. J. Prod. Econ. **224**, 1–17 (2020). https://doi.org/10.1016/j.ijpe.2019.107546
7. Burganova, N.; Grznár, P.; Mozol, Š.: Challenges of Factory of Future in the Context of Adaptive Manufacturing (2021)
8. Shafiq, S.I., Sanin, C., Szczerbicki, E.: Knowledge based virtual modeling and simulation of manufacturing processes for Industry 4.0. Cybernet. Syst. **51**(2), 84–102 (2020). https://doi.org/10.1080/01969722.2019.1705546
9. Lai, X., Shui, H., Ding, D., Ni, J.: Data-driven dynamic bottleneck detection in complex manufacturing systems. J. Manuf. Syst. **12**, 662–675 (2021). https://doi.org/10.1016/j.jmsy.2021.07.016
10. Lin, L.: A systematic-theoretic analysis of data-driven throughput bottleneck detection of production systems. J. Manuf. Syst. **47**, 43–52 (2018). https://doi.org/10.1016/j.jmsy.2018.03.001
11. Kovalčík, J., Balog, M.: Automatic identification systems for management - material flow control and stock status. Acta logistica. **8**(1), 55–64 (2021). https://doi.org/10.22306/al.v8i1.202
12. Zupancic, D., Buchmeister, B., Aljaz, T.: Reducing the time of task execution with existing resources –comparison of approaches. Int. J. Simulat. Model. **16**(3), 484–496 (2017). https://doi.org/10.2507/IJSIMM16(3)10.394
13. Stefanik, A., Grznar, P., Micieta, B.: Tools for Continual Process Improvement - Simulation and Benchmarking. In: Katalinic, B. (ed.) Annals of DAAAM for 2003 & Proceedings of the 14th International DAAAM Symposium: intelligent manufacturing & automation: focus on reconstruction and development, pp. 443–444. Daaam Int Vienna, Wien (2003). https://www.webofscience.com/wos/woscc/full-record/WOS:000245326200218
14. Grznar, P., Gregor, M., Mozol, S., Krajcovic, M., Dulina, L., Gaso, M., Major, M.: A system to determine the optimal work-in-Progress inventory stored in interoperation manufacturing buffers. Sustainability. **11**(14), 3949 (2019). https://doi.org/10.3390/su11143949
15. Gabajová, G., Krajčovič, M., Matys, M., Furmannová, B., Burganová, N.: Designing virtual workplace using unity 3D game engine. Acta Tecnología. **7**(1), 35–39 (2021). https://doi.org/10.22306/atec.v7i1.101

Implementation of the RTLS Localization System in the Micro Logistics Laboratory

Peter Trebuňa, Marek Mizerák, and Tomáš Švantner

1 Introduction

A real-time location system (RTLS stands for Real-Time Location Systems) refers to any system that accurately determines the location of an item or person. RTLS is not a specific type of system or technology but rather a goal that can be achieved using a variety of asset location and management systems. An important aspect of RTLS is the time that assets are tracked, and this data can be used in different ways depending on the application. For example, some applications only need timestamps when an asset moves through an area, while other RTLS applications require much more detailed visibility and require constant updating of time data [1].

The ideal real-time location system can accurately locate, track, and manage assets, inventory, or people, helping companies make informed decisions based on collected location data. Furthermore, it is possible to say that the RTLS system serves to identify the tag and locate it and track its movement in real time. The system determines the location using small devices placed on the objects we track, active radio frequency identification (RFID) tags. RTLS technology is mainly intended for tracking and determining the location of objects indoors or outdoors (the premises of a manufacturing company, etc.). RTLS is used in many industries with specific employee and asset-tracking applications. These applications can be found in the manufacturing and mining industries but are most prominent in the healthcare industry [2].

The accuracy of this system ranges from meters to tens of centimeters, depending on the technology used.

P. Trebuňa · M. Mizerák (✉) · T. Švantner
Department of Industrial and Digital Engineering of the Faculty of Mechanical Engineering, Technical university of Košice, Košice, Slovakia
e-mail: marek.mizerak@tuke.sk

M. Balog et al. (eds.), *EAI International Conference on Automation and Control in Theory and Practice*, EAI/Springer Innovations in Communication and Computing, https://doi.org/10.1007/978-3-031-31967-9_10

Each RTLS system uses a certain infrastructure. For the system to work smoothly and correctly, the monitored area must be covered by the wireless network so that each label has a sufficient signal. Each active RFID tag in the network then independently transmits data from that network to a server, where the data is further processed. The position or trajectory of the marked movement is evaluated via the application. Labels are small electronic devices that are attached to any object or individual that needs to be tracked. The tags serve as a transmitter and transmit flashes that are captured by anchors and sent to a location server to calculate the tag positions [3].

2 Hardware and Software Support

Sewio UWB RTLS consists of two types of hardware: signal transmitters, called tags, used to track entities, and signal receivers, called anchors, used to receive signals from tags.

Tags are small electronic devices that are attached to any object or individual that needs to be tracked. The tags act as a transmitter and emit flashes that are received by the anchors and sent to a location server to calculate tag positions.

Anchors are electronic devices that detect UWB pulses emitted by UWB tags and then forward them to a location server to calculate tag positions. To cover an area with an indoor tracking system, a set of anchors must be installed over the area to create a location infrastructure [4].

Sewio built its vision on the idea of "all in one," where it created a software platform called RTLS studio. This software is designed for seamless deployment, maintenance, and complete control of the internal tracking platform.

The contents of this package include several modules, namely:

- RTLS Manager,
- RTLS Planner,
- RTLS Monitor,
- RTLS Player,
- Sensmap (Fig. 1).

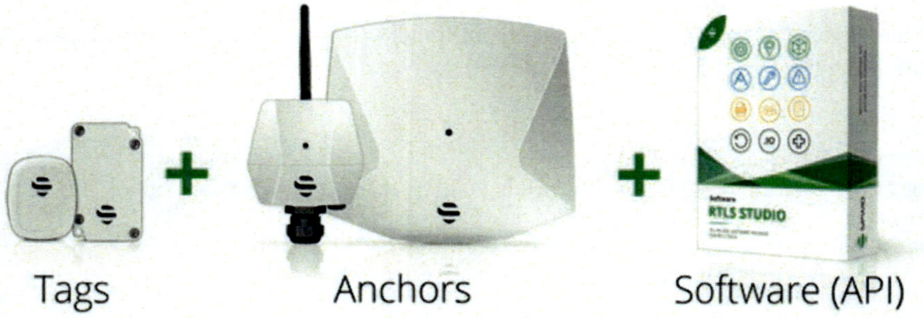

Fig. 1 Sewio hardware and software platform

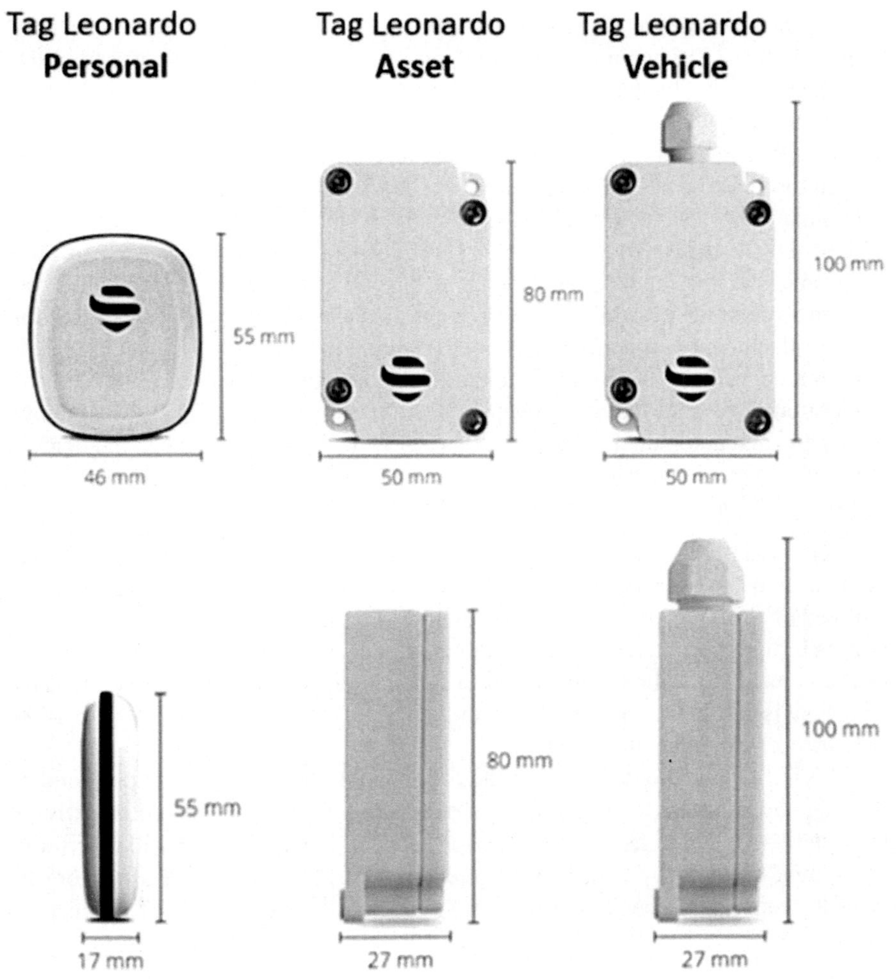

Fig. 2 RTLS tags

The picture shows three types of tags from the company Sewio, which deals with the technology of tracking property and people. To this end, it has developed the following three types of tags, which you can see in Fig. 2:

- Leonardo personal tag – the Leonardo personal tag is intended for identifying and localizing employees and other persons. In addition to industry, it is often used in healthcare, sports, retail, and museums. Its task is to increase the safety of employees, navigate people within the monitored area, analyze their movement, and replace the classic methods of recording employee attendance. It is equipped with UWB radio, near field communication (NFC), and Bluetooth technologies, LED indicators, allows wireless recharging, fast reconfigurability via RTLS studio, and an accelerometer and barometer for determining the Z-axis coordinates. It weighs 30 g and is equipped with a Li-Pol battery with a capacity of 300 mAh. The monitored persons most often wear it on their wrists in the manner of a wristwatch, and it is also possible to fasten it on clothes or a safety helmet [5].

- Leonardo Asset tag is designed to track and identify any asset. It is used for monitoring and analyzing material flows and in the e-Kanban system. It provides companies with a quick return on investment. Just as the personal tag is equipped with UWB radio, Bluetooth, NFC technology, and an accelerometer, the barometer is missing. It is powered by a coin cell battery with a capacity of 1000 mAh without the possibility of recharging. It can be extended with a magnet for easy attachment to monitored objects. Its weight is 70 g,
- Leonardo Vehicle tag does not contain a battery and is therefore intended for monitoring devices that can power it with their own battery. These are mainly forklifts and automated guided vehicle (AGV) trucks or milk runs in the field of intralogistics and storage. It is equipped with accelerometer, barometer, UWB radio, NFC, and Bluetooth. It weighs 104 g [6].

Anchors are electronic devices that detect shortwave pulses and then forward them to a location server to calculate tag positions. These devices are usually mounted on the walls or columns of production areas, providing an optimal view of various business processes. To cover the area with an internal monitoring system, anchors must be installed above the area to create the location infrastructure. The advantage is that the technology is not limited in this respect; therefore, up to 1000 such anchors can be mounted (Fig. 3).

All RTLS applications consist of the following basic components: transponder, receiver, and software for interpreting data from each of them. The complexity of the system, the chosen technology, and the scope of the application will determine the amount of hardware and software needed to create an ideal RTLS network to cover the required space for the collection and subsequent analysis of production

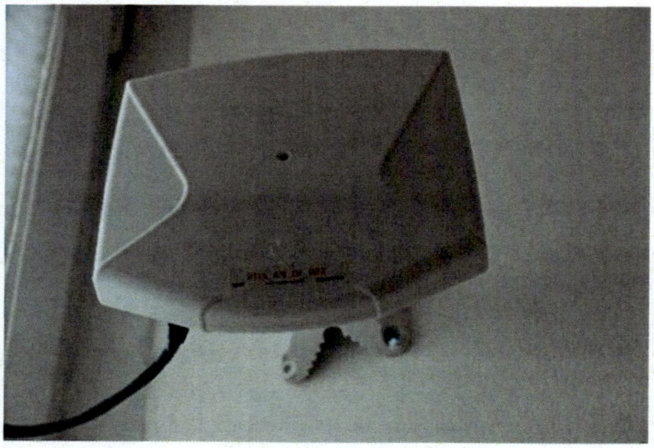

Fig. 3 UWB anchor for indoor location tracking used in laboratory

data. We want to follow this idea when comparing collected production data and results when performing simulations in a software environment [7].

3 Technical Realization of the Implementation

In the first step, the issue of the placement of anchors in the laboratory was solved. In this step, care is taken above all to achieve the greatest possible accuracy in the localization of tags. It was necessary to observe the following two principles:

- All anchors must be mounted in one horizontal plane around the perimeter of the laboratory,
- The anchors must be fixed on the brackets to avoid connection with the walls or ceiling (the brackets are included in the box of each anchor) [8].

Four anchors were chosen for the best possible network coverage of the laboratory. For a better imagination, the layout of the anchors can be seen in Fig. 4.

In order to be able to receive data and information from the anchors, as a result, a network had to be created using UTP cables for a secure connection with the server (Fig. 5).

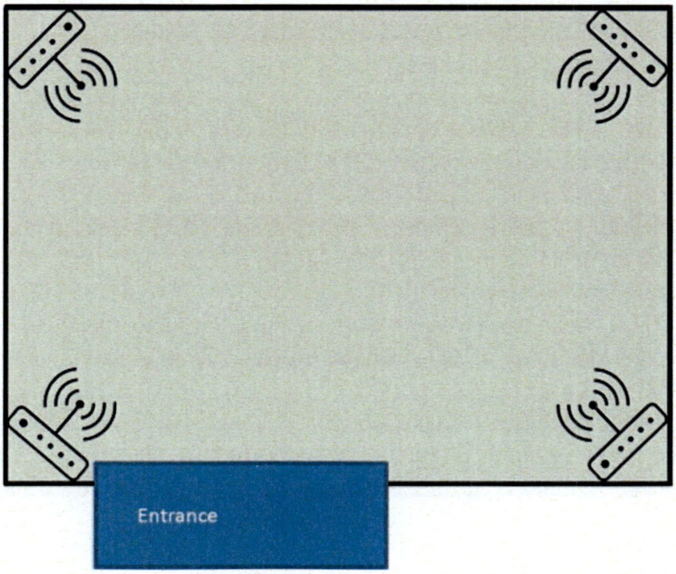

Fig. 4 Drawing of the placement of anchors in the micro logistics laboratory

Fig. 5 Cisco switch used in laboratory

4 Configuring Anchors and Tags

In our case, the connection of the anchors with the switch and the computer assembly can also be done wirelessly, but the decision was made for a more reliable option, which is the mentioned UTP cable. The unique IP address of each anchor and switch is used for configuration. After the hardware side of the preparation is done, the software part comes next. The configuration itself takes place in the software portfolio from the technology manufacturer itself. In our case, it is from Sewio, namely Sewio – RTLS studio. The program must be installed according to the instructions, and then it will start [9] (Fig. 6).

In the next step, anchor synchronization is started using the "Anchor Initiatlization" module, which is extremely necessary for the overall correct operation. This module sets the range of IP addresses and the anchors' initialization. If this process is carried out correctly, the software offers to select the anchor that is most adequate and, at the same time, will be designated as the main anchor. The choice of such an anchor depends on its position, where it is the one that has the greatest "supervision" of the others [10].

Sewio – RTLS studio has another module, namely Sensmap, thanks to which the user can set the positions of the anchors. We can do this setting manually or automatically.

The advice for configuring tags comes in the step when we have configured anchors. The tags – summary function is used for this operation. All tags are configured wirelessly and according to parameters selected by the user, such as:

- Refresh rate,
- Channel,
- RF Profile,
- Sleep mode and so on.

#	Anchor Mac	Update clickable	Alias editable	Status	IP	FW
	All Anchors	UPDATE ALL	N/A	N/A	N/A	N/A
1	d8803961fe3c	UPDATE	SE12	✔	N/A	1.024
2	d88039621805	UPDATE	SE07	✔	N/A	1.024
3	d8803961fd3c	UPDATE	SE14	✔	N/A	1.024
4	d88039625495	UPDATE	SE13	✔	N/A	1.024
5	d88039623ec5	UPDATE	SE04	✔	N/A	1.024
6	d8803961ed86	UPDATE	SE08	✔	N/A	1.024
7	d8803962053e	UPDATE	SE05	✔	N/A	1.024
8	d8803962794e	UPDATE	SE06	✔	N/A	1.024
9	d880396230c4	UPDATE	SE52	✔	N/A	1.024
10	d88039623772	UPDATE	SE38	✔	N/A	1.024
11	d8803962934c	UPDATE	SE42	✔	N/A	1.024
12	d88039625905	UPDATE	SE40	✔	N/A	1.024
13	d88039628a85	UPDATE	SE50	✔	N/A	1.024

Fig. 6 RTLS Manager

Fig. 7 RTLS Tag used in laboratory

After successfully setting all selected parameters, it is necessary to choose an anchor, thanks to which the wireless configuration will be transferred. However, this rule mostly applies to larger objects such as production halls. This rule does not apply in our case, as our laboratory does not have large dimensions. For that reason, in our case, a random anchor was selected, which was intended as a configuration anchor and had the best communication reach to tags. After the configuration itself, tags must be restarted. The next phase is testing, when the set parameters are verified and the modules are checked (Fig. 7).

5 Use of the RTLS Localization System in Laboratory Conditions and Verification of the State of the Simulation Model

As part of the use of the RTLS localization system in laboratory enterprises at the Department of Industrial and Digital Engineering, we want to focus not only on localization as such but also on the creation of a mini production line and the use of robots and conveyors, which would result in the creation of a space not only for students and their work but also for beneficial scientific articles and publications. Furthermore, another important aim would be to create a system and methodology for verifying simulation models using RTLS localization technology [11].

The idea of state verification based on simulations is based on the nature of the properties, parameters, and state of RTLS technology, which can research a given area. There are two variants of using RTLS technology in cooperation with simulation models.

The first variant estimates the possibility of use in monitoring and localization in verifying the veracity of the simulation results. This is an examination of quantities that should confirm or refute the simulation outputs in different areas. In practice, we can imagine this as monitoring production and logistics flows, people, machines, and other various equipment for a specified period of time, which would be sufficient to verify the simulation. Furthermore, it can be argued that based on these monitoring measurements, it would be possible to create a superstructure for superstructure optimization and the like [12].

The second variant is the opposite method of measurement. This method would be used in companies with no data for the creation of simulations, or it is a problem to create such data, respectively, measured in conventional ways. Based on this, the localization RTLS system would function as a data bus to monitor this predetermined data and then evaluate it in a software environment. Thus, it would be possible to create tables as a basis for entering the simulation software Technomatix plant simulation [13].

As a form of these analyses, we can imagine the above-mentioned spaghetti flow diagrams or so-called heatmaps, with which we can even track specific employees or forklifts. Of course, RTLS localization technologies are a form of optimization technologies in their own right, but nowadays, the idea arises when these two technologies could be controlled or supplemented with each other [14].

This method of verification and cooperation of two different technologies has not yet been physically verified, but there is a precondition for their closer cooperation [15].

6 Conclusion

Today, localization technologies are becoming an integral "collaborator" for executives and managers, who ultimately help them with a general overview of the operation of a manufacturing company. In such cases, it is possible to build on other optimization options, both material and production flows. As part of the creation of the localization network infrastructure, there is an interest in cooperation with practice on case study solutions and cooperation with students on their final theses.

Acknowledgments This article was supported by projects VEGA 1/0438/20 Interaction of digital technologies to support software and hardware communication of an advanced production system platform, KEGA 020 TUKE-4/2023 Trebuňa (2023–2025) Systematic development of the competence profile of students of industrial and digital engineering in the process of higher education, APVV- 17-0258 Digital engineering elements application in innovation and optimization of production flows, APVV-19-0418 Intelligent solutions to enhance business innovation capability in the process of transforming them into smart businesses. VEGA 1/0508/22 (2022–2025) Innovative and digital technologies in manufacturing and logistics processes and systems. 020 TUKE-4/2023 Systematic development of the competence profile of industrial and digital engineering students in the process of higher education.

References

1. Mozolova, L., Grznár, P., Mozol, Š., Furmanová, B.: Development of Digital Support Tools (2021)
2. Sinko, S., Marinic, E., Poljanec, B., Obrecht, M., Gajsek, B.: Performance-oriented UWB RTLS decision-making approach. Sustainability. **14**(18), 11456 (2022)
3. Fedorko, G., Vasil, M., Bartosova, M.: Use of simulation model for measurement of MilkRun system performance. Open Eng. **9**(1), 600–605 (2019)
4. Burganova, N., Grznár, P., Mozol, Š.: Challenges of factory of future in the context of adaptive manufacturing, 1st edn, pp. 32–35. InvEnt 2021: Industrial Engineering – Invention for Enterprise (2021)
5. Stefanik, A., Grznar, P., Micieta, B.: Tools for continual process improvement - simulation and benchmarking. In: Katalinic, B. (ed.) Annals of DAAAM for 2003 & Proceedings of the 14th International DAAAM Symposium: Intelligent Manufacturing & Automation: Focus on Reconstruction and Development, vol. 2003, pp. 443–444. Daaam Int, Wien (2018)
6. Masek, P., Sedlacek, P., Ometov, A., et al.: Improving the precision of wireless localization algorithms: ML techniques for indoor. In: 43rd International Conference on Telecommunications and Signal Processing (TSP), pp. 589–594. Electr Network (2020)
7. Boulos, M.N.K., Berry, G.: Real-time locating systems (RTLS) in healthcare: a condensed primer. Int. J. Health Geogr. **11**(1), 25–32 (2012)
8. Choi, H., Baek, Y., Lee, B.: Design and implementation of practical asset tracking system in container terminals. Int. J. Precision Eng. Manuf. **13**(11), 1955–1964 (2012)
9. Li, H., Chan, G., Wong, J.K.W., Skitmore, M.: Real-time locating systems applications in construction. Autom. Constr. **63**, 37–47 (2016)
10. Trebuna, P., Petrikova, A., Pekarcikova, M.: Influence of physical factors of working environment on worker's performance from the ergonomic point of view. Acta Simulat. **3**(3), 1–9 (2017)
11. Straka, M., Khaouri, S., Rosova, A., Caganova, D., Culkova, K.: Utilization of computer simulation for waste separation design as a logistics system. Int. J. Simulat. Model. **17**(4), 583–596 (2018)
12. Kovalčík, J., Balog, M.: Automatic identification systems for management - material flow control and stock status. Acta Logistica. **8**(1), 55–64 (2021)
13. Gabajová, G., Krajčovič, M., Matys, M., Furmannová, B., Burganová, N.: Designing virtual workplace using unity 3d game engine. Acta Tecnología. **7**(1), 35–39 (2021)

14. Grznar, P., Gregor, M., Mozol, S., Krajcovic, M., Dulina, L., Gaso, M., Major, M.: A system to determine the optimal work-in-Progress inventory stored in interoperation manufacturing buffers. Sustainability. **11**(14), 3949 (2019)
15. Stefanik, A., Grznar, P., Micieta, B.: Tools for continual process improvement - simulation and benchmarking. In: Katalinic, B. (ed.) Annals of DAAAM for 2003 & Proceedings of the 14th International DAAAM Symposium: Intelligent Manufacturing & Automation: Focus on Reconstruction and Development, pp. 443–444. Daaam Int, Vienna: Wien (2003)

Theoretical Basis of Application of the Parameter of Dielectric Permeability of Hydrocarbon Feedstock during its Processing

Nabil Abdel Sater, Andrey Grigorov, Alona Tulska, Mikhail Nahliuk, and Peter Lazorik

1 Introduction

Reducing the costs associated with the production of a commercial petroleum product is the main task of any oil refinery, the solution of which the cost of the product and, as a consequence, its competitiveness will significantly depend. Basic technological expenses consider raw materials, logistic research, and energy for raw material processing. One of the main mechanisms, generally recognized in the world practice of oil refining, for reducing production costs is the timely forecasting of the most rational direction of processing of raw materials and the flexibility (controllability) of the technological process. The implementation of these mechanisms is based on objective information about the group and chemical composition of crude oil and distillate fractions obtained from it. Information about the composition of all material flows, in order to intensify the process (e.g., to minimize fluctuations in the composition of raw materials) of petroleum goods production, must be obtained in real time. To achieve this, it is necessary to use a certain integral parameter that will promptly provide the plant operator with information about the composition of material flows. On the basis of this information, an understanding of the efficiency of oil raw material processing at individual sites (preparation, processing, and compounding) and the energy efficiency of production as a whole will be formed in the future.

N. A. Sater · A. Grigorov · A. Tulska (✉)
National Technical University, Kharkov Polytechnic Institute, Kharkov, Ukraine
e-mail: alona.tulska@khpi.edu.ua

M. Nahliuk
Kharkiv National Automobile and Highway University, Kharkiv, Ukraine

P. Lazorik
Faculty of Manufacturing Technologies with the seat in Prešov, Technical University of Košice, Prešov, Slovakia

M. Balog et al. (eds.), *EAI International Conference on Automation and Control in Theory and Practice*, EAI/Springer Innovations in Communication and Computing, https://doi.org/10.1007/978-3-031-31967-9_11

2 Literature Review

Electrophysical indicators are often used as indicators of certain qualitative characteristics of oil and petroleum products such as dielectric permittivity, tangent of the dielectric loss angle, and electrical conductivity [1]. Devices like quality analyzers and sensors based on the measurement of these indicators are widely used for determining the quality of oil and petroleum products [3]. Due to the ease of measurement and the insignificant temperature coefficient, the use of dielectric constant has become the most widespread in the field of oil refining [4].

The possibility of obtaining objective factual information on the composition of feedstock and products of its processing gives the use of electrophysical indicators, particularly the relative dielectric constant (ε). We have chosen this indicator in accordance with the fact that oil and oil products, products of destructive processing of polymers, belong to nonpolar dielectrics, and the measure of their electrical properties is the index ε [6]. At the same time, the index ε is one of the so-called integral indicators and has proven itself well in laboratory practice when determining the quality of oil and oil products [8]. Its value depends both on the hydrocarbon composition of oil and the products of its processing and on various impurities that can accumulate in them when the technological regime of processing, storage, and transportation are violated. Aromatic (average value 2.4105), cyclic (average value 2.1206), and naphthenic (average value 2.0501) hydrocarbons have the biggest impact on the value of dielectric constant of oil prepared for processing [9]. Paraffin hydrocarbons perform not have such a considerable effect on dielectric constant (average value 1.9760). Also, the value of the dielectric constant of oil and its fractions will significantly depend on the content of oxygen-, nitrogen-, and sulfuric compounds, which are considered as undesirable impurities that complicate the processing of raw materials and pass into the composition of products – distillate and residual fractions, worsening their quality [7]. Besides, it should be considered that as the boiling point of petroleum distillates increases, their dielectric constant also increases. Thus, among the products of primary oil processing, gasoline has the minimum dielectric constant value (boiling point −180 °C), and fuel oil has the maximum value (boiling point above 360 °C). Additives in petroleum products have a significant impact on the dielectric constant of commodity oil products [8]. Due to their polarity, they tend to increase the dielectric constant of the commodity product compared to the corresponding oil fraction. Thus, applying the index ε directly in the conditions of an oil refinery plant at processing units of crude oil and polymer feedstock as part of the global monitoring and correction system of the technological process, it becomes possible to significantly increase its controllability and flexibility. This obstacle, in its turn, is a very important task that should be solved. There should be mentioned that a solution of this problem is being provided by such companies as Chevron Corporation, Royal Dutch Shell, KMG International, and British Petroleum [10].

3 Materials and Methods

The main sections of the primary oil refining unit and destructive processing of polymer feedstock which is monitored by global monitoring and correction system (GMCS). The GMCS realize control by placed capacitive sensors that collecting primary information. Structural block diagram of a typical GMCS is presented in Fig. 1.

Main structural elements of the global monitoring and correction system (see Fig. 2) are a primary information collection unit, which consists of sensors with a two- or three-electrode design; a unit for converting and processing the received primary information and a unit for light indication and light signaling of a technological mode violation; database; correction block.

The principle of operation of the given above global monitoring and correction system is based on obtaining primary information from the primary information collection unit by means of capacitive sensors (D1-D9) combined into blocks that are located in the areas of raw material preparation (Sensor block 1), raw material processing (Sensor block 2) and compounding products (Sensor block 3). Then, information from the capacitive sensors blocks will go to the information conversion unit (the capacitance measured by the sensors is converted into dielectric constant) and the data comparison unit. Finally, the data is compared with the values in the data bank.

In case of violation of the technological regime, which entails a change in the composition of the product (prepared raw materials, distillate fractions, or commercial oil products), the information is transmitted to the signaling unit (light indication and sound signaling are provided) and the correction unit, where the technological parameters of the process change. Then, off-test products are sent back to the process.

The global monitoring and correction system also provides blocks of level gauges, which are placed in tanks at the storage areas for raw materials (Block 1 of

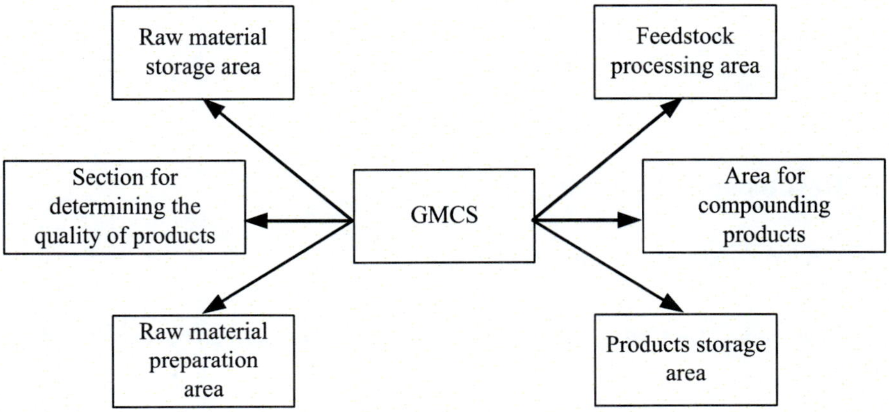

Fig. 1 Structural block diagram of the GMCS at the oil refinery plant

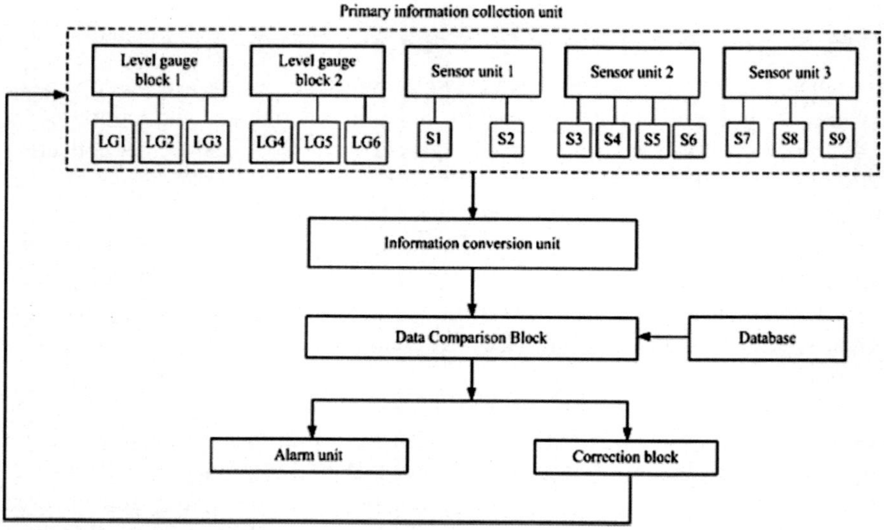

Fig. 2 Block diagram of GMCS

level gauges) and finished products (Block of level gauges 1). An example is wave radar level transmitters, the principle of operation of which is based on the use of pulses of electromagnetic energy transmitted through a waveguide (probe). When the pulse reaches the surface of a liquid with a higher dielectric constant than that of air ($\varepsilon r = 1$), in which it propagated before, the pulse is reflected. The fast response timing circuitry accurately determines the pulse propagation time and provides accurate measurement of the liquid (or bulk material) level. The amplitude of the reflected pulse depends on the dielectric constant of the product. The higher the dielectric constant is, the stronger the reflection is [11].

The presented global monitoring and correction system carries out continuous monitoring of the technological process at the unit for the primary processing of crude oil.

Further, we will take a closer look at the work of the global monitoring and correction system at each of the sections of the primary processing unit for crude oil.

4 Results and Discussion

4.1 Raw Material Preparation Area

The value of the exponent εc.fe. for crude petroleum feedstock can be represented as

$$\varepsilon_{\text{c.fe.}} = \varepsilon_{\text{pol.}} + \varepsilon_{\text{c.c.}} \tag{1}$$

where ε_{pol}. - is the value of ε, which depends on the presence of water in the raw material, mechanical impurities, and water-soluble chloride salts;

$\varepsilon_{c.c.}$ - is the value of ε, which depends on the chemical composition of the petroleum feedstock.

When removing contaminants at the site of preliminary preparation of petroleum feedstock, for example, at an electric desalting dewatering unit, the value of ε_{pol}. Will decrease significantly, which will lead to a decrease in the value of $\varepsilon_{c.fe.}$ to a certain level $\varepsilon_{p.o}$.

Considering this, the value $\varepsilon_{p.o.}$ can be used to determine the degree of readiness of crude oil for processing. To do this, it is necessary to create a databank of the value εp.o., petroleum feedstock, depending on the content in it of various concentrations of water, mechanical impurities, and chloride salts.

The operation of the global monitoring and correction system is ensured by capacitive sensors S1–S2, combined into a block of sensors 1 (see Fig. 2), which are located in the pipelines after the electric dehydrators of the stages I, II.

After the raw material passes the preliminary preparation stage, the value εp.o. will be compared with the value from the data bank εDbp.o, provided that $\varepsilon_{Dbp.o} < \varepsilon_{p.o.}$, the system will signal violations of the technological regime and correct the technological parameters of the process.

Taking into account the fact that after preliminary preparation εpol. = const, and the value of the parameter εp.o., will be determined mainly due to the term εc.c. with then it can be used at the preliminary stage of research to determine the potential content of light (fuel) fractions of various oils and gas condensates [12]. This, in turn, makes it possible to timely determine the most rational direction for processing raw materials: fuel oil or fuel oil.

At the sections of preparation of polymer raw materials during its thermodestructive processing, sensors can be used to monitor the residual moisture content in dried feedstock.

A study of the influence of different types of polluting impurities on the value of the indicator (ε) of the BC was carried out at a laboratory by performing so-called model environments. Model environments imitate real (natural) objects, artificially created for laboratory research to study certain processes of effects of different factors. Some samples are "water in oil." Other samples are suspensions of oil or gas condensate with solid admixtures (Fe_2O_3 and SiO_2) of 1–3 μm in size dispersed in the liquid area. Dispersion of admixtures in oil and gas condensate was carried out by an ultrasonic dispersant brand UZD-22/44 for 60 minutes at a frequency of 44 kHz.

The results obtained are experimentally presented in Figs. 3 and 4.

The conducted laboratory studies has proved that in the studied range of values water as an oil impurity component has the most significant effect on the ε value ($\Delta\varepsilon$ = 0.51). A smaller effect ($\Delta\varepsilon$ = 0.18) has Fe_2O_3 particles, which is the most obvious case of corrosion of tanks, pipelines, etc. This effect is because water is a polar substance ($\varepsilon \approx 80$), and dehydrated oil is nonpolar dielectrics ($\varepsilon_{max} \approx 2.7$). Fe_2O_3 particles are semiconductors, and at this point, their presence in oil can also

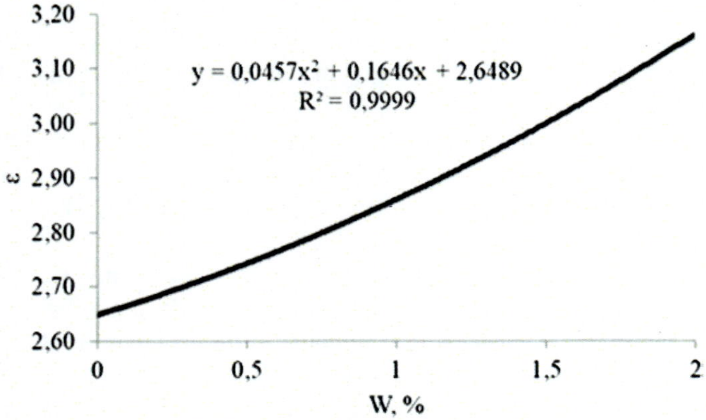

Fig. 3 Effect of water content (*W*) on the (ε) value

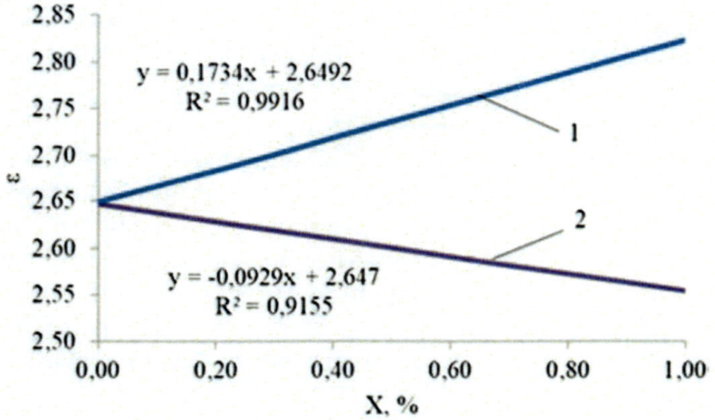

Fig. 4 Effect of Fe_2O_3 and SiO_2 concentration (X) on the (ε) value: 1– oil +Fe_2O_3; 2 – oil +SiO_2

increase increase the overall (ε) value of oil. Conversely, SiO_2 is a dielectric material. In an electric field they contribute to the improvement of the dielectric properties of the oil (decrease in value by $\Delta\varepsilon = 0.09$). But the influence of SiO_2 on the (ε) value is usually insignificant due to its low concentration.

4.2 Raw Material Processing Area

It is known that the value of the exponent ε of distillate fractions depends on their boiling point, which means that the following expression is valid

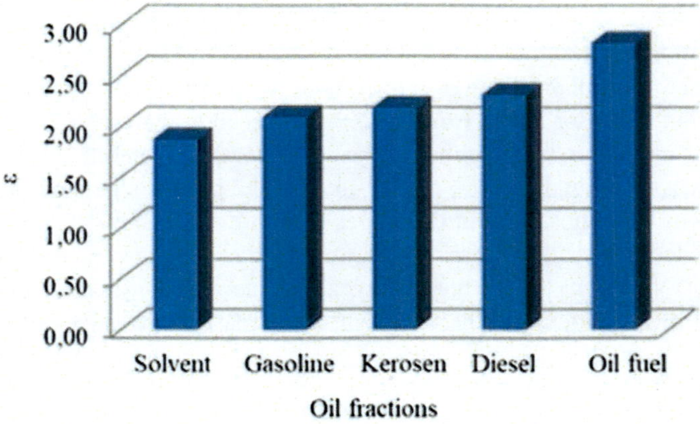

Fig. 5 Values of (ε) for different oil fractions

$$\varepsilon_{\text{p.f.}} < \varepsilon_{\text{:.f.}} < \varepsilon_{\text{d.f.}} < \varepsilon_{\text{o.f.}} \tag{2}$$

where $\varepsilon_{\text{p.f.}}$ – the value of ε, for the gasoline fraction (b.p.-180 ° C);
$\varepsilon_{\text{κ.f.}}$ – the value of ε, for the kerosene fraction (180–240 ° C);
$\varepsilon_{\text{d.f.}}$ – ε value for the diesel fraction (240–360 ° C);
$\varepsilon_{\text{o.f.}}$ – ε value, for fuel oil fraction (> 360 ° C).

The expression (2) can be illustrated by the diagram based on the results of laboratory studies (see Fig. 5).

Taking into account the effect of overlapping fractions, which is a common phenomenon at primary oil refining plants and depends mainly on the design features of the column equipment and technological parameters of the process (vapor velocity, amount of reflux, etc.) [13], the value of the index ε for each faction can be represented as

$$\left\{\begin{array}{l} \varepsilon_{\text{p.f.}} = \varepsilon^{\text{Imp}}{}_{\text{h.g.}} + \varepsilon_{\text{m.f.}} \\ \varepsilon_{\text{:.f.}} = \varepsilon^{\text{Imp}}{}_{\text{p.f.}} + \varepsilon_{\text{m.f.}} \\ \varepsilon_{\text{d.f.}} = \varepsilon^{\text{Imp}}{}_{\text{:.f.}} + \varepsilon_{\text{m.f.}} \\ \varepsilon_{\text{o.f.}} = \varepsilon^{\text{Imp}}{}_{\text{d.f.}} + \varepsilon_{\text{m.f.}} \end{array}\right. , \tag{3}$$

where $\varepsilon^{\text{Imp}}{}_{\text{h.g.}}$ – value of ε, impurity of liquefied hydrocarbon gases;
$\varepsilon^{\text{Imp}}{}_{\text{p.f}}$ – value of ε, impurities of the gasoline fraction;
$\varepsilon^{\text{Imp}}{}_{\text{k.f.}}$ – value of ε, impurities of the kerosene fraction;
$\varepsilon^{\text{Imp}}{}_{\text{d.f.}}$– value of ε, admixture of diesel fraction;
$\varepsilon^{\text{Imp}}{}_{\text{o.f.}}$ – value of ε, impurities of the fuel oil fraction;
$\varepsilon_{\text{m.f.}}$ – value ε for the main fraction, which, depending on the boiling point, can be gasoline, kerosene, diesel, or fuel oil.

The value $\varepsilon_{\text{Impp.f}}$, $\varepsilon_{\text{Impk.f.}}$, $\varepsilon_{\text{Impd.f.}}$, $\varepsilon_{\text{Impo.f.}}$ will be determined by the concentration of impurities of the corresponding fraction in the main distillate fraction.

Note that to control the depth of extraction of fractions, capacitive sensors S3-S6, combined into a sensor unit 2 (see Fig. 2), are placed at the outlet from the main atmospheric column in the pipelines of the corresponding material flows.

The values of the indices $\varepsilon_{\text{p.f.}}$; $\varepsilon_{\text{k.f.}}$; $\varepsilon_{\text{d.f.}}$; $\varepsilon_{\text{o.f.}}$ of the products passing through the pipelines will be compared with the values from the data bank $\varepsilon^{\text{Db}}{}_{\text{p.f.}}$; $\varepsilon^{\text{Db}}{}_{\text{k.f.}}$; $\varepsilon^{\text{Db}}{}_{\text{d.f.}}$; $\varepsilon^{\text{Db}}{}_{\text{o.f.}}$, provided that $\varepsilon^{\text{Db}}{}_{\text{p.f.}} < \varepsilon_{\text{p.f.}}$, $\varepsilon^{\text{Db}}{}_{\text{k.f.}} < \varepsilon_{\text{k.f.}}$, $\varepsilon^{\text{Db}}{}_{\text{d.f.}} < \varepsilon_{\text{d.f.}}$, $\varepsilon^{\text{Db}}{}_{\text{o.f.}} < \varepsilon_{\text{o.f.}}$ the system will signal about violations of the technological regime and adjust the technological parameters of the process in the direction of decreasing the values of $\varepsilon^{\text{п}}{}_{\text{p.f.}}$; $\varepsilon^{\text{п}}{}_{\text{j.f.}}$; $\varepsilon^{\text{п}}{}_{\text{d.f.}}$; $\varepsilon^{\text{п}}{}_{\text{o.f.}}$

4.3 *Compounding Products Area*

The use of the index ε in this area is reduced to the control of the dosage of the obtained distillate fractions with basic components and additives that improve the operational properties of commercial fuels. Often, the refinery uses an automatic system of dosed injection (by weight) of components and additives. An additional indicator of the control of this input can be the value of the change in the index ε, relative to the initial value.

In this case, we have the following

$$\varepsilon_{\text{c.f.}} = \varepsilon_{\text{b.f.}} + \Delta\varepsilon_{\text{add.}}, \tag{4}$$

where $\varepsilon_{\text{c.f.}}$ - the value of ε, commercial fuel;

$\varepsilon_{\text{b.f.}}$- the value of ε, the base fraction of the fuel;

$\Delta\varepsilon_{\text{add.}}$ - the value of ε change due to the addition of additives (components).

And the value of $\Delta\varepsilon_{\text{add.}}$ can be represented in the form of.

$\Delta\varepsilon_{\text{add.}} = f(X)$, (5).where X - is the concentration of the added additive (component), % wt.

Nowadays alcohols like methanol, ethanol, propane, and isobutane [9] are widely as the components of gasoline. It decreases the harmful environmental effect of automobile gasoline by decreasing the concentration of CO, NOx, and solid C in the waste gases [10]. Besides, alcohols makes motor gasoline resistant to detonation by 6–8 points [12].

The value (ε) of alcohols varies from 17.7 to 33.1 units. [14], and added to a gasoline fraction they increase the total value of (ε_{sum}) as shown at Fig. 6.

Alcohols can be arranged in the following order according to the (ε) value: methanol > ethanol > propanol > iso-butanol. So at the maximum alcohol content in the mixture at 15%, the (ε) values are: for methanol $\Delta\varepsilon = 2.46$; for ethanol $\Delta\varepsilon = 2.33$; for propanol $\Delta\varepsilon = 2.27$; for iso-butanol $\Delta\varepsilon = 1.93$.

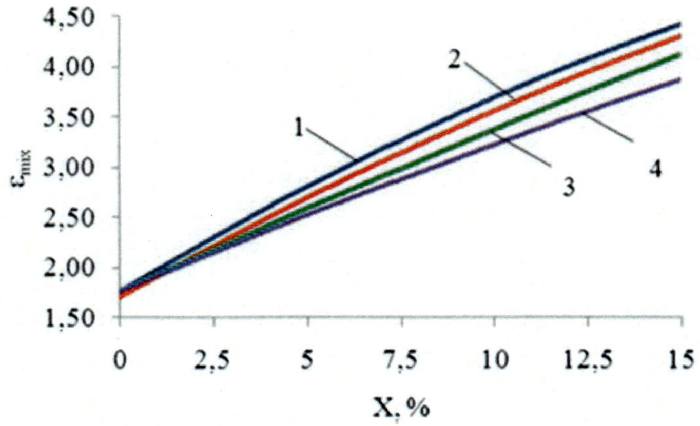

Fig. 6 Effect of the concentration of alcohols (X) on the (ε) value: 1 – methanol; 2 – ethanol; 3 – propanol; 4 – iso-butanol

In addition to alcohols, ethers (dimethyl ether (DME), methyl tert-butyl ether (MTBE)), and N-methylaniline (MMA) are used to increase the resistance to detonation of automobile gasoline.

In this section of the installation, a block of sensors 3 is located, consisting of capacitive sensors S7–S9 (see Fig. 2), placed on the material flows of commercial gasoline, aviation kerosene (jet fuel), and diesel fuel.

An example of the use of the parameter (ε) to determine the quality of petroleum products, e.g., motor gasoline, is their resistance to detonation, expressed in points of octane number (ON) depending on (ε). Based on the data given in [20], this dependence has the character shown at Fig. 7.

This diagram is used for the development of portable express analyzers of the quality of commercial motor fuels – octanometers, which are widely used at gas stations (gas stations), motor vehicle enterprises (ATP), etc.

4.4 *Section for Determining the Quality of Products*

At the oil refinery plant, this section is represented by the central plant laboratory and is not included in the global monitoring and correction system scheme presented by us. However, in the central plant laboratory, for the operational control of the quality of the fuel produced (identification, determination of contamination, pour point, resistance to detonation), methods and instruments (quality analyzers) can be used, the principle of which is based on the measurement of the index ε [21].

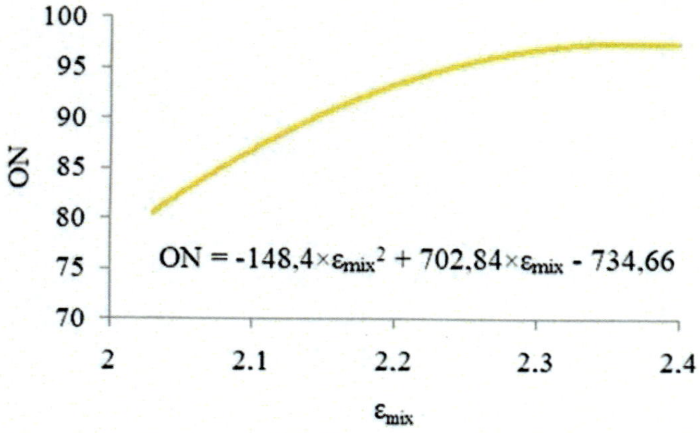

Fig. 7 Dependence of the (ε) value on the ON of commodity motor gasoline

5 Conclusions

Considering the current research results based on the indicator of the relative dielectric constant of hydrocarbon feedstock and its fractions, a global monitoring and correction system has been proposed. This system will enable operational control of the technological process at all stages of fuel production at installations for the primary processing of crude oil and plant for thermo-destructive processing of polymer feedstock.

A structural diagram of the main elements of the global monitoring and correction system is proposed. It contains primary information on a collection unit based on capacitive sensors and level gauges, information on a conversion unit, a data comparison unit, and a correction unit. The principle of operation of this system is based on a comparison of the obtained values of the relative dielectric constant with the values in the data bank and, on the basis of this comparison, in cases where it is necessary to correct the technological parameters of the fuel production process.

This is primarily important for the control of the technological process at the compounding stage. It happens because, in recent years, motor fuels can often be mixed with alcohols (up to 10%) to improve these fuels' operational and ecological properties. The dielectric constant of alcohols is in the range of 17.7–33.1 units, which is significantly higher than the dielectric constant of distillate oil fractions (1.8–2,2 on average). Therefore dielectric constant is considered a suitable unit for the evaluation of alcohol percentage in motor fuels. Also, the value of the dielectric constant of the mixture allows to calculate the octane number of compound gasoline. And all measurements are carried out continuously, in a flow, which allows to increase efficiency of refineries significantly.

The presented information is only the first step toward developing and implementing the global monitoring and correction system at primary oil refining units. A further study of this area involves a deeper analysis of the existing information, which should be supported by a significant amount of experimental research.

Acknowledgments This work was supported by the Slovak Research and Development Agency under contract No. APVV-19-0590, and also by the projects VEGA 1/0700/20 and KEGA 022TUKE-4/2023 granted by the Ministry of Education, Science, Research, and Sport of the Slovak Republic

References

1. Xia, Y., Sun, X.: Xinjie zhao at all: In-depth characterization of nitrogen heterocycles of petroleum by liquid chromatography-energy-resolved high resolution tandem mass spectrometry. Talanta. **249**, 123654 (2022)
2. Barman, B.N., Cebolla, V.L., Membrado, L.: I: Chromatographic techniques for petroleum and related products. Crit. Rev. Analyt. Chem. **30**(2–3), 75–120 (2000)
3. Rakhmatullin, I.R., Sidenev, D.M., Akhmetov, A.F.: at all: Development of a method for predicting the quality of petroleum coke. Chem. Technol. Fuels Oils. **58**(3), 439–443 (2022)
4. Yu, Z., Wang, Z., Jiang, Q., at all.: Analysis of factors of productivity of tight conglomerate reservoirs based on random Forest algorithm. ACS. Omega. **7**(23), 20390–20404 (2022)
5. Ayobami, A.O.: An assessment of trace metal pollution indicators in soils around oil well clusters. Petrol. Res. **7**(2), 275–285 (2022)
6. Punase, A., Hascakir, B.: Stability determination of asphaltenes through dielectric constant measurements of polar oil fractions. Energy Fuels. **31**(1), 65–72 (2017)
7. Vrålstad, H., Spets, Ø., Lesaint, C., Lundgaard, L., Sjöblom, J.: Dielectric properties of crude oil components. Energy Fuels. **23**(11), 5596–5602 (2009)
8. Dedong, T., Liguang, W.: Study of the petroleum products micro water online detemination method. J. Chin. J. Sci. Instru. **25**(4), 34–36 (2004)
9. Mingfang, W., Dedong, T.: The oil product moisture meter based on the electromagnetic resonance. J. Instru. Techniq. Sensor. **4**, 16–18 (2008)
10. Arik, E., Altan, H.: Okan Esenturk; dielectric properties of diesel and gasoline by terahertz spectroscopy. J. Infrared Millimeter Terahertz Waves. **35**, 759–769 (2014)
11. Sumant Juvekar: Guided Wave Radar Level Measurement Reduces. Costs, Improves Accuracyin Wastewater Applications. Express Water Magazine Issue June 2018, 46–47 (2018).
12. Dobrotvorskiy, S., Basova, Y., Dobrovolska, L., Sokol, Y., Kazantsev, N.: Big challenges of small Manufacturing Enterprises in Industry 4.0. In: Ivanov, V., Trojanowska, J., Pavlenko, I., Zajac, J., Peraković, D. (eds.) Advances in Design, Simulation and Manufac-turing III. DSMIE 2020 Lecture notes in mechanical engineering, pp. 118–127. Springer, Cham (2020)
13. Zatonskiy, A.V., Tugashova, L.G., Alaeva, N.N., Gorshkova, K.L.: Controlling the oil rectification process in a primary oil refining unit using a dynamic model. Petroleum Chemistry. **57**, 1121–1131 (2007)
14. Dobrotvorskiy, S., Balog, M., Basova, Y., Dobrovolska, L., Zinchenko, A.: Concept of the Software for Materials Selection Using .NET Technologies. In: Advanced Manufacturing Processes. InterPartner 2019. Lecture notes in mechanical engineering, pp. 32–43. Springer, Cham (2020)
15. De Souza, J.E., Scherera, M.D., Cáceresb, A.S., Cairesa, R.L., Peko, C.M.: A close dielectric spectroscopic analysis of diesel/biodiesel blends and potential dielectric approaches for biodiesel content assessment. Fuel. **105**, 705–710 (2013)
16. Guan, L., Feng, X.L., Li, Z.C., Lin, G.M.: Determination of octane numbers for clean gasoline using dielectric spectroscopy. Fuel. **88**(8), 1453–1459 (2009)
17. Shatokhina, E.V.: Fast analysis fo the quality and environmental safety of motor fuels. Chem. Technol. Fuels Oils. **43**, 242–247 (2007)

18. Kutia, M., Mukhin, N., Petrova, H., et al.: Sensor for the evaluation of dielectric properties of sulfur-containing heteroatomic hydrocarbon compounds in petroleum based liquids at a microfluidic scale. AIP Advances. **10**, 025006 (2020)
19. Lapuerta, M., Rodríguez-Fernándeza, J., et al.: Determination of optical and dielectric properties of blends of alcohol with diesel and biodiesel fuels from terahertz spectroscopy. Fuel. **274**, 117877 (2020)
20. Khodabakhshikoulaei, A., Sadrnia, H., et al.: Bioethanol fuel quality assessment using dielectric spectroscopy. Biofuels. **13**, 1894002 (2022)
21. Pavlenko, I., Piteľ, J., Ivanov, V., Berladir, K., Mižáková, J., Kolos, V., Trojanowska, J.: Using regression analysis for automated material selection in smart Manufacturing. Mathematics. **10**(11), 1888 (2022)
22. Berladir, K., Trojanowska, J., Ivanov, V., Pavlenko, I.: Materials selection in product development: Challenges and quality management tools. In: Hamrol, A., Grabowska, M., Maletič, D. (eds.) Advances in Manufacturing III. MANUFACTURING 2022 Lecture notes in mechanical engineering, pp. 72–86. Springer, Cham (2022)

Part III
Augmented Reality, Virtual Reality, Computer Modelling

Description of the Basic Tools of Augmented Reality in the Design of Computer Models in Unity

Stella Hrehova and Darina Matisková

1 Introduction

Augmented reality (AR) is now a recognized term, but it yet begins in the 1960s. The concept of AR was only formalized in 1994 as a relationship among real space, virtual space, and all intermediate forms of mixed space [1]. In augmented reality, the virtual information generated by the computer is over-covered in the real scene. It is the process of displaying digital information in a realistic way so that it appears as part of the real environment [2]. The availability and hardware simplicity of this technology give the possibility of its use to an increasing number of ordinary users and give more opportunities for use in various areas as well [3]. With the advent of information technology, the creation of simulation and computer models moved to a new level. We can understand simulation as a special approach to the study of models, which is essentially experiential or experimental. The simulation includes:

- Creation of a model that imitates the observed behavior,
- Working with a model to gain observations of this behavior,
- Attempts to understand, summarize, and/or generalize these behaviors.

In many applications, simulation also includes testing and comparing alternative designs and validating, explaining, and supporting simulation results and study recommendations [4]. One of the tasks of the created model is to enable the analyst to predict the impact of changes on the system. On the one hand, the model should approximate the real system and include most of its most important properties. On the other hand, it should not be so complex that it cannot be understood and

S. Hrehova (✉) · D. Matisková
Faculty of Manufacturing Technologies with the seat in Prešov, Technical University of Košice, Prešov, Slovakia
e-mail: stella.hrehova@tuke.sk; darina.matiskova@tuke.sk

M. Balog et al. (eds.), *EAI International Conference on Automation and Control in Theory and Practice*, EAI/Springer Innovations in Communication and Computing, https://doi.org/10.1007/978-3-031-31967-9_12

experimented with. A good model is a reasonable compromise between realism and simplicity.

2 Literature Review

Currently, the development of new technologies has led to the expansion of augmented reality applications in various areas of industry. From the analysis of publications dedicated to AR applications, they found applications in the following areas (Fig.1) [5]:

- Maintenance,
- Assembly,
- Robot-human collaboration,
- Manufacturing,
- Education and logistics [5].

In AR environment, the real world is augmented with virtual content displayed by devices such as tablets, smartphones, or smart glasses. The user has the opportunity to interact with the real world as well as with the virtual world and manipulate virtual content [6]. The implementation of augmented reality tools in the creation of models can also be seen in earlier works [7], where several applications for the use of augmented reality in production planning are presented [8]. In work [9] a system is described that uses augmented reality to simulate real production processes machines using virtual objects. The works [10, 11] show the use of AR in production, maintenance, and assembly. Other works describe the application of AR applications, e.g., in warehouse management [12], training and education [13], and last but not least, in the field of tourism and health [14] but also in other areas.

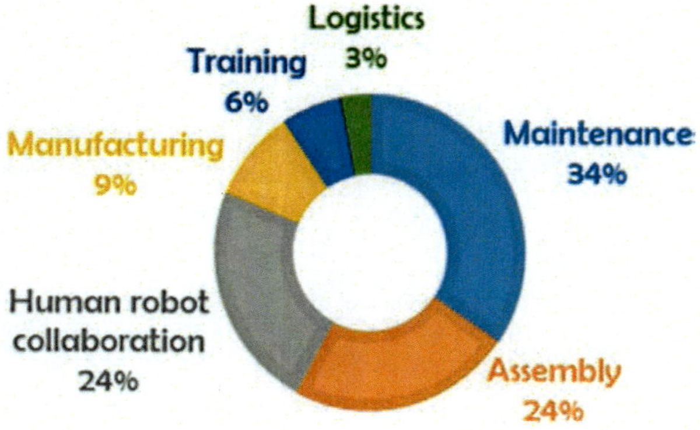

Fig. 1 Distribution of the type of AR application according to the area of its implementation

Augmented reality systems can be characterized by the following properties [14]:

- Mixing virtual objects and objects from the real world in a real environment.
- Synchronization between real and virtual objects.
- They are highly interactive and run in 3D in real time [14].

AR applications are considered to be significantly helpful in improving efficiency and effectiveness in industrial operations. Figure 2 shows the design of AR layers, from the user interface to the display associated with AR applications.

Tracking systems for AR can be:

- Marker-based AR.
- Markerless.
- Objects placement.

Marker-based AR: Marker-based AR uses image recognition technology using a camera and some visual marker types, for example, QR/2D code, that only provides a result when scanned by a reader. Marker-based applications using a camera can distinguish between markers and other real-world objects. Markers are generally based on simple pictures, which can be easily defined, requiring comparatively low processing power to read. In these cases, the AR scene is affected by the quality of the markers, the maturity of the AR SDK, and the complexity of the virtual models [15], as well as the distance from the virtual object to the marker, the size, and the spatial arrangement of the markers. However, the analysis of published contributions on this topic showed that systems based on markers are the most frequently adopted solutions in the development of applications [16].

Markerless AR: uses miniature versions of GPS, a digital compass, and a speedometer or accelerometer built into the device to provide data based on the device's precise location and orientation. The availability of smartphones with location detection functions anywhere allows greater application of the markerless technology solution. However, these apps are mostly used to map routes, find nearby businesses, or other location-based mobile apps.

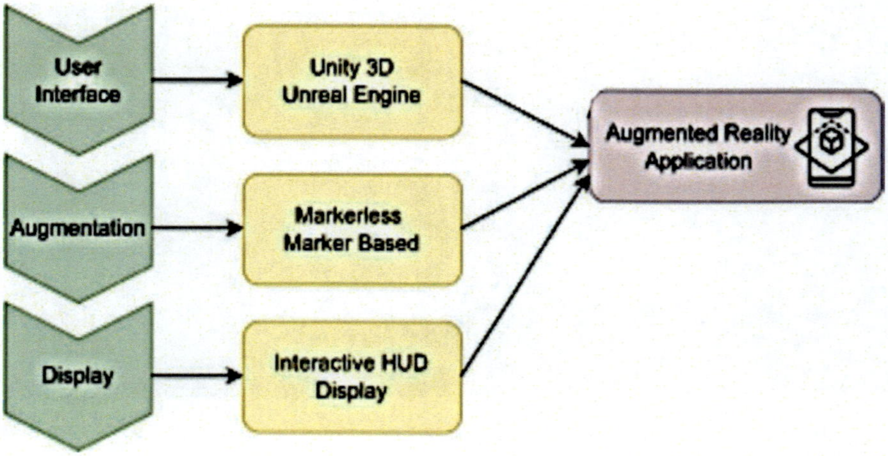

Fig. 2 AR layers [2]

3 Tools and Methodology

In the following parts of this paper, the chosen basic tools for creating computer models based on the use of augmented reality will be described.

3.1 *Brief Description of the Unity Environment*

Unity Technologies develop this software and provides the ability to develop both 2D and 3D applications. The first version of Unity was presented at Apple's conference in 2005 and supported only their OS X platform. In addition to 2D and 3D development, Unity also provides an environment for creating scripts in the Java and C programming languages. Today, Unity supports more than 25 platforms in mobile, desktop, console, TV, VR, and web applications. Unity software is the development tool of choice for most VR creators. More than two-thirds of all applications for virtual reality are created in it. Based on research [17], compared to Unreal Engine, Unity is more suitable for beginners because:

- Has a simpler user interface;
- Provides many tutorials and examples;
- Many assets are available [17].

Moreover, its installation does not require high-performance hardware. When it comes to Android app development, the process is quite simple, and exporting is easier. The basic environment of this engine is shown in Fig. 3.

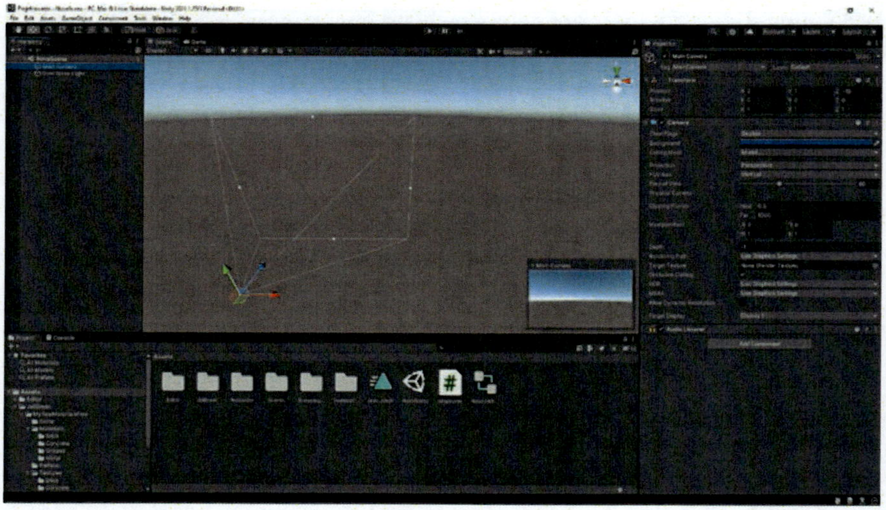

Fig. 3 The basic screen of the Unity environment

The basic element of each scene is the "camera," which is actually a separate object. It is possible to use standard transformation tools on it. The properties of each selected object are displayed in the right part of the screen, labeled "Inspector."

3.2 *Brief Description of the Vuforia Environment*

The Vuforia platform offers robust and efficient image recognition technology based on computer vision and includes a number of features that facilitate the development of mobile applications (Fig. 4).

This platform includes a target management system (Target manager), cloud target database, target device database, and Vuforia engine [4]. The tracked target is uploaded as an input image, and the mobile application then uses a cloud connection or local storage to access the required resources. AR application based on Vuforia SDK includes camera, image converter, and internal monitoring. The SDK enables native development for iOS and Android and the creation of AR applications in Unity that are easily portable among multiple platforms. In addition, the SDK includes features such as virtual buttons for localized occlusion identification, real-time target image selection, and the ability to reconfigure and create target sets based on circumstances. Important features of the Vuforia environment [2]:

- Tracking is effective even when the subject is partially obscured in low light.
- Provides fast local target recognition with the ability to simultaneously monitor up to five targets.
- Provides VuMarks, a new type of barcode, to create a unique design with an emphasis on marking and encoding data that serves the purpose of AR [2].

In order for the "markers" in the applications to work correctly, it is necessary to test them in the Vuforia environment (Fig. 5). The most ideal condition is if the system shows five stars. If there are only three, the marker needs to be modified.

This platform is supported by all major platforms and software, and it offers many services for free (image target, object scanner, QR code scanner, etc.), which allows you to create applications in many fields.

4 Result and Discussion

In this section, we will focus on the possibilities of using basic Unity tools to create a model without using a marker. In the given environment, we can insert basic geometric objects, such as a sphere, cuboid, or cube, i.e., 3D objects, which are available directly in the Unity program (Fig. 6). Almost the entire application environment can be created with these elements. Working with objects in this program is easy and intuitive, although sometimes time-consuming, especially when creating more demanding objects.

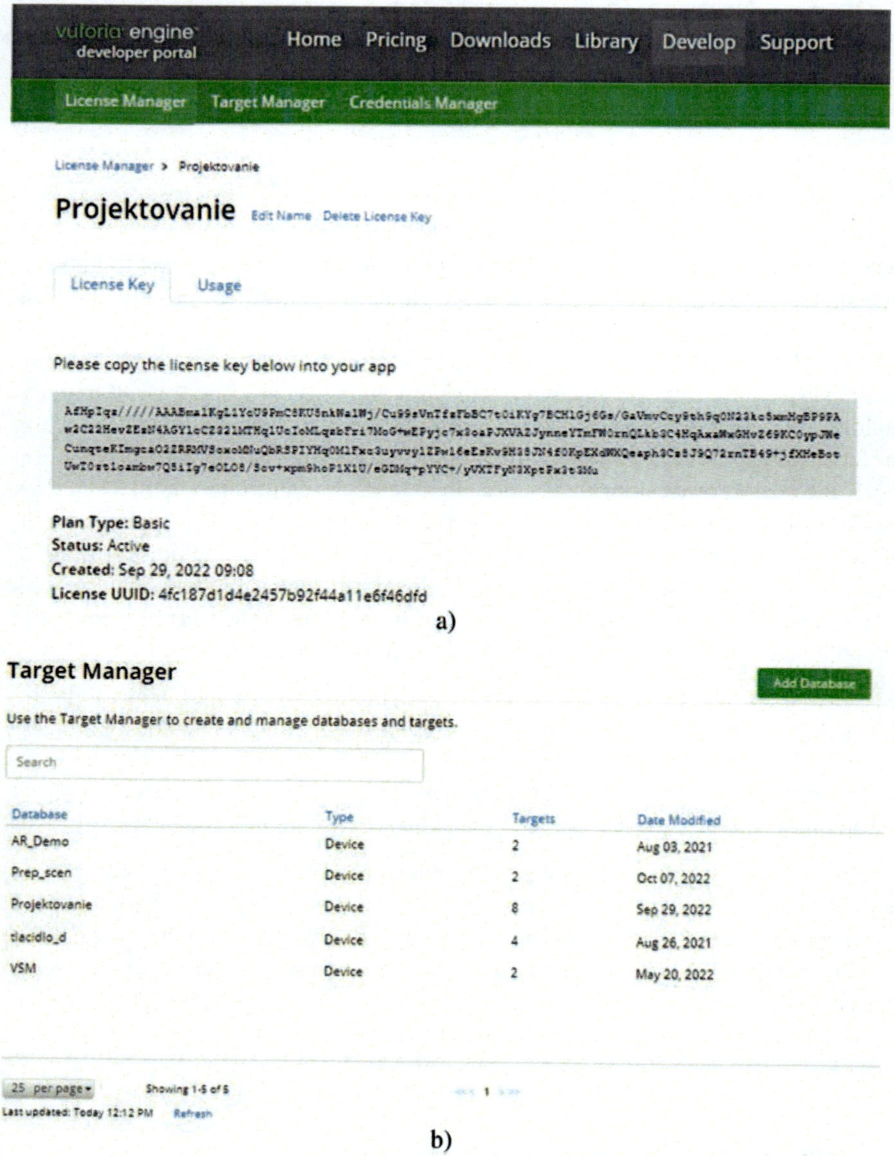

Fig. 4 Licence manager (**a**) and Target manager (**b**) of Vuforia environment

The advantage of this environment is that we can use ready-made objects created by other software applications, including Inventor, which is used to create drawing documentation. The following figure shows the Yumi ABB robot, which was in .obj format. Fig. 7 shows the robot object in Unity along with the camera view.

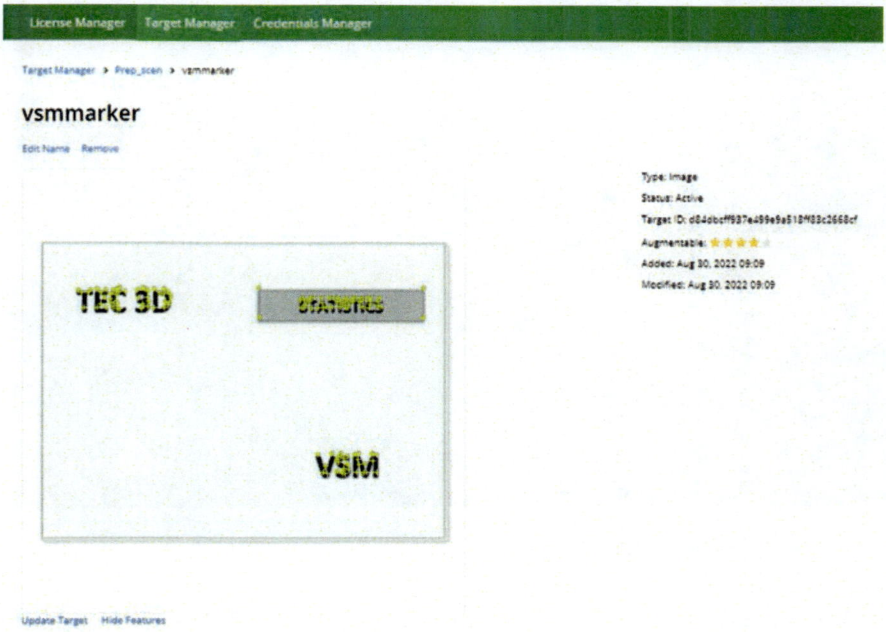

Fig. 5 Marker resolution test in the Vuforia environment

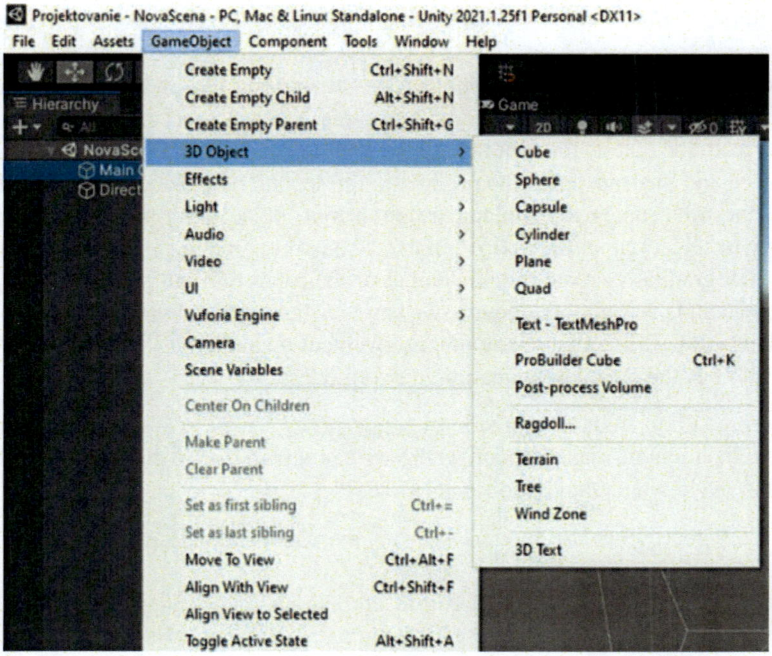

Fig. 6 3D objects of Unity environment

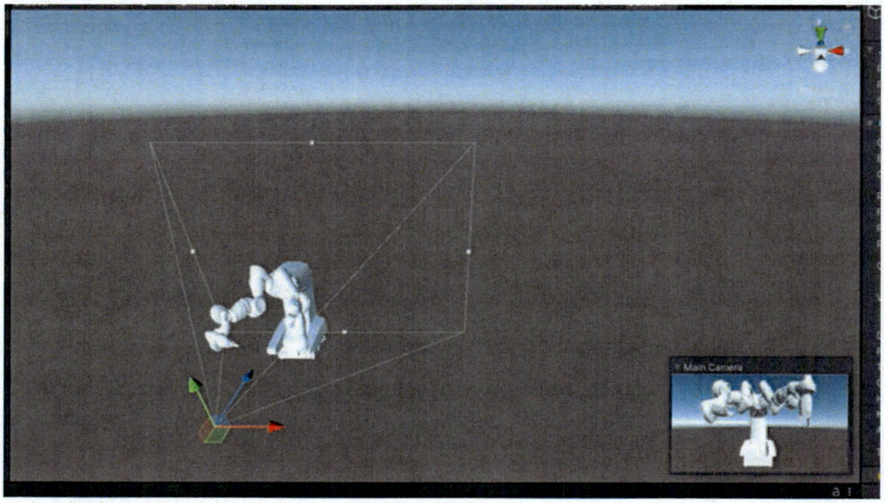

Fig 7 A preview of the imported robot object

How the object will behave in the Unity environment, and its other formatting options will depend on how it was created in its source environment (e.g., Inventor). Unity's formatting and animation tools can then be applied to the entire object or only to selected parts of it. Inserting external objects into the given environment is very simple. This is done by "dragging" files into a selected folder, usually "Resource."

Another tool for creating an application is the possibility of animation without scripting. In Fig. 8, we have created a scene that can represent the removal of the product from the belt by the robot to the work table, which can serve as an assembly workplace and the transfer of the product to the second belt. We can set the animation of the robot using the standard rotation tool. Select the object and use Add Property to select one of the options in the "Transform" menu.

We will gradually create the movement of the robot through manual filming. If the movement is not smooth enough, we can view and edit individual curves for all axes. This allows us to ensure smoother movement manually.

We can test the created application in two modes:

- Pressing "Play" in basic mode,
- In the "Simulator" environment, where we can set the device that will be the final output device (tablet and phone).

Figure 9 represents the sequences from the robot animation in the simulator environment

Another possibility is the use of command buttons, the functionality of which is provided by a program created in the C# language in the Visual Studio development environment. We then assign the given script to individual buttons via the *OnClick* property.

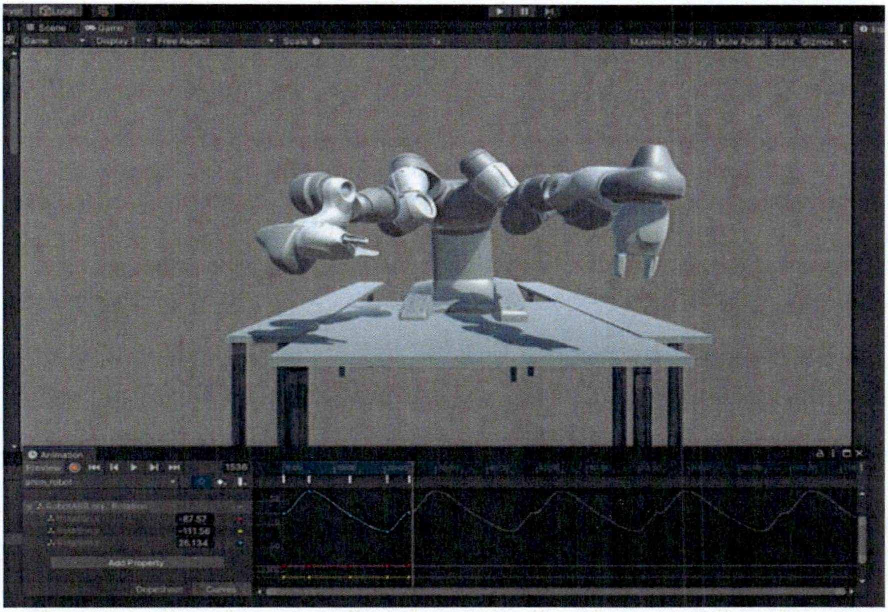

Fig. 8 Creating an object rotation using the "Animation" option

Fig. 9 Individual sequences after running in the simulator view

4.1 Connection of Unity and Vuforia – Marker-Based Application

The connection of the Unity tools environment with Vuforia is carried out via "AR camera" from the Unity package tools menu – Vuforia Engine AR (Fig. 10).

The basic element is the "Image Target," which is defined in the database in the Vuforia environment (Target manager). We import this database into the Unity environment, and then we have the opportunity to choose the appropriate "marker" – an image (scheme) over which the desired object will be displayed. Various images and videos are most often used. These are transferred to the Unity environment using Drag and Drop, and we can join them to the appropriate markers. For example, the following image (Fig. 11) shows a video of the laboratory virtual tour, which starts above the corresponding marker. The given scene also contains a "Back" button, which allows us to return to the main scene.

The merge of the development environment Unity and Vuforia enables the implementation of AR applications for almost any system - Android, universal Windows platform, iOS, Mac, Linux, and other game platforms (Fig. 12).

When using both methods for creating a computer model and its testing (e.g., also on a simulator), we can develop this model in the form of an application (.apk)

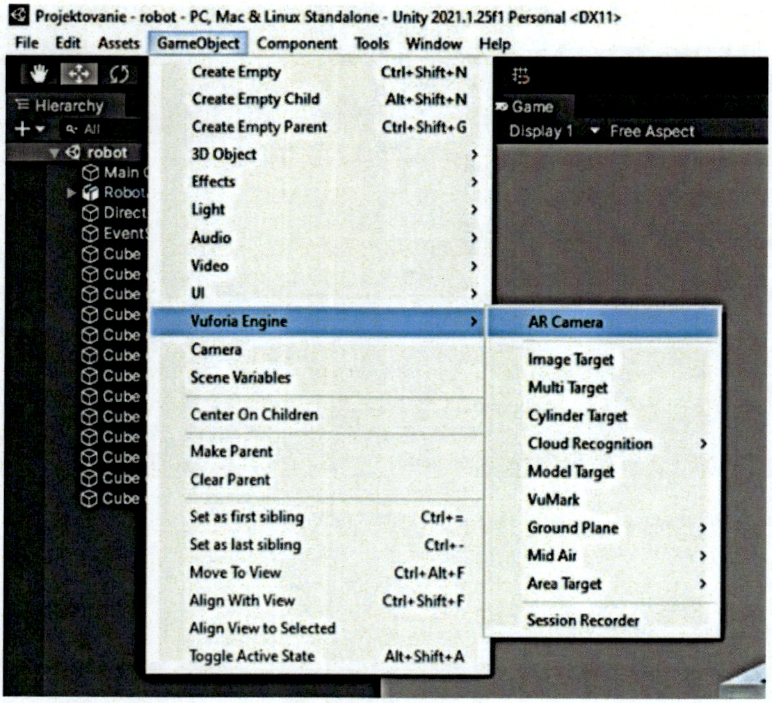

Fig. 10 Required objects selection

Fig. 11 Display the video above the marker

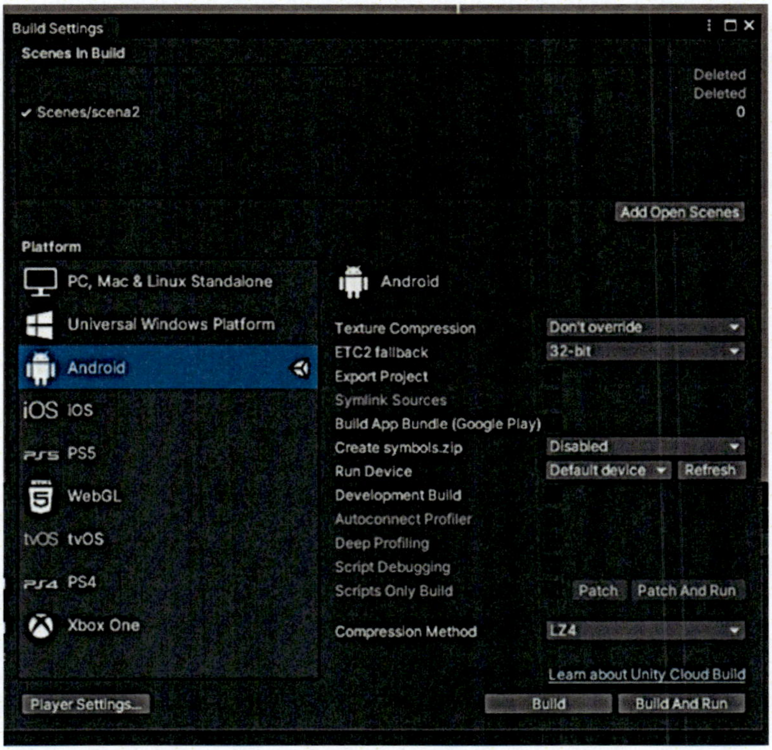

Fig. 12 Selecting a platform for model export

for the platform of our choice. The given application, together with the marker, thus becomes widely portable, providing a real view of the monitored process, environment, product, etc.

5 Conclusion

Augmented reality is currently a very advanced modern medium, while its availability and hardware simplicity make it much more accessible to common users and have the potential to be applied in every phase of industrial product life cycle management, from conception to disassembly [18]. Augmented reality improves perception and user interaction with the real world. Models created using augmented reality tools in the industrial field can be considered relevant because they have the potential to significantly improve communication in product design and production development, help identifies and prevent errors in individual process phases, reduce the number of physical prototypes and, last but not least, save time, and costs for businesses. We can consider AR as an effective tool for improving and accelerating the development of products and processes in many industrial applications. In the presented contribution, we pointed out the use of the basic tools of the Unity development environment in creating models without using markers. The combination of the Unity and Vuforia environments makes it possible to create computer models using markers, which further enhances the impression of reality, and the models thus provide additional information needed to solve problem situations.

Acknowledgments This work was supported by the Slovak Research and Development Agency under contract No. APVV-19-0590, and also by the projects VEGA 1/0700/20, and KEGA 014TUKE-4/2023 granted by the Ministry of Education, Science, Research, and Sport of the Slovak Republic.

References

1. De Pace, F., Manuri, F., Sanna, A.: Augmented Reality in Industry 4.0. Am. J. Computer Sci. Informat. Technol. **6**(1:17) (2018). https://doi.org/10.21767/2349-3917100017
2. Devagiri, J.S., Paheding, S., Niyaz, Q., Yang, X., Smith, S.: Augmented reality and artificial intelligence in industry: Trends, tools, and future challenges. Expert Sys. Applicat. **207**, 118002 (2022). https://doi.org/10.1016/j.eswa.2022.118002
3. Kaščak, J., Telišková, M., Török, J., Baron, P., Zajac, J., Husár, J.: Implementation of Augmented Reality into the Training and Educational Process in Order to Support Spatial Perception in Technical Documentation, pp. 583–587. IEEE 6th International Conference on Industrial Engineering and Applications (ICIEA) (2019). https://doi.org/10.1109/IEA.2019.8715120
4. Fernández, E.R., Delgado, M.L.: Augmented reality as a didactic resource for teaching mathematics. Appl. Sci. **10**, 2560 (2020). https://doi.org/10.3390/app10072560
5. Reljić, V., Milenković, I., Dudić, S., Šulc, J., Bajči, B.: Augmented Reality Applications in Industry 4.0 Environment. Appl. Sci. **11**, 5592 (2021). https://doi.org/10.3390/app11125592

6. Buń, P., Gapsa, J., Husár, J., Kaščak, J.: Mixed Reality Training in Electrical Equipment Operating Procedures. In: Advances in Manufacturing III. MANUFACTURING 2022. Lecture Notes in Mechanical Engineering. Springer, Cham (2022). https://doi.org/10.1007/978-3-030-99310-8_24
7. Doil, F., Schreiber, W., Alt, T., Patron, C.: Augmented reality for manufacturing planning. In proceedings of the workshop on virtual environments 2003, EGVE '03, 71–76. ACM, New York, NY, USA (2003)
8. Karlsson, I., Bernedixen, J., Ng, A.H.C., Pehrsson, L.: Combining augmented reality and simulation-based optimization for decision support in manufacturing, pp. 3988–3999. Winter Simulation Conference (WSC) (2017). https://doi.org/10.1109/WSC.2017.8248108
9. Pehrsson, L.: Manufacturing management and decision support using simulation-based MultiObjective optimisation. Volvo Car Corporation, Sweden University of Skovde, Sweden (2013)
10. Esengün, M., İnce, G.: The Role of Augmented Reality in the Age of Industry 4.0. In: Industry 4.0: Managing The Digital Transformation. Springer Series in Advanced Manufacturing. Springer, Cham (2018). https://doi.org/10.1007/978-3-319-57870-5_12
11. Iakovets, A., Balog, M., Hrehova, S.: Employee Training Programs forSmall-and Medium-Sized Manufacturing Enterprises. In: 5th EAI International Conference on Management of Manufacturing Systems. EAI/Springer Innovations in Communication and Computing. Springer, Cham (2022). https://doi.org/10.1007/978-3-030-67241-6_6
12. Husár, J., Knapčíková, L.: Possibilities of using augmented reality in warenhouse management: a study. Acta Logistica 2021. **8**(2), 133–139 (2021). https://doi.org/10.22306/al.v8i2.212
13. Shyshkina, M.P., Marienko, M.V.: Augmented reality as a tool for open science platform by research collaboration in virtual teams. In: Proceedings of the 2nd International Workshop on Augmented Reality in Education, Ukraine, ISSN1613-0073 (2019)
14. Anand, N., Bandana, M., DacNhuong, L., Suseendran, G.: Virtual Reality (VR) & Augmented Reality (AR) technologies for tourism and hospitality industry. International Journal of. Eng. Technol. **7**(2.21), 156–160 (2018)
15. El Barhoumi, N., Hajji, R., Bouali, Z., Ben Brahim, Y., Kharroubi, A.: Assessment of 3D models placement methods in augmented reality. Appl. Sci. **12**, 10620 (2022). https://doi.org/10.3390/app122010620
16. Bottani, E., Vignali, G.: Augmented reality technology in the manufacturing industry: A review of the last decade. IISE Transactions. **51**(3), 284–310 (2019). https://doi.org/10.1080/24725854.2018.1493244
17. Christopoulou, E., Xinogalos, S.: Overview and Comparative Analysis of Game Engines for Desktop and Mobile Devices. Int. J. Serious Games. **4**(4), 10.17083/ijsg.v4i4.194 (2017)
18. Durchon, H., Preda, M., Zaharia, T., Grall, Y.: Challenges in Applying Deep Learning to Augmented Reality for Manufacturing. In: Web3D '22: Proceedings of the 27th International Conference on 3D Web Technology, No: 13,1–4 (2022). https://doi.org/10.1145/3564533.3564572

Preparation of an Assembly in Virtual Reality

Eduard Franas , Petr Baron , and Marek Kočiško

1 Introduction

The use of virtual reality tools in solving the creation of CAD models can provide a new perspective on problem-solving and speed up the detection of errors already during the creation of models that may not be detected during normal creation. This approach helps to identify design errors and prevent them in the initial stages of the development process, reduces the number of physical prototypes, and saves companies time and costs. Virtual reality tools are a suitable technology for implementation in Industry 4.0. They offer several functions, such as real-time digital twin display, workplace presentation, training of new employees in the company, display, and interaction with the simulation of the production process. Virtual reality aims to replace the real world with a computer-generated environment. In this environment, the engineer can interact and manipulate a realistic display of the product in real time. Unlike VR, augmented reality (AR) merge the real world with virtual objects that are placed at the right time and place [1]. VR is an important technology in the industry today, for example, in product development, engineering, and service. It makes it easier for the engineer to understand complex design concepts and enables more effective interaction between the engineer and the computer. This saves time and costs and ultimately increases the quality of the product [2].

The advantage of using virtual reality in CAD systems is the possibility of collaboration between several users on one project, where the environment is displayed in real time and in the dimensions of the model. [3]

E. Franas (✉) · P. Baron · M. Kočiško
Faculty of Manufacturing Technologies with the seat in Prešov, Technical University of Košice, Prešov, Slovakia
e-mail: eduard.franas@tuke.sk

M. Balog et al. (eds.), *EAI International Conference on Automation and Control in Theory and Practice*, EAI/Springer Innovations in Communication and Computing, https://doi.org/10.1007/978-3-031-31967-9_13

CAD (computer-aided design) software are systems that uses computer systems in the creation of 2D and 3D models, analysis, or design optimization. The purpose of CAD systems is to streamline and optimize work processes, increase productivity, improve the quality of design details, and improve communication in the field of documentation. A combination of CAD/CAM systems is often used, where the necessary procedures for production can be created [4]. CAD systems are divided into small, medium, and large. Small CAD systems are mainly used to create drawings and sketches, medium CAD systems are used to create simple models and drawings, and large CAD systems also include various options for motion, deformation, and computational fluid dynamics (CFD) analysis [5]. VR/AR are valuable tools for improving and accelerating product and process development in many industrial applications, but there are still many unsolved requirements that leave potential for further development and improvement of VR/AR-based tools and methods [2].

2 Preparation of Models by Using the VR

2.1 A Hardware Solution for Modeling in VR

To create a model in VR, in addition to software, hardware (VR headset) is also required. A suitable device is the HTC Vive Cosmos headset. The reason for choosing this device was its compatibility with Mindesk and Rhinoceros 3D software and its availability at the Department of Computer-aided Manufacturing Technologies. It is the fourth generation VR headset from HTC. HTC Vive Cosmos offers a resolution of 2880x1700px (1440x1700px for one eye), a 90 Hz refresh rate, and a 110° field of view. The headset also includes the possibility of changing the interpupillary distance (IPD) and basic sensors such as the Gyroscope and G-sensor. Two external sensors are used to monitor the location of the device in space. It also requires a powerful computer to run [6].

Minimum computer requirements needed to run HTC Vive Cosmos:

- Graphics card: Nvidia Geforce GTX 970 4GB / AMD Radeon R9 290 4GB,
- Processor: Intel Core i5-4590/ AMD FX 8350,
- Operating memory: 4GB RAM,
- Display port 1.2, USB 3.0 [6].

Figure 1 shows the resources that were used for creating models in VR.

2.2 A Software Solution for Modeling in VR

There are many softwares enabling the creation of virtual reality, such as the Unity 3D engine, Unreal Engine, SimLab, Vizard, or CAD systems Rhinoceros and Solidworks, in combination with the Mindesk module.

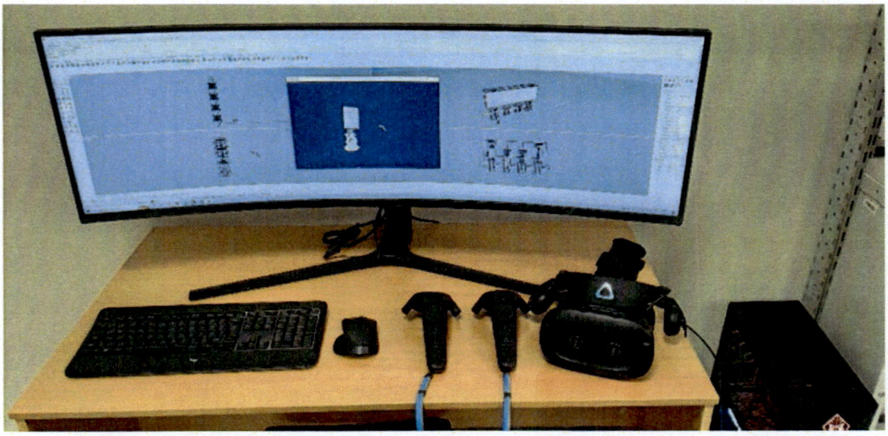

Fig. 1 Resources used for modeling in the VR

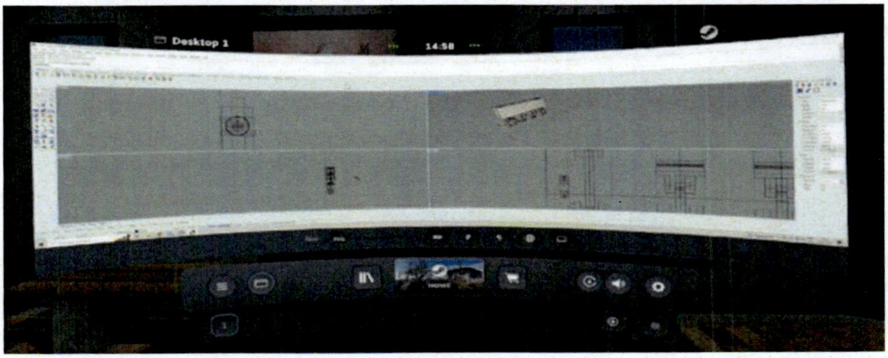

Fig. 2 A view of the desktop that is displayed in VR

For proper VR functionality, you need to have SteamVR installed, which is available for download in the Steam digital store, from Valve Corporation. When SteamVR is launched for the first time, the boundaries of the room in which the user will move, the height of the floor and the location of the monitor are set. In addition to running VR-supporting applications, SteamVR also allows displaying the desktop of the monitor, which is shown in Fig. 2. The desktop display is suitable for use if the running program does not contain some functions and it is necessary to switch to the classic display, or if necessary, start another program that does not support VR [7].

The Mindesk Environment Mindesk is a 3D platform that can be connected to the graphics engine Unreal Engine or CAD programs such as Solidworks and Rhinoceros. This means that Mindesk uses the functions used by the program to which it is connected. The functions can be opened using the button on the left controller, where a

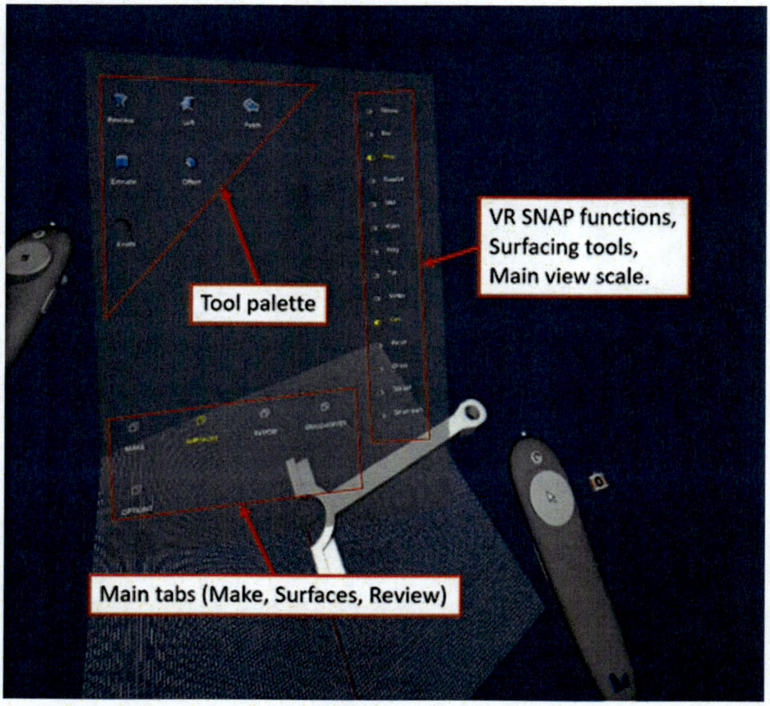

Fig. 3 Mindesk program environment

menu with options for selecting tools will appear after clicking. These functions are divided into four categories (make, surfaces, review, grasshopper) [8].

The make tab contains tools that enable the creation of 2D sketches (circle, poly, rectangle). Also, this tab provides tools for creating shapes (cylinder, sphere, box). Furthermore, some tools allow various manipulations with objects (rotate, move, copy). Surfaces are created in the surfaces category. Among the most frequently used tools are extrude, which enables the creation of a 3D model from a 2D sketch, and the revolve function, which enables the creation of rotating surfaces. The review category contains tools that allow, for example, to measure distances on the part and create sections using the section tool. This category also includes the snapshots tool, used to create screenshots directly in VR. The last category is the grasshopper function, which is used to create algorithms for parametric systems (Fig. 3).

2.3 A Description of the Model Being Created

A simplified assembly of a four-cylinder engine was chosen for modeling, where all components were developed in VR using Rhinoceros and Mindesk. Subsequently, a functional assembly of this engine was created using PTC Creo. Several methods of

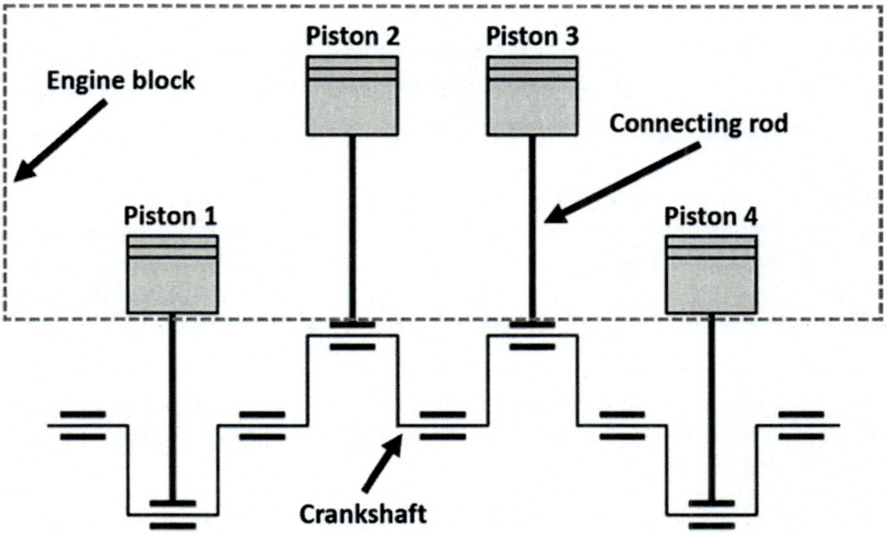

Fig. 4 Diagram of a four-cylinder engine

computer modeling are known, which differ in some ways but can achieve similar results. We know parametric, direct (explicit), and generative modeling. [5] The direct modeling method was used to create the model, which allows for working directly with the geometry of the model. An assembly consisting of five different parts (engine block, piston, connecting rod, pin, and crankshaft) is shown in Fig. 4.

2.4 Preparation of Models in a VR Environment

The workflow of creating an engine assembly in VR:

- Creating a crankshaft model,
- Creating a piston model,
- Creating a connecting rod model,
- Creating a connecting pin,
- Creating a model of the engine block,
- Creating a four-cylinder engine assembly.

Crankshaft Model The crankshaft model is one of the main parts of the assembly being created. It is the mechanical part of the engine that converts the rectilinear reciprocating motion of the pistons into rotary motion. The process of creating the shaft model began with the creation of a sketch of the crank arm. The advantage of working in the Rhinoceros program and VR environment is feedback, which means that changes made in VR will be displayed on the program monitor and vice versa. This sketch was created directly in the Rhinoceros program in the SteamVR virtual environment due to its complexity and the unavailability of the trim tool in the Mindesk program, which is necessary to create this sketch. The trim tool is found in the Rhinoceros application and removes unnecessary entities in the model sketch.

In the next step, the extrude tool is used in the surfaces tab, and the crank arm is created. Finally, the cylinder tool was used to create the part that connects the individual crank arms (Fig. 5).

After creating these parts, the assembly of the crankshaft begins. Copying, moving, and rotating tools were used for composing. Since the parts are duplicated and identical, just copy them with the copy tool and move them to the required place with the move tool. The rotate tool is used to rotate the crank arms.

Piston Model Pistons perform the basic movements of a four-cylinder engine. There are four identical pistons in the entire engine assembly. The modeling process begins with the creation of a sketch. Next, sketches are created using the poly tool, which is used to create simple lines (Fig. 6).

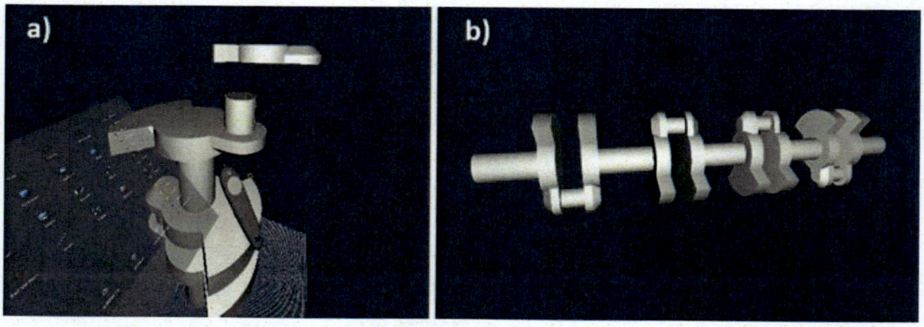

Fig. 5 (**a**) Creation of the crankshaft, (**b**) Final model of the crankshaft

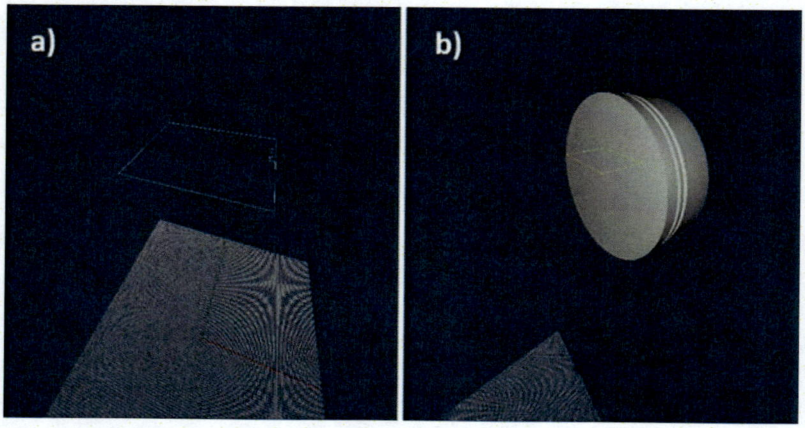

Fig. 6 (**a**) Sketch of the piston, (**b**) Use of the revolve tool

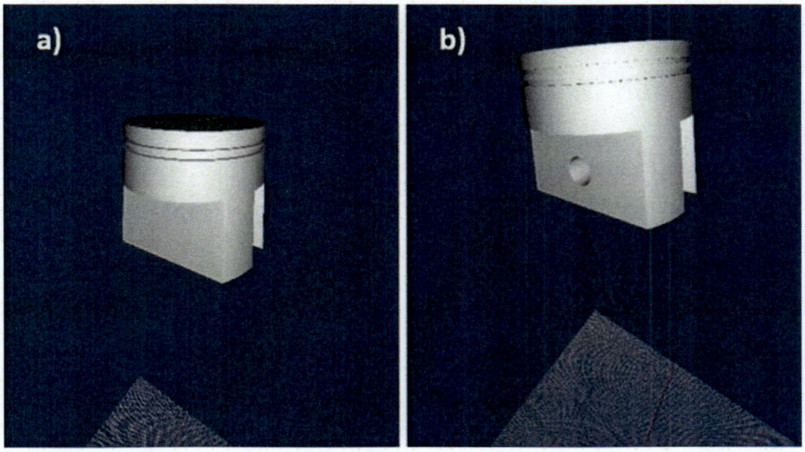

Fig. 7 (**a**) Using the extrude tool, (**b**) The final piston model

The first step is the creation of a simple sketch, from which the basic shape of the piston is created using the revolve function. Subsequently, it is necessary to create a part where the piston will be connected to the engine connecting rod. This part was created using the poly and circle tools. Next, a circle was created that describes this part, and four lines that divide the circle so that two parts can be created to connect the connecting rod to the piston. Excess parts of the sketch were removed using the trim function. Finally, the resulting shape of the piston was created using the extrude function.

The last step is to create a hole for the pin that connects the piston to the connecting rod. This part was created using the circle tool, where a sketch of the hole was made and then the necessary hole was drawn into the model using the MakeHole tool in the Rhinoceros program (Fig. 7).

Creating a Connecting Pin This pin connects the piston, and connecting rod of the assembly is created. The circle tool was used for its creation, where two circles were created, one with Ø8 mm and the other with Ø4 mm. Subsequently, a surface was created, which was pulled into the space to a length of 31 mm using the extrude tool. Finally, in the Rhinoceros program, chamfers were created on both ends of the study, using the chamfer tool.

Creating a Connecting Rod Model The first part of the model is the connecting rod sleeve, which is used to attach the connecting rod to the crank pin that connects the crank arms. Next, the circle and poly tools were used to create the sketch, and unwanted lines were removed using the trim tool. Subsequently, the extrude function was used, and a 3D model of the part was created. The last step is to create a hole through which the screw will go. These holes are created using the circle tool and then the material is removed from the part using the extrude tool to create a hole for the screws (Fig. 8).

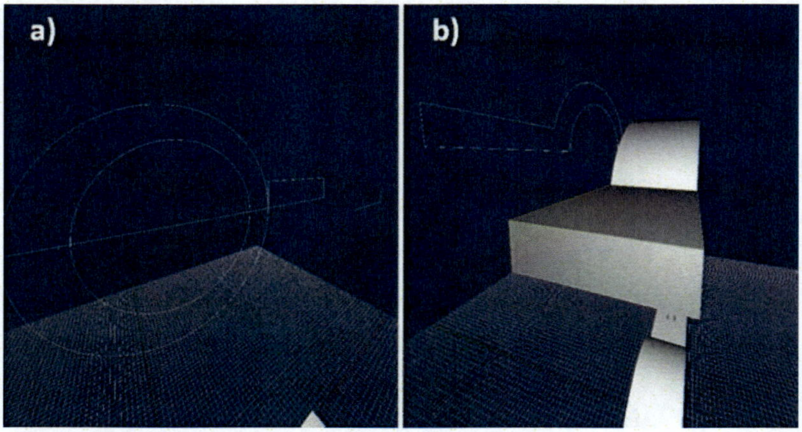

Fig. 8 (**a**) Sketch of a part of the connecting rod, (**b**) Connecting rod cap

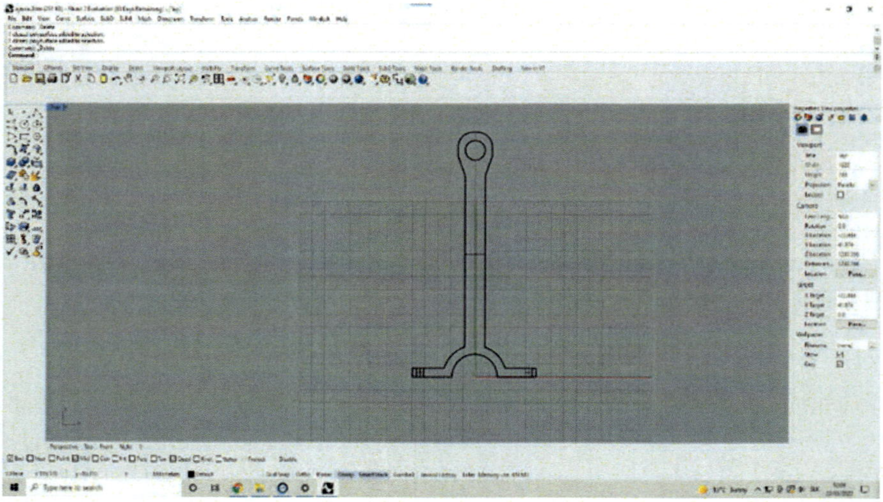

Fig. 9 Sketch of the entire connecting rod model

The connecting rod connects the piston and the crankshaft. The modeling started by creating the bottom part, which has the same shape as the sleeve, so the process of creating this part was the same. First, a sketch is created using the circle and poly tools, and then unnecessary lines are removed in the Rhinoceros software environment using the trim function. Subsequently, the upper part was created, where the circle function was used to create a part that serves to connect the connecting rod and the piston. Finally, the two parts were joined together using the poly tool (Fig. 9).

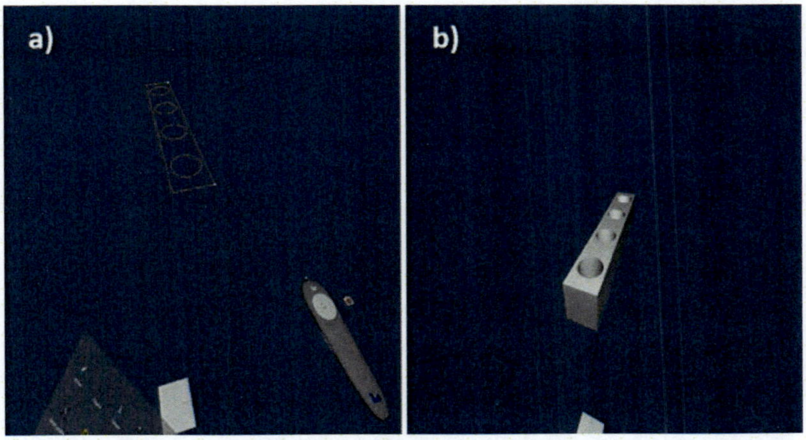

Fig. 10 (**a**) Sketch of the engine block, (**b**) Final model of the engine block

Creating a Model of the Engine Block This engine block is very simplified and serves mainly to illustrate and guide the movement of the cylinders. The creation of the engine block began with the creation of a sketch. First, the rectangle function was used, and a rectangle with dimensions of 65 x 392 mm was created. Subsequently, holes were created for the movement of the pistons. These holes are oriented to the center of the width of the model and are 98 mm apart. Subsequently, using the extrude function, the sketch was drawn into a space with a height of 100 mm, and the final model of the engine block was created (Fig. 10).

Final Engine Build in VR Figure 11 shows the final engine assembly in VR. The models of the assembly are positioned in VR in the Mindesk environment, where when the model is marked with the controller, the coordination system of the part is displayed, which can be used to change the position of the models in the assembly.

3 VR for Multiple Users (Multiuser Link)

To use the multiuser interface, each member must have a VR device. The CAD software is only needed for the administrator, and connected users use the Viewer application. For the connection to work properly, it is necessary to use the same LAN (Local Area Network), either using a network cable, wireless connection, or using a virtual private network (VPN). The administrator must manually enter the IP addresses of the other members when setting up the connection, and if they are set correctly, the "Linking to Rhinoceros" loading window will be displayed to all connected participants [3] (Fig. 12).

All participants can use most of the available Rhinoceros and Mindesk tools (navigation, modeling, creation, and surface modification). Each modification of the

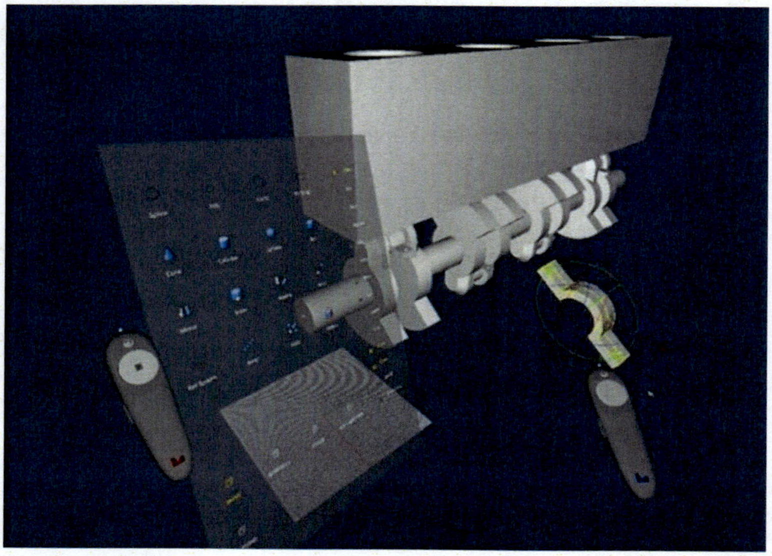

Fig. 11 Final engine build in VR

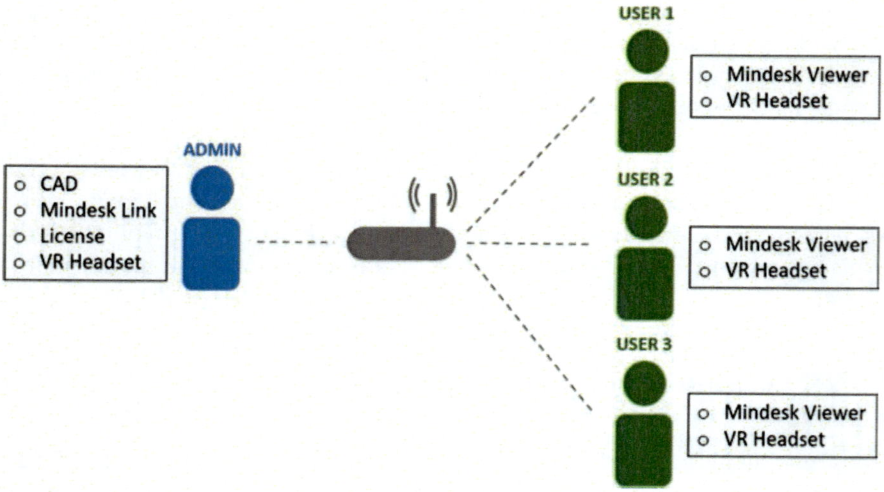

Fig. 12 Multiuser connection diagram [9]

model made by the participants of the connection directly affects the CAD database stored on the administrator's workstation [3].

Figure 13 shows an example of the use of several users, where with the help of the 3D pen tool, it is possible to sketch a new design of the model freehand, which is subsequently modified in the Rhinoceros environment.

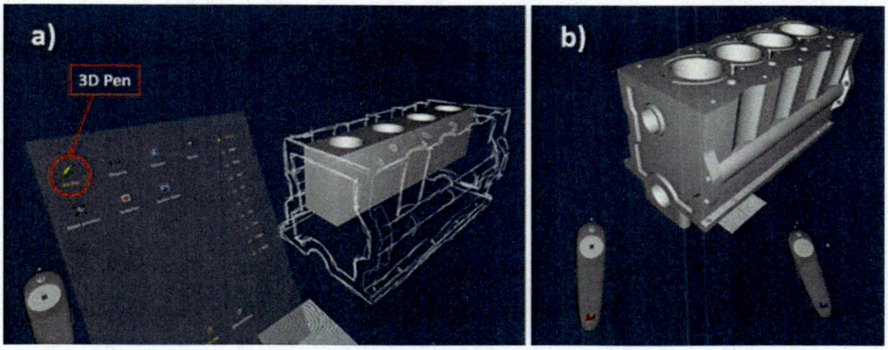

Fig. 13 (**a**) Freehand sketch, (**b**) New model

4 Creating an Engine Assembly in PTC Creo

The program Rhinoceros does not offer the creation of assemblies, but since it is an open program and supports various extensions, the possibility of creating animations that are similar to the creation of assemblies exists. An extension called Bongo allows you to create various animations, such as engine animation, but this is not a suitable solution for this case, so PTC Creo and its assembly module were used. The PTC Creo program was opened and controlled in the SteamVR environment using the desktop view in VR. Initially, all models need to be saved in the correct format, as PTC Creo does not support the format of files saved in Rhinoceros. For these purposes, the .stp file format was chosen. Subsequently, the open function is used to open these components, we choose which component we want to open, and then the component model is saved in the .prt format, which PTC Creo supports, and it simplifies the work when creating assemblies. A new assembly file is created using the new function (Fig. 14).

The first step in building an assembly is to import the engine block model. This model is set as the default model, and the next creation of the assembly will be based on it. The next step in the creation is to insert the crankshaft model. The shaft will perform a rotating movement, so when creating connections, the connection type is set as pin. The crankshaft is located exactly in the center of the engine block and 100 mm away from the top of the block. Subsequently, the connecting rod model is added to the assembly, and the pin connection is set. Next, a link is created that connects the connecting rod and the crankshaft, and another link determines the correctness of the location. These steps are repeated four times so that four connecting rods are in the assembly. Subsequently, a connecting rod sleeve model is used so that the connecting rod can be secured from the other side with it. Finally, the function in the tools tab, under the name screw, is used to create a joint. This function will make the work easier when modeling the bolt and nut. ISO 7380-1 screws with M3 x 16 mm thread, ISO 7040-5 nuts, and DIN EN ISO 7092 washers were used. These screws are used throughout the assembly. Subsequently, the piston

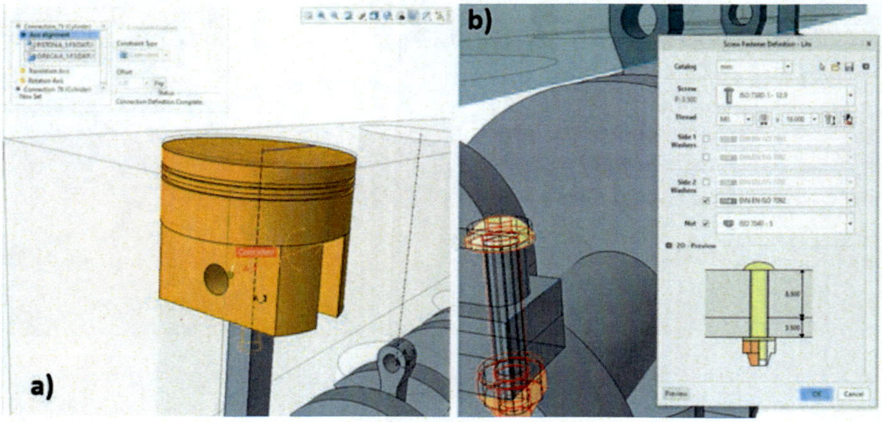

Fig. 14 (**a**) Creating links for the engine piston, (**b**) Creating screws using the screw tool

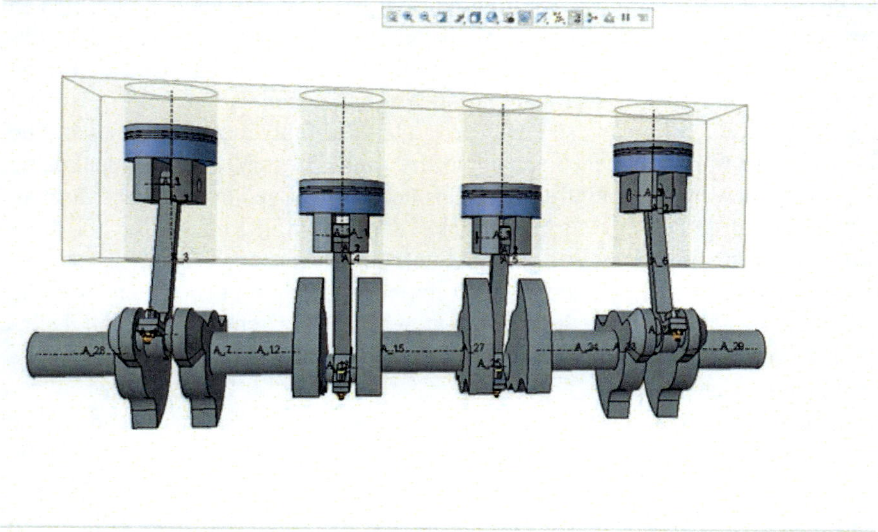

Fig. 15 The final model of the four-cylinder engine assembly

is inserted into the assembly using two cylinder ties, which determine the connection of the connecting rod with the piston and the place for the movement of the piston in the engine block. The last step is inserting the connecting pin to hold the piston and connecting rod. The final assembly is shown in Fig. 15.

5 Conclusion

This paper focuses on the description of computer modeling tools in VR for the needs of preparing an assembly. In the practical part, a model of a four-cylinder engine was created, where all the necessary components were developed in a VR environment. It is during the creation of individual components that various shortcomings are described, where it was necessary to choose an alternative method of creating some models. A frequent problem was the unavailability of tools in the Mindesk program. However, the advantage of creating models in VR is the constant connection with the Rhinoceros application, so if the tool is not available in Mindesk, then this tool can be found in the Rhinoceros program, and the modification of the model will be displayed in real time in the VR environment. Finally, an assembly was created from the created components in the PTC Creo program since Rhinoceros and Mindesk do not offer assembly creation. Currently, modeling in virtual reality does not replace computer modeling but can serve as its complement. The main reason is precisely the difficulty of creating models, where it is often necessary to use tools outside of VR in the application with which Mindesk is connected, so modeling is still dependent on creating models in a classic way. Furthermore, it is difficult to create an accurate 2D sketch due to the nonavailability of tools during creation, such as trim, chamfer, and round, which would facilitate the work. On the other hand, if it is necessary to create models where the same parts are repeated, such as the creation of a crankshaft, it is easy to copy these repeating components and assemble them into the final model. A suitable solution to these shortcomings could be the creation or implementation of VR in parametric CAD software, where there would be a greater selection of tools and an easier process of creating sketches for model creation. VR can serve in the process of checking the created models, where the model is displayed in its real size, and it is easier to diagnose various shortcomings and errors of the model compared to the classic display. The advantage of using VR in computer modeling is the possibility of sharing the project with several users and changing the model in real time.

In the practical part, all known possibilities of VR were used, from interaction with the virtual environment, to its display itself, where the innovative part can be considered the complex use of VR for the creation of the entire model, using a 2D display of the desktop in VR, which provides the use of all functions of the CAD program.

Acknowledgments The chapter was created with the support of the project VEGA: No. 1/0026/22 Augmented reality and research of its application possibilities in the process of assembly in mechanical engineering, and KEGA: No. 004TUKE-4/2022.

References

1. Baron, P., Kocisko, M., Franas, E.: Application of augmented reality tools to the design preparation of production. MM Sci. J., 4428–4433 (2021)
2. Ma, D., Gausemeier, J., Fan, X., Grafe, M.: Virtual reality & augmented reality in industry. Springer, Berlin, Heidelberg (2011)
3. Franas, E., Kočiško, M., Kačšák, J., Hlavatá, S., Vodilka, A.: Virtual reality and its possible integration into the process of distance learning focused on technically oriented subjects. 2021 IOP Conf. Ser.: Mater. Sci. Eng. **1199**, 012028 (2021)
4. Vijay, D., Primer, C.: A general guide to computer aided design and drafting-Cadd. MailMax Publishing, New York (2000)
5. Kuric, I., et al.: Počítačom podporované systémy v strojárstve. Edis, Žilina (2002)
6. HTC Corporation. Vive Cosmos, https://www.vive.com/us/product/vive-cosmos/specs, Last accessed November 3, 2023.
7. Murray, J.W.: Building virtual reality with Unity and Steam VR. Taylor and Francis Group, New York (2017)
8. Ververidis, D., Nikolopoulos, S., Kompatsiaris, I.: A review of collaborative virtual reality Systems for the Architecture, engineering, and construction industry. Architecture. **2**(3), 476–496 (2022)
9. Mindesk. Multiuser VR in Rhinoceros, https://mindesk.freshdesk.com/support/solutions/articles/33000255190-multiuser-vr-in-rhinoceros, Last accessed November 3, 2023.

Part IV
3D Printing, Composite Materials, Fused Deposition Modelling

FDM Printing and Parameter Setting in the KISSlicer Program

Ján Kopec, Miriam Pekarčíková, and Jozef Trojan

1 Fused Deposition Modeling (FDM)

3D printing and modeling a part of several bachelor theses and diploma theses, while they represent the method of rapid prototyping, the possibility of testing models for production processes, or is possible to create a miniature of the production hall, in the optimization of production or the introduction of a new production line or warehouse. Print parameters are influenced by several factors. The first factor is the filament. The storage method, age, and the printer itself affect the filament. This brings us to the second factor, which is the 3D printer. Printers range from hundreds of euros to tens of thousands. It depends on the manufacturer, the size of the printer, the materials used for production, and the closure of the printing area (there are open printers and printers with a protective cover to better maintain the temperature of the extruder). The last factor is the software used to create the .gcode.

With most 3D printers using the FDM process, selecting several parameters to achieve more accurate printing is possible. Controllable parameters are, e.g., printing speed and filament extrusion speed. The temperature of the nozzle and the temperature of the work surface are optional due to the different typologies of the filament and, therefore also, the different melting temperatures of the filament and the different solidification temperatures on the work surface. We corrected the nozzle's size and the required layer's height at the initial level. All parameters define the dimensional accuracy of the print; only the smaller diameter of the nozzle and the lower height of the layer determines the solution for parts where a smooth surface and a high level of detail must be achieved [1].

J. Kopec (✉) · M. Pekarčíková · J. Trojan
Faculty of Mechanical Engineering, Technical University of Košice, Košice, Slovakia
e-mail: jan.kopec@tuke.sk; miriam.pekarcikova@tuke.sk; jozef.trojan@tuke.sk

M. Balog et al. (eds.), *EAI International Conference on Automation and Control in Theory and Practice*, EAI/Springer Innovations in Communication and Computing, https://doi.org/10.1007/978-3-031-31967-9_14

The extruded material in this type of 3D printing is in the form of a thermoplastic polymer, the so-called filament. Using the extruder, the filament is fed toward the heated nozzle, where it is melted. Subsequently, the molten string is applied to the working surface of the printer through the nozzle along a predetermined path. On the surface, the string solidifies and cools, and the desired object is created by successive layering. The filament in the form of a string is wound on spools of various sizes. The coil is inserted into the printer, and then, using the extruder, it is guided in the channel toward the heated nozzle, where the filament is melted. Calibration in 3D printing technologies is very important because the printers work at the micron level. Calibration is performed manually, or the 3D printer performs it independently before each print run. After reaching the correct filament melting temperature and heating the work surface (if the printer provides it), the 3D printer starts moving the print head and applies the melted filament to a precisely determined place where it solidifies. After each layer is completed, the moving platform (either the print head or work surface) moves to a new position and continues the printing process layer by layer.

1.1 Materials for 3D Printing Using the FFF/FDM Method

- ABS: 3D models printed from this type of filament have high hardness and impact resistance. They are also characterized by good resistance to high temperatures and scratching. The printing temperature varies between 235–265 °C, and the substrate temperature must be at the level of 90 °C [2].
- ABS+: This material is offered in a wide range of colors on the market. ABS+ material is a suitable and versatile material from which many models can be created, either according to construction or models intended for general use. Models printed from this material should not be exposed to the weather and are also not intended for printing components for medical purposes. The recommended temperature of the nozzle is in the range of 230–240 °C, and the recommended temperature of the pad is 90–100 °C.
- ABS-T: ABS-T material is used when highly detailed 3D printing is required. The use is like ABS+, where a heating printing surface is also required. ABS-T is stronger compared to ABS material, but a higher melting temperature is required. The percentage of model shrinkage is smaller than ABS, which means printing models of larger dimensions from this material is possible. Prints from the material are glossy to transparent. On the other hand, ABS is tougher and offers matte final prints. The recommended nozzle temperature for the printing surface is 230–250 °C and 90–110 °C.
- ASA: This is a very durable thermoplastic sought after for its properties. It is a material suitable for outdoor models, functional prototypes, or spare parts. ASA material withstands weather conditions and does not lose its properties even after exposure to UV radiation. The printing temperature is in the range of 200–240 °C, and the substrate temperature must reach 40–60 °C.

- HIPS: HIPS material is a quality material that offers trouble-free printing. The HIPS material is so versatile that it can be used as a building material and as a support material when the main building material is another type of filament. The model has a light pattern with a matte surface. Secondary surface treatment can be done by sandblasting or painting with acrylic paint. It is a filament from the category of thermoplastics (a mixture of rubber and polystyrene that is modified). Acetone or limonene acts as a solvent for the HIPS material. In terms of properties, HIPS is sufficiently strong and tough. To achieve the best properties, choose a combination of two or more head 3D printers (one head prints support from HIPS, and the other forms a model structure from, for example, ABS). The printing temperature when using filament made of HIPS material is 235–250 °C, and a heated substrate is also required at 80–100 °C [3, 4].
- PLA: This is one of the most widespread materials used in the FFF/FDM method. It offers high-quality printing, precise details, and excellent lamination. As one of the few materials intended for 3D printing, it meets the requirements for contact with food. The printing string is made of natural materials and easily recyclable. The printing temperature ranges from 195 °C to 220 °C and the plate temperature must reach 50–70 °C [5].
- PLA-ST: This material is characterized by high impact resistance and high toughness. The PLA-ST material is suitable for 3D printing of prototypes of mechanical parts, where it is necessary to maintain the accuracy and strength of the print. PLA-ST achieves minimum tolerance deviations and small shrinkage in the final model. A closer comparison of PLA, PLA+, PLA-ST materials is shown in Table 1.

- PET-G: For this string, it is necessary to have a 3D printer with a heating printing surface. Printing from PET-G filament is not difficult, and the resulting model is glossy with good color reproduction. The strength compared to ABS is better. It is a strong and tough material. PET-G material is characterized by good thermal stability. It is also characterized by a small shrinkage of the final model, which is why it is also suitable for printing large models. Compared to other printing materials, PET-G is highly resistant to acids and solvents and is also resistant to temperatures, be it high or low temperatures. The printing temperature ranges between 230–250 °C, and the heating surface must have a temperature of 90 °C [7].

Table 1 Comparison of PLA-based material [6]

	PLA	PLA+	PLA-ST
Tensile strength	**78 MPa**	63 MPa	34.3 MPa
Elongation to breaking point	14%	20%	**90%**
Bending strength	**90 MPa**	74 MPa	43 MPa
Modulus of strength in bending	**1984 MPa**	1934 MPa	1477 MPa
Notch toughness	5.6	9	**63**

2 Working with KISSlicer Printing Software

KISSlicer is a program that offers users a simple environment and the ability to set several 3D printing parameters so that the printout meets the quality requirements. KISSlicer is free software. The program serves as an export part to obtain the .gcode file format. The .gcode format instructs the 3D printer where to move, at what speed, and what path to follow.

The environment of the KISSlicer program is clear and simple (Fig. 1). After downloading and subsequent successful installation, it is necessary to set the print parameters. The setting option is available in three levels (classic, advanced, and expert). According to the selected option of the setting level, the parameters that the given level allows to be set become available. This is the size of the printing surface and the size of the filament from which we will print the model nozzles. It is necessary to go to the Profile Settings section and select the printer option on the bar. An option to set the printer's name will appear, as multiple printers can be configured [8].

The most basic settings that must be defined include the following:

- Number of Extruders: that is, the number of extruders, since some printers work with several printing systems,
- Loop/Solid Infill Overlap: this is the degree of overlap and fullness of the fill,
- Z Step: adjustment of the height displacement of the extruder in millimeters,
- Min Segment Length: minimum segment length,
- Bed Size: specified size in millimeters, dimensions of the printer in the X, Y, and Z axes,
- Bed Roughness: the thickness of the printing surface, also given in mm (Fig. 2).

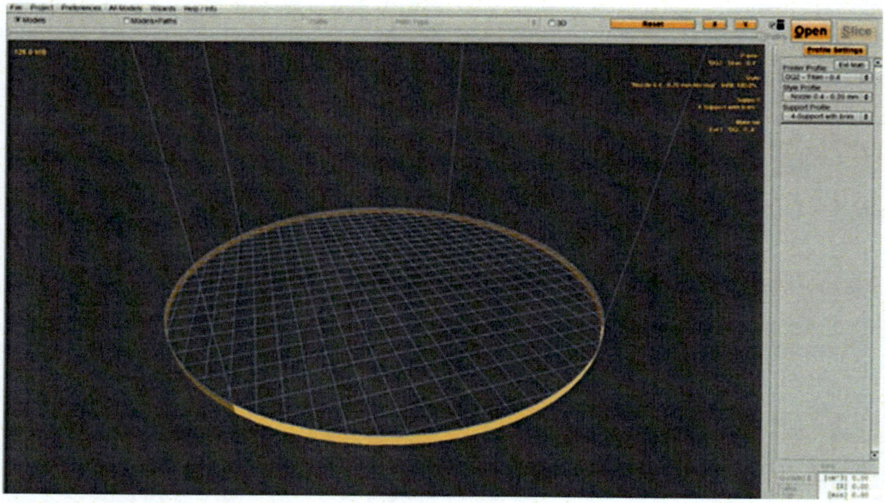

Fig. 1 Environment of the KISSlicer program

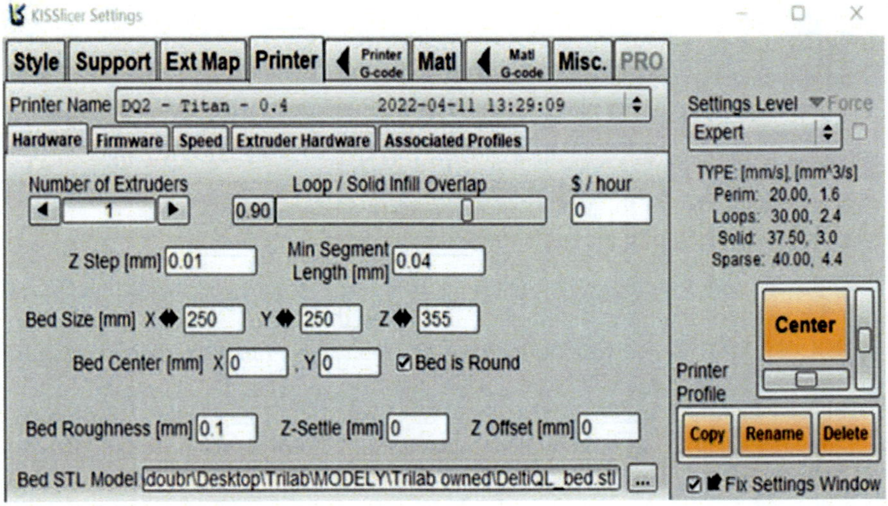

Fig. 2 Setting the initial parameters of the printer

Setting style parameters:

- Layer Thickness: For faster printing with low-quality requirements, 75% of the nozzle diameter is chosen (that is, if we have a nozzle with a diameter of 0.4 mm, then 0.3 mm is chosen). For better quality printing, at least 50% of the diameter is chosen (in our case, it is 0.2 mm). However, if there is a requirement for high-quality details but a long printing time, it is possible to use up to 20% or even 10% of the average.
- Extrusion Width: the extrusion width should always be greater than or equal to the diameter of the nozzle. A larger extrusion width will speed up printing but at the cost of a lower detail of the printed model. If you choose smaller values, the model is more detailed but fragile.
- Skin Thickness: this parameter should correspond to the product of the thickness of the layer and the width of the extrusion or come as close as possible to the multiple (in our case, it is 0.7 mm).
- Inset Surface: this value is usually left at zero. This parameter makes it possible to shrink the model by the selected amount. It is defined in case of insufficient clearance for holes or shafts [8].
- Stacked Layers: the possibility to define this parameter is only for PRO versions of the program. If left at the default level of 1, the support and padding for each layer of the model will be printed. If the data is already changed to 2, one layer is printed, but the support is printed only for every second layer, which causes the speed of printing at the cost of reducing strength.
- Speed Slider: the push speed assigned to the given style.
- Number of Loops: this parameter should be used to calculate the thickness of the edge layer.

- Loop Order: if this option is enabled, the model will have better-looking circuits but slightly more imprecise outer dimensions. If exact models are required, it is recommended to disable this option.
- Infill Slider: selection of the amount of infill. Moving the slider to the left (100% padding) and moving the option to the right will set zero padding and no top padding. Using the fill percentage setting option also makes it possible to precisely define the number of fill percentages [9].
- Infill Extrusion Width: the infill width can be defined at three levels:
  - thinner extrusion: less deformation at the expense of a weak model,
  - the width of the filling is the same as the width of the extrusion: the pressure in the print nozzle will be constant, a good appearance of the printout is achieved,
  - larger than the extrusion width: the model will gain more strength, but the appearance will deteriorate due to the definition of the perimeter, and the risk of deformation in the printing process will increase (Fig. 3).
- Infill: the infill style can have three forms:
  - Straight: fast printing, but not too firm,
  - Octagonal (Octagonal filling): stronger model than with rectilinear filling, slower printing,
  - Rounded (Rounded filling): slightly weaker than the octagonal filling, but faster printing. Very similar to the octagonal print, but upon closer examination, the difference can be observed in the form of rounded corners (in the octagonal print, the print string is placed in sharp corners),

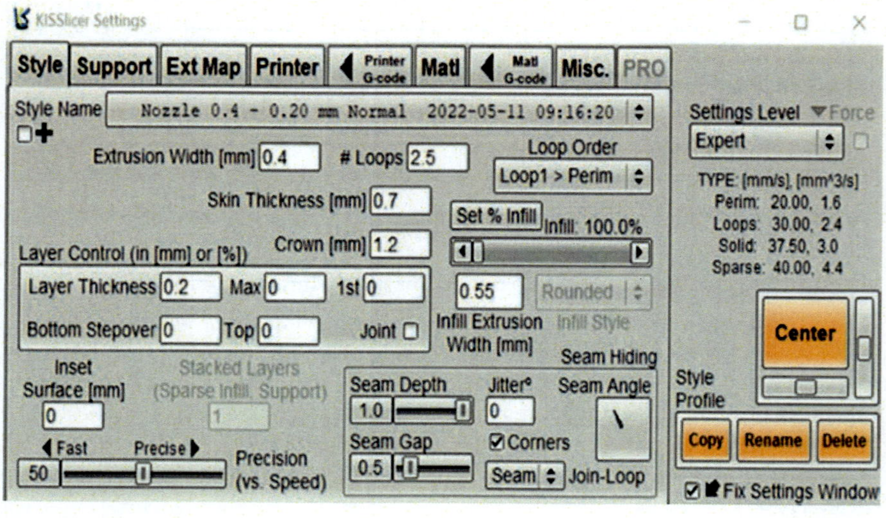

Fig. 3 Style setting

- Cellular (Cellular infill): the base is like the octagonal infill, with the difference that with cellular infill, an expansion is pressed onto each octagon of the infill with increasing layers until a connection occurs (Fig. 4).

- Support: support can be set at several levels (Fig. 5):
  - Off: the support is turned off, so no support material is used during printing,
  - Coarse: used for basic parts with low quality,
  - Rough: this is a medium-low carrier density,
  - Medium: as the name already indicates, it is a medium-density,
  - Dense: dense support used when printing round objects,
  - Fine: provides a high support density, but the print will increase significantly,
  - Ultra: very high density, used for demanding geometries,

- Support/Object Interface: This option must be enabled if a perfectly flat bottom layer is required. However, it should be noted that this option is only relevant for printers with a double extruder, or with the use of soluble support material,
- Z Band: control element of the last stage of support,
- Support degree: this element gives the possibility to set the angle to which support will not be needed,
- XY gap: represents the horizontal distance between the support and the object. We try to achieve low values (20%) if the print is round and high values (70%) for vertically flat surfaces,
- Zgap: this is the vertical distance between the support and the part; it should correspond to a value equal to a fraction of the thickness of one layer,

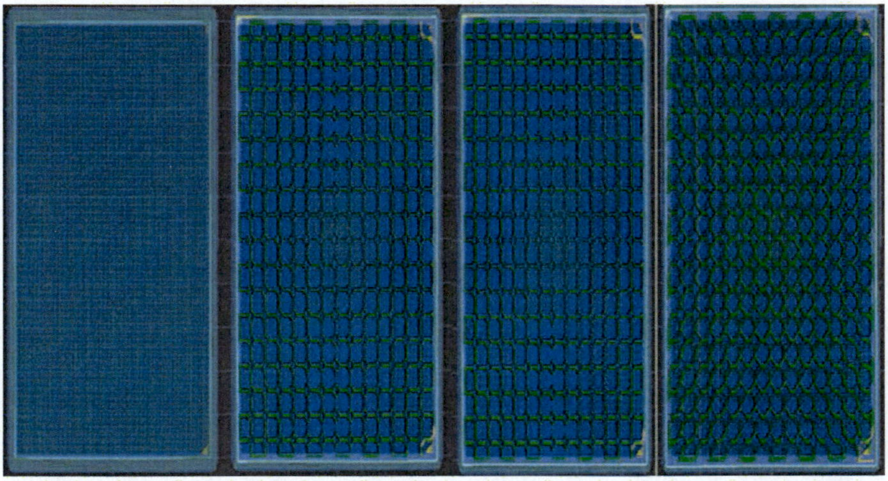

Fig. 4 Filling (Straight, Octagonal, Rounded, and Cellular)

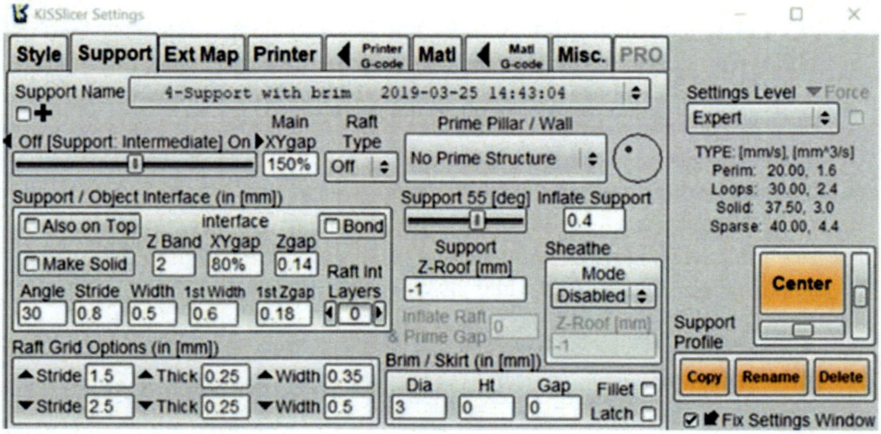

Fig. 5 Setting the filling parameters

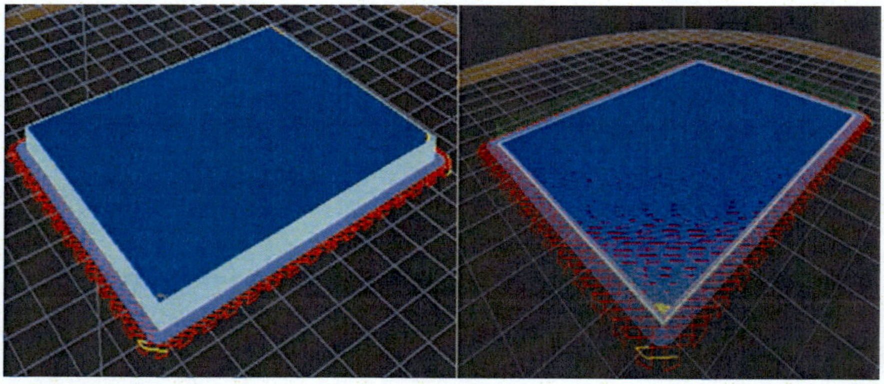

Fig. 6 Support in the form of raft-type Grid and Pillar

- Support z-roof: the default value is −1, and the program will automatically create support where necessary. If the value is overwritten to any value, support printing will stop for parts of the object whose height is greater than the specified value,
- Sheath: this option allows for creating a perimeter around the supporting structure. The final support part thus gains strength but is more difficult to remove [7]:
  - Raft type:
- Grid: a grid is printed between the model and the printing area,
- Pillar: almost identical to the grid option, but a layer of micropillars is added between the grid and the printing surface (Fig. 6).
- Raft Grid Options: each grid is made of two layers, but the properties can be changed. By adjusting it, it is possible to achieve a wide, thick, and properly

distributed bottom layer, which has good adhesion and thus compensates for errors on the surface of the print,

- Skirt: useful parameter when applying a layer at the very beginning. No other filling is performed during printing,
- Brim Dia: when adhesion is difficult, the rim will make this parameter larger to ensure better adhesion (obviously at the expense of increased time spent on pre-printing treatment). The diameter indicates how far the support surface should be to improve adhesion between the filament and the print surface,
- Brim Ht: left at zero in most cases. However, if it is changed to another numerical data, the surface to improve adhesion will increase and be rough according to the given number,
- Brim Fillet: If the BRIM HT is large enough, a fillet is applied to the print surfaces to increase adhesion and minimize distortion [10],

3 Conclusion

3D printing, as already mentioned in the introduction, is influenced by several factors. The article describes filaments, their properties for the user, working with the KISSlicer software, and the effect of changes in the program on the final print. The benefit is a summary of the work in the program so that even a layman who does not know how to work with the KISSlicer program can set up the 3D printer in the best possible way so that the final printout meets the necessary requirements.

3D printing as an additive technology is becoming an important part of prototyping. As a tool for rapid production, 3D printing is used in engineering, medicine, construction, and the like. The FDM method is the most basic method of creating 3D models. Several materials are used for this printing, which is described in the article. Each material has its own properties, and according to the determination of the print, it is necessary to choose a filament with the necessary properties. The KISSlicer program is designed to convert a 3D model in .stl format to a .gcode program. The program is clear and easy to set up, but appropriate experience with 3D printing is required.

The next continuation will be the dissertation and the measurements of the test prints and their comparison in the SolidEdge software.

Acknowledgments The implementation of the grant projects created this chapter: APVV-19-0418 Intelligent Solutions to enhance business innovation capability in the process of transforming them into smart businesses. VEGA 1/0438/20 Interaction of digital technologies to support software and hardware communication of the advanced production system platform. KEGA 020TUKE-4/2023 Systematic development of the competence profile of students of industrial and digital engineering in the process of higher education. VEGA 1/0508/22 "Innovative and digital technologies in manufacturing and logistics processes and system."

References

1. Anakhu, P., Bolu, C., Abioye, A.: Fused deposition modeling printed patterns for sand casting in a nigerian foundry. Rev. Internat. J. Appl. Eng. Res. **13**(7), 5113–5119 (2018) © Research India Publications. http://www.ripublication.com
2. HUBS: What is FDM (fused deposition modeling) 3D printing? https://www.hubs.com/knowledge-base/what-is-fdm-3d-printing/.
3. MakeShaper: Your Guide to 3D Printer Filaments; https://www.makeshaper.com/3d-printer-filament-comparison/
4. SIMPLIFY3D: HIPS, KISSlicer© v1.5 User manual; https://www.kisslicer.com/u loads/1/5/3/8/15381852/kisslicer_manual_en_201015.pdf
5. Uysal, R., Stubbs, J.B.: A new method of printing multi-material textiles by fused deposition modelling (FDM). Tekstilec, Web of Science. **62**, 248 (2019)
6. NA3D: ePLA-ST filament. https://www.na3d.sk/p/4420/epla-st-filament-175-mm-1-kg-esun-pla-super-tough?gclid=CjwKCAjw9-KTBhBcEiwAr19ig7Kvgz2aJbWf2e6FMTKk58ci9APUxpNfhaXyKw4r1S_5Tb-WdelEABoChvQQAvD_BwE
7. Tay, Y.W.D., Li, M., Tan, M.J.: Effect of printing parameters in 3D concrete printing: Printing region and support structures. Journal of Materials Processing Technology, Web of Science. **271**, 261 (2019)
8. KISSlicer© v1.5 User manual; https://www.kisslicer.com/u loads/1/5/3/8/15381852/kisslicer_manual_en_201015.pdf
9. Urbonaite, G., Kibirkstis, E., Miliunas, V.: 3D print technologies analysis. In: 18th International conference on Mechanika, web of science (2014)
10. Son, J., Choi, S.: Orientation selection for printing 3d models. In: 2015 International Conference on 3D imaging (IC3D). Web of Science (2015)

Application of Composite Materials Natural Fibers in Automotive Industry – Short Review

Zuzana Mitaľová, Dušan Mitaľ, and Vladimír Simkulet

1 Introduction

Compared to 1990, there is currently a greater than 25% increase in greenhouse gas emissions from transport (transport accounts for one-fifth of total emissions in the EU). Light commercial vehicles (cars and vans) account for about 15% of total CO_2 emissions. Average CO_2 emissions in the period from 2015 to 2019 amounted to 130 g CO_2 / km, in 2019, the value decreased to 122.3 g CO_2 / km [1]. In the period between 2020 and 2024, the following values were regulated for: passenger cars: 95 g CO_2 / km, vans: 147 g CO_2 / km. The target is to minimize emissions below 50 g CO_2 / km [2]. In order to comply with EU regulations and directives, manufacturers are trying to reduce the amount of CO_2 by reducing the weight of vehicles (1 kg of vehicle saves the environment about 20 kg of carbon dioxide, according to Martin Goede et al.: 100 kg mass reduction in a car weight could result in a 0.3–0.5 dm^3 / 100 km). The weight reduction of a motor vehicle can be achieved by replacing the parts with lighter materials. Plastics are also an alternative, either in virgin or recycled form / and so are the NFRP materials – composite plastic materials reinforced with natural fibers (benefits of natural fibers: biodegradation, fiber modification, thermal stability, crystallinity). The use of recycled plastic is claimed to be able to reduce 80% of greenhouse gas emissions [3]. The present short review discusses composite materials reinforced with natural fibers and recycled plastics in the automotive industry. According to Jato Dynamics – total average CO_2 emissions in Europe indicate a Fig. 1.

Z. Mitaľová (✉) · D. Mitaľ · V. Simkulet
Faculty of Manufacturing Technologies with the seat in Prešov, Technical University of Košice, Prešov, Slovakia
e-mail: zuzana.mitalova@tuke.sk

M. Balog et al. (eds.), *EAI International Conference on Automation and Control in Theory and Practice*, EAI/Springer Innovations in Communication and Computing, https://doi.org/10.1007/978-3-031-31967-9_15

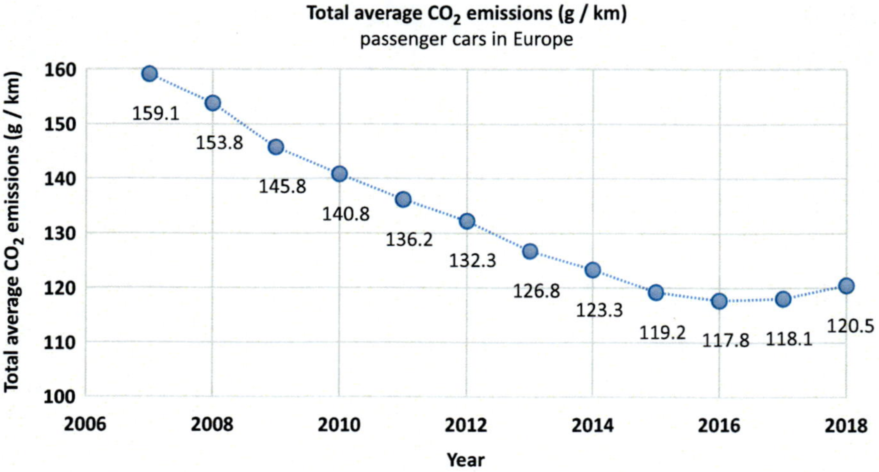

Fig. 1 Total average CO_2 emissions (g/km) for passenger in Europe. (Source: agency Jato Dynamics)

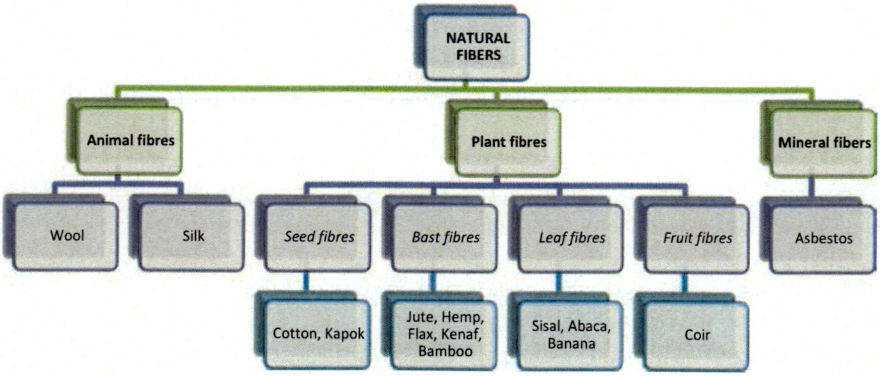

Fig. 2 Classification of natural fibers

1.1 Classification of Natural Fibers, History of the NFRP Application

Interest in the NFRP materials is growing – both in terms of industrial applications and the basic research. Advantages of the NFRP materials: low specific weight, renewability (production of an NFRP composite requires low energy inputs), production at low input costs, (complete or partial) recyclability, biodegradability, reduced wear of tooling, healthier working conditions + good insulation, thermal and acoustic properties [4]. The term natural fibers refer to fibers of vegetable or animal origin (Fig. 2). Only cellulose fibers of plants obtained from various parts of these plants are used as reinforcement for the NFRP composites (see Fig. 2 classification of natural fibers and Table 1 describing selected mechanical properties of

Table 1 Mechanical properties of bast fibers [10–12]

Type of fiber	Tensile strength [MPa]	Elongation [%]	Young's modulus [GPa]	Study
Jute	300–700	1.2–3.0	20–50	Raja, et al., 2017
Hemp	350–800	1.6–4.0	30–60	Raja, et al., 2017
Flax	800–1500	2.7–3.2	60–80	Shahria, 2019
Kenaf	350–600	2.5–3.5	20.6	Khalid and Nor, 2017

Note: values can be different in relation to its type, origin, humidity, etc.

natural fibers). Natural reinforcements can be used in the form of long fibers / short fibers or crushed grains (for example – wood flour in the WPC products). In addition, fibers can be spun into threads or ropes (see below for powerRibs™ threads from Bcomp Ltd.). A disadvantage of natural fibers is their poor durability – moisture absorption, subsequent swelling, susceptibility to UV light (and the subsequent changes in molecular structure), poor fire resistance, and insufficient compatibility between the fiber and the matrix (i.e., the resulting composite). However, the disadvantages can be "suppressed" by appropriate treatment methods, thus modifying the fibers' properties [5]. The modification can be carried out by applying physical techniques (like plasma treatment), chemical techniques (modifying their crystalline structure – by removing weak components or by adding appropriate binders – for example silane), or through biological treatments [6–8]. Mechanical characteristics of the NFRP depend on several factors [9]:

- the strength of the fiber itself and its physical properties, moisture absorption, impurities,
- volume fraction and fiber orientation in the resulting composite; and,
- interfacial adhesion property of fiber (and others).

2 Application of NFRP in Automobile Industry

The application of natural fibers in the automotive industry dates back to the first half of the Twentieth century. In 1941, Henry Ford developed the first prototype of a composite car referred to as a Soybean Car = ... "soybean fiber in a phenolic resin with formaldehyde used in the impregnation" Davis, 1941 (cutting weight to 907 → 454 kg lighter than a steel car) [13, 14]. *However, a breakthrough in the field of the NFRP application occurred later - in 1957 Trabant 601 was made in East Germany with a body made of duroplast (hardened plastic made of waste fibers of cotton+ phenolic resins)* [15]. *Followed by Fiat punto (1982) with wood flour-filled door panels. Since the 1990s, there has been intense research in the field of replacement of original parts (whether metal or composite containing glass fibers) by composites reinforced with natural fibers (Table 2). The automakers, auto suppliers and research academies play an important role in the NFRP development*

Table 2 LCA (Life cycle Analysis) of NFRP in production [16]

Type of product	NFRP type	Traditional material	Weight reduction [GPa]	Change in greenhouse gas emission [%]
Car side door	Hemp + EP	ABS	−27	−15
Under-floor pan	Flax + PP	GF + PP	0	NA
Rotor blade	Flax + EP	CF + EP	0	−45
Car interior	Bagasse + PP	Talc + PP	−20	−21

Bcomp Ltd. (Fribourg, Switzerland) was nominated as one of the six candidates for the 'Biocomposites of the Year 2019' award in 2019. It is dedicated to developing reinforcements for the NFRP composite materials (e.g., *ampliTex™ - reinforcement made of flax fibers – for applications with thermoplastics /thermosets - for the RTM technology, in the form of prepreg for processing in an autoclave, or to be processed by the compression molding technology). Their reinforcements were applied by automotive companies in the production of parts as outlined below* [17–19]

- VOLVO cars: powerRibs™ (source: hemp) with recycled ocean plastic for XC60 passenger car parts (Fig. 3);
- McLaren: application of flax fibers AmpliTex™ in the production of racing car seats (advantages of composite material with flax fibers: reduction of vibration, increased safety);
- PORSCHE: replacement of traditional monolithic carbon fibers with powerRibs™ fibers, reducing the carbon footprint in the production of a particular part by 85% and improving the safety conditions in the case of an impact (ductile crash behavior sharp shattering), Fig. 3 - front bumper;

Faurecia - a part of Forvia Group (Nanterre, France) – has been the market leader in sustainable vehicle interiors since the 1980s. Since 2003, it has focused on the application of the NFPP (Natural Fiber-reinforced Polypropylene) composite materials with more than 50% share of natural fibers (e.g., NAFILean – lightweight natural, recyclable fiber material with hemp fibers, 2011) – Fig. 4 [22]. Currently, recycled polymers (modified NAFILean R – using 100% recycled plastic) and the so-called rCF NFPP materials – recycled carbon fiber (composite material for compression technology consisting of 40% kenaf fibers / 50% polypropylene and 10% recycled carbon fiber as a nonwoven reinforcement layer) for smart and sustainable interiors are also being used in their production. rCF NFPP achieves a weight reduction of 50% compared to injected technologies and 25% compared to standard compression technologies [23]. NAFILean bio composites currently equip 17 vehicles - for example: Renault Clio, the Alfa Romeo Giulia, Land Rover Velar, Peugeot 508, etc.

Johnson Controls GmbH & Co. KG, a global leader in automotive seating (interiors and electronics), introduced a nonvisible door panel elements for the BMW 3 series in 2012 made of wood fiber and molded plastic matrix. This innovation won

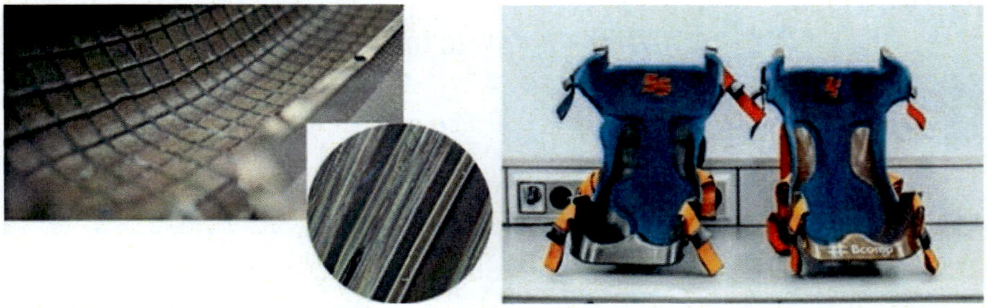

Fig. 3 Left side: Front bumper with powerRibsTM reinforcement for Porsche (micrograph of hemp fiber, Nikon), right side: racing car - seat McLaren [17, 20, 21]

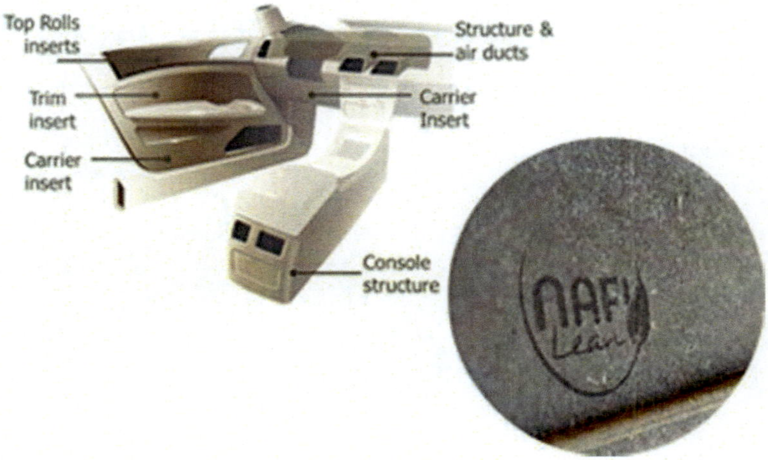

Fig. 4 Application of NAFILean: for covered or nonvisible applications for door panels, center consoles [24]

it the Automotive Division Award from the Society of Plastics Engineers (SPE) since the 'innovative production process makes the door panel 20% lighter than conventional components' [25]. The NFRP materials with abacus reinforcement (type of banana fibers) were applied in automotive product by Rieter Automotive Systems [26]. Green dot BIOPLASTICS applies Terrater® NFRP - a bio-composite material reinforced with jute, sisal, hemp, American bamboo, and flax fibers in conjunction with a plastic matrix as a substitute for composite materials with glass fibers (Fig. 6). For comparison, the density of glass fibers is approx. 2.6 $g.cm^{-3}$, while the flax fiber accounts for 1.5 $g.cm^{-3}$ – see the graphic representation below – Fig. 5. They use the material in the production of components from door and trunk liners to under-floor pans to floor panels (and for sport equipment, too).

The company is also working on the production of composite material from corn cob with green PE (Terratek CC200500, for injection molding) and 100% biodegradable fabric – plant-based synthetic leather as an alternative to leather car seat upholstery. Similarly, it is possible to apply Piñatex (a material referred to as Vegan

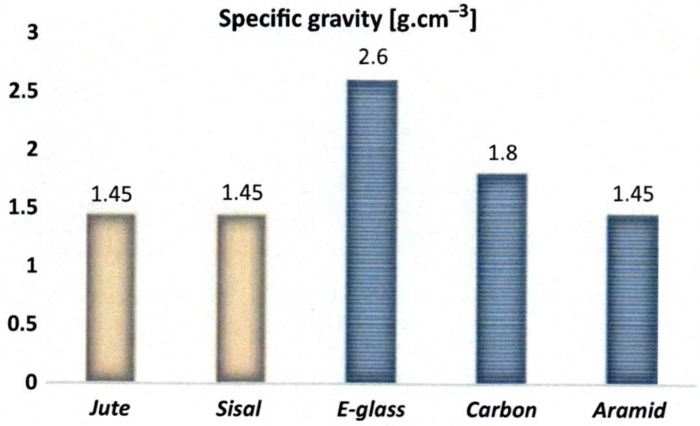

Fig. 5 Comparison of specify gravity of different fibers type [32]

Fig. 6 Presentation model – John Deere EcoRigs made with corncob fibers and green PE (Green dot BIOPLASTICS) [28]

Leather), made by combining waste pineapple fibered with PLA and a petrol-based resin coating. The material was patented by a Spanish entrepreneur Carmen Hijosa, who became a finalist for the European Inventor Award 2021 [27–30]. The German company Quadrant Natural Fiber Composites GmbH (Lambrecht, Germany) has established a strong position in the natural fiber composites market. Quadrant's Nafcoform material is a mixture of 50% PP and 50% natural fibers (kenaf, hemp, and flax). Among the vehicles using Quadrant's natural fiber composite are the Audi A8, Mitsubishi Space Star minivan, BMW 7-series, Iveco heavy truck line, Volkswagen Touareg, Porsche Cayenne, Audi Q7 sport utility vehicles. The components are sandwiches of expanded PP foam covered on each side with natural fiber / PP composites and topped with PET carpet [31].

Currently, car manufacturers in Europe are trying to incorporate the NFRP materials into the production process (as a substitute for original materials). European manufacturers integrate these materials in the interiors of passenger cars/ commercial vehicles. A 2004 study shows that the BMW group used 10 kilotons of natural fibers in its production of the NFRP materials [33]. From the BMW 3 series to → the BMW X5 Series, the company has achieved savings of over 300 kg in weight by applying the NFRP composite materials [34]. Volkswagen used hemp, flax, and kenaf fiber for a composite part – the door trim for its Golf model. In 2021, they

Fig. 7 Left side: Oryzite material applied in the production of discs (aluminum replacement); right side: rice husks [36, 37]

included materials from SEAQUAL YARN in their production – material made from recycled polymer fibers (from collected plastic waste). SEAT used rice husk composite material (the secondary waste in harvesting rice) in combination with thermoplastics (PUR / PP / PA) as a substitute for plastic (for flooring and roof upholstery). The pilot project started in 2019. A material with the trade name of Oryzite was created (modified rice husks). The composition of rice husks is: 20% silica (increases resistance), 40% cellulose, 30% lignin and 10% fatty acids (Fig. 7). Almost 40% of this material is carbon dioxide captured from the atmosphere during the plant life cycle. This composition is preserved once it is turned into Oryzite and even when applied in the final vehicle part [35]. The economic appreciation and potential benefits of the material are described in detail in the study by Ilijevic, Colet and Ganduxé in the ATZ worldwide journal (2021).

The leader in applying sustainable materials is Ford Motor Co. Plastic reinforcement using rice husks is used in its F-150 pickup truck‘s wiring harnesses. Fibers from wheat straw are used in the Flex full-size SUV to reinforce storage bins. The Lincoln MKX armrests are made of cellulose fibers (replacing the original glass fibers, which results in a weight loss of 10% compared to the original version of the part). In addition, the company seeks to apply fiber found in the vicinity of its production sites. E.g., "waste" – fiber from the agave plant (pulp is used for the production of alcoholic beverages) in Mexico constitutes the input raw material of ecological Ford composites for their plants in that territory. They are used in the production of bioplastic parts – such as storage boxes, HVAC units, etc. [38, 39]. A mixture of kenaf and flax has gone into package trays and door panel inserts for Opel Vectras [40]. Biomimetic materials – where the resulting structure of the composite material is inspired by biological patterns – animals, are also possible. The study by Hoffmann et al. dealt with the production process of an NFRP composite material suitable for car interior parts.

The material had a sandwich structure with thin, unidirectional flax fiber layers surrounding a needle felt based core (with the resulting reduced weight of the part compared to the original one). The idea was inspired by the carapace structure of the red-eared slider Trachemys scripta elegans [41] (Table 3).

Table 3 LCA (Life cycle Analysis) of NFRP in production [38]

Manufacturer	Model of automobile	Model applications
AUDI	A2 / A3 / A4 / avant / A6	Seat backs, door panel, boot lining, hat truck, spare tire linning
BMW	3 / 5 / 7 series	Door panels, headliner panel, noise insulation panels
FORD	Mondeo CD 162 / FOCUS	Door panels and boot liner, B-pillar
Mercedes-Benz	Trucks	Internal engine cover, engine insulation, sun visor, bumper, wheel box
Toyota	Brevis / harrier / Celsor / RAUM	Door panels, seat backs, spare tire cover
Volkswagen	Golf / Passat / variant / bora / fox / polo	Door panels, boot liner
VOLVO	C70 / V70	Seat padding, natural foams, cargo floor tray

3 Future Trends

The global market for NFRP materials has declined due to the economic recession caused by the coronavirus pandemic. However, it is currently recovering and is expected to reach an estimated USD 0.6 billion by 2026 (according to a large study: Trends, Opportunities and Competitive Analysis of the Natural Fiber Composite Market) [42]. The nature of the development of the European market is similar. According to the German Nova-Institute, up to 80,000 tons of wood / natural fibers are used for the production of 150,000 tons of composite parts for cars and trucks produced in Europe. Fibers of European origin (wood, cotton flax) are predominant, but imports (Asia - kenaf) are present, too [43]. In relation to their advantages and modification of properties (methods for "dampness suppression"), their further application in the short term is understandable. In large cities, however, the use of passenger cars is in decline due to the congestion of local infrastructures. Public transport serves as a substitute for the transport of inhabitants. It is, therefore, a question of how this aspect will be reflected in demand for composite materials with natural fibers on the part of car manufacturers.

4 Conclusion

The use of composite materials in the automotive industry is currently growing at an enormous rate. The replacement of original materials by "lighter" versions has an impact on the actual consumption of the car and ultimately on the elimination of CO_2 emissions into the air (reduction of the carbon footprint). Composite materials based on natural fibers have some excellent properties, such as: low specific weight, renewability, biodegradability, and reduced wear of tooling. Fibers of composite materials are of vegetable origin and, in many cases, constitute waste from the primary production process (e.g. the mentioned agave fiber as a succulent product in the production of alcoholic beverages or rice husks). Their disadvantage is the retention of moisture (stemming from their hydrophilic origin), but chemical, physical methods for modification exist. The paper describes the current NFRP materials applied in the automotive industry, where we witness apparent diversity of applied plant fibers in conjunction with polymeric matrices. Another challenge, in relation to current trends, is the creation of substitutes for polymeric matrices through biopolymers. Vehicle manufacturing is constantly innovating and giving impetus to other sectors of the economy in relation to the development and research of new ultra-lightweight and "greener" materials.

Acknowledgment This work was funded by the Slovak Research and Development Agency under contract No APVV-21-0228 and the projects VEGA1/0080/20, KEGA 018TUKE-4/2021 were granted by the Ministry of Education, Science, Research and Sport of the Slovak Republic and this publication is the result of the Project implementation: Development of excellent research capacities in the field of additive technologies for the industry of the twenty-first century, ITMS: 313011BWN5, supported by the Operational Program Integrated Infrastructure funded by the ERDF.

References

1. Európsky parlament, https://www.europarl.europa.eu/news/sk/headlines/society/20180920STO14027/nove-limity-na-emisie-co2-z-aut-a-malych-dodavok-v-skratke, last accessed December 10, 2022.
2. Climate action, https://climate.ec.europa.eu/eu-action/transport-emissions/road-transport-reducing-co2-emissions-vehicles/co2-emission-performance-standards-cars-and-vans_sk, last accessed November 10, 2022.
3. Gu, F., Hall, P., Miles, N.J.: Development of composites based on recycled polypropylene for injection moulding automobile parts using hierarchical clustering analysis and principal component estimate. J. Cleaner Product. **137**, 632–643 (2016)
4. Chandramohan, D., Marimuthu, K.: A review on natural fibers. Internat. J. Res. Rev. Appl. Sci. **8**(2), 194–206 (2011)
5. Cruz, J., Fangueiro, R.: Surface modification of natural fibers: A review. Procedia Eng. **155**, 285–288 (2016)
6. Shahidi, S., Wiener, J., Ghoranneviss, M.: Surface modification methods for improving the Dyeability of textile fabrics. Eco-Friendly Textile Dyeing and Finishing. **10**, 53911 (2013)
7. Xie, Y., Hill, C.A., Xiao, Z., Militz, H., Mai, C.: Silane coupling agents used for natural fiber/polymer composites: A review. Composit. Part A: Appl. Sci. Manufact. **41**(7), 806–819 (2010)
8. Pommet, M., Juntaro, J., Mantalaris, A., Lee, A.F.: Surface modification of natural fibers using bacteria: Depositing bacterial cellulose onto natural fibers to create Hierachical fiber reinforced nanocomposites biomacromolecules. **9**(6), 1643–1651 (2008)
9. Mohammed, L., Ansari, M.N.M., Pua, G., Jawaid, M., Islam, M.S.: A review on natural fiber reinforced polymer composite and its applications. Internat. J. Polym. Sci., 1–15 (2015)
10. Raja, T., Anand, P., Karthik, M., Sundaraj, M.: Evaluation of mechanical properties of natural fibre reinforced composites: A review. Internat. J. Mech. Eng. Technol. **8**(7), 915–924 (2017)
11. Shahria, S.: Fabrication and property evaluation of hemp–flax fiber reinforced hybrid composite. Cellulose. **77**(56.5), –64–1 (2019)

12. Khalid, S.N.A., Nor, N.H.M.: A review on the perforated impact energy absorption of kenaf fibres reinforced composites. In J. Physics Conference Ser. **914**(1), 012044 (2017)
13. The Henry Ford Homepage, https://www.thehenryford.org/collections-and-research/digital-resources/popular-topics/soy-bean-car/, last accessed November 11, 2022.
14. Chen, K.I., et al.: Soyfoods and soybean products: From traditional use to modern applications. Appl. Microbiol Biotechnol. **96**(1), 9–22 (2012)
15. New Atlas, https://newatlas.com/fraunhofer-cotton-composite-cars/35445/, last accessed October 10, 2022.
16. Duflou, J.R., Dang, Y., Van Acker, K., Dewulf, W.: Materials for sustainable development. MRS Bulletin. **37**(4), 374–382 (2012)
17. Innovation in Textiles, https://www.innovationintextiles.com/first-for-flax-from-mclaren-and-bcomp/, last accessed October 10, 2022.
18. CW Composites World, https://www.compositesworld.com/news/six-candidates-nominated-for-biocomposite-of-the-year-2019, last accessed October 10, 2022.
19. McLaren, https://www.mclaren.com/racing/sustainability/natural-fibre-sustainable-composite-racing-seat/, last accessed Septeber 10, 2022.
20. Renewable carbon, https://renewable-carbon.eu/news/powerribs-enables-ocean-plastic-in-automotive-interior-parts/, last accessed October 11, 2022.
21. Microscopy U, https://www.microscopyu.com/gallery-images/hemp-fibers, last accessed November 10, 2022.
22. Forvia Faurecia Homepage, https://www.faurecia.com/en/newsroom/spotlight-sustainable-materials-nfpp-and-next-generation-vehicle-interiors, last accessed November 10, 2022.
23. JEC Composites Connect, https://www.jeccomposites-connect.events/essential_grid/recycled-carbon-fiber-nfpp-rcf-nfpp/, last accessed November 12, 2022.
24. APM Homepage, https://www.apm-planet.com/products/nafilean/what-is-nafilean/, last accessed October 10, 2022.
25. Osta Hompeage, https://o-sta.si/en/9248/society-of-plastics-engineers-spe-central-europe-honors-johnson-controls-for-innovative-door-panel-made-of-natural-fibers, last accessed September 12, 2022.
26. Scherübl, B.R.: The realization of natural fibre-reinforced plastics in the automotive exteriors. In: Documentation of the Third International Conference of the European Industrial Hemp Association (EIHA), pp. 1–28. Nova Institut GmbH, Hürthm (2005)
27. Green Dot Bioplatics Homepage, https://www.greendotbioplastics.com/materials/natural-fiber-reinforced-composites/, last accessed October 12, 2022.
28. Plant Based Composite Materials for Automotive, https://speautomotive.com › uploads › 2020/12, last accessed October 9, 2022.
29. European Pattent Office, https://www.epo.org/news-events/events/european-inventor/finalists/2021/hijosa.html, last accessed December 1, 2022.
30. Institut of Making, https://www.instituteofmaking.org.uk/materials-library/material/pinatex, last accessed October 11, 2022.
31. Renewable Carbon News, https://renewable-carbon.eu/news/natural-fiber-composites-slowly-take-root/, last accessed September 9, 2022.
32. Pandita, S.D., Yuan, X., Manan, M.A., Lau, C.H., Subramanian, A.S., Wei, J.: Evaluation of jute/glass hybrid composite sandwich: Water resistance, impact properties and life cycle assessment. J. Reinforced Plast. Composit. **33**(1), 14–25 (2014)
33. Holbery, J., Houston, D.: Natural-fiber-reinforced polymer composites in automotive applications. JOM. **58**(11), 80–86 (2006)
34. Pervaiz, M., Panthapulakkal, S., Sain, M., Tjong, J.: Emerging trends in automotive lightweighting through novel composite materials. Mat. Sci. Appl. **7**(01), 26 (2016)
35. Ilijevic, S., Colet, J., Ganduxé, I.: Renewable raw Material from Rice husks for vehicle interior parts. ATZ Worldwide. **123**(4), 44–47 (2021)
36. ABC Economia Homepage, https://www.abc.es/economia/insospechadas-vidas-paralelas-arroz-20220911130449-nt.html, last accessed September 12, 2022.

37. Montsia Homepage, https://montsia.es/en/tags/oryzite, last accessed October 12, 2022.
38. Assembly Homepage, https://www.assemblymag.com/articles/94029-natural-fibers-may-hold-the-key-to-lightweighting, last accessed October 10, 2022.
39. Continental tires, https://www.continental-tires.com/sk/sk/b2c/stories/sustainable-materials-in-cars.html, last accessed October 10, 2022.
40. Linkedin Homepage, https://www.linkedin.com/pulse/natural-fiber-reinforced-polymer-composites-automobile-harikrishnan-c, last accessed October 12, 2022.
41. Hoffmann, K.G., Haag, K., Müssig, J.: Biomimetic approaches towards lightweight composite structures for car interior parts. Materials & Design. **212**, 110281 (2021)
42. Globe News Wire, https://www.globenewswire.com/news-release/2021/11/09/2330290/0/en/Trends-Opportunities-and-Competitive-Analysis-of-the-Natural-Fiber-Composite-Market.html, last accessed December 12, 2022.
43. Wilson, A.: Vehicle weight is the key driver for automotive composites. Reinforced Plastics. **61**(2), 100–102 (2017)

Part V
Small and Medium-sized Enterprises, Management, Manufacturing, Industry 4.0, Digitalization

Increasing the Sustainability of Manufacturing Processes in the Conditions of SMEs Enterprises by Predicting, Their Information Intensity

Sergey Dobrotvorskiy, Michal Balog, Andrew Ruzmetov, Yevheniia Basova, Serhii Hrdzelidze, and Ludmila Dobrovolska

1 Introduction

In the conditions of small and medium-sized enterprises (SMEs), due to the wide technological nomenclature, there is a big problem in predicting the accuracy of the duration of the secondary process and repairing processes. This prevents the correction of the composition of accessory tooling and increases the level of the basic equipment loading. The least labor-intensive and reliable structure of the preparatory process will favor the equipment loading, taking into account the possibility of planning multiple-machine supervision. An essential and defining load in compiling the structure of a secondary process is a significant amount of information that must be processed by a human controller. Therefore, the efficiency and reliability of its operation depend on the thoroughness and correctness of taking into account the current characteristics of the technological system.

This paper presents a model for determining the employee's information tension as part of an automated system of microelement synthesis of a secondary process.

S. Dobrotvorskiy (✉) · A. Ruzmetov · Y. Basova · S. Hrdzelidze · L. Dobrovolska
National Technical University "Kharkiv Polytechnic Institute", Kharkiv, Ukraine
e-mail: sergiy.dobrotvorskyy@khpi.edu.ua

M. Balog
Faculty of Manufacturing Technologies with the seat in Prešov, Technical University of Košice, Prešov, Slovakia

M. Balog et al. (eds.), *EAI International Conference on Automation and Control in Theory and Practice*, EAI/Springer Innovations in Communication and Computing, https://doi.org/10.1007/978-3-031-31967-9_16

2 Literature Review

According to the source [1], there are more than 22 million SMEs in the EU countries, which handle 67% of the total number of employees. In most of these enterprises (especially small ones), only the processing cycle is automated [2]. This requires the participation of a human controller when performing adjustments and secondary processes. At the same time, the number of tasks that the human controller solves in the conditions of micro- and small enterprises reaches 27 different items, while work in the conditions of large enterprises usually consists only of certain jobs and routines [3]. Therefore, machine operators of micro and small enterprises perform much more complex work than operators of large enterprises [2].

Optimization during the planning of the front and the sequence of work, both on the scale of the workshop and a separate workplace using lean manufacturing technologies (MES – Manufacturing Execution System), allows for guaranteeing the human controller's performance [4]. According to studies [5, 6], to estimate the manual work time spent on preparing the basic equipment for the process, the planning concept MTM (Methods-Time Measurement) can be used. It consists in using a standard system of elementary actions that provide a consistent and reproducible description of the manufacturing process. As a rule, time is determined based on MTM standards and PMTS (Predetermined motion time system), used to assess the complexity of manual assembly production steps [7–9]. However, the MTM's manual work planning system, determining the labor intensity by the quantity and quality of the elements involved in it, does not take into account the energy and information components of the process.

The wide range and complexity of products in such industrial areas as aerospace production [10, 11], assembly of internal combustion engines [12], automobile production [13], and production of industrial equipment and tools create tasks that cannot be solved without special requirements for cognitive human controller's abilities [14, 15]. It was noted that improving the efficiency of strategies for the implementation of manufacturing processes is based both on the use of modern digital technologies [16, 17] and on the identification and analysis of factors affecting their performance and on the reliability of human controller's actions [18].

There are two approaches to determining human reliability in manual work: the development of methods for human error identification (HEI) [19] and the use of classical methods HRA (Human Reliability Assessment) to assess the impact of human controller's actions in operating the system [12, 20, 21]. It should be noted that the use of HEI and HRA methods will make it very difficult for the information potential objective assessment of the manufacturing complex under study since these methods consider not so much the technological environment properties as the reaction to them from a certain human. As a result of research [22–24], it was found that the human controller's performance is influenced by such factors as the number of features of the objects of the manufacturing complex, the complexity of the manufacturing process algorithm, and the level of order in the organization of the working environment. At the same time, the bulk of information that the human controller must process is associated not so much with the manufacturing technology but with the search and identification of a certain design and the principle of tooling operation, which allows intensifying the functionality required for work. Based on the identified functional series, the determination of the sets of microelements – actions (BSM-1 – The basic system of microelements [25]), which make up the methods of work, in the appropriate organizational and technological conditions [26]. The operator is forced, to master information about the tooling characteristics, to control their condition, and control is always the solution to a certain complex of logical tasks of varying degrees of complexity. Each solved logical task, especially in conditions of limited time, increases the neuro-emotional stress of the employee and provokes a slowdown in the performance of the sensorimotor reaction [26]. Thus, the employee's information tension contributes to a slowdown in the manufacturing process as a whole, which affects production cycle losses, especially in conditions of regular changeover.

With digitalization within the framework of Industry 4.0, there has been a tendency to create a typological model of the human controller – Operator 4.0 [27]. According to this model, a human-centered production system will combine the planning and implementation of the manufacturing process [28]. It is expected that the human controller (Operator 4.0) will control and adjust both the manufacturing process and the technology, which will maximize the use of human competencies.

The purpose of the work is to create a model for calculating the coefficient of employee information tension, depending on the characteristics of tooling and equipment used in the conditions of SMEs.

3 Materials and Methods

Practically on each microelement complex of the secondary process, corresponding to the manufacturing method, the process structure can be developed alternatively within the framework of the requirements of the processing technology. In conditions of regular change of technology and tooling used (due to the heterogeneity of products), such variability significantly reduces the guarantees of timely completion of work.

Information Certainty of the Manufacturing Complex State A manufacturing complex (MC) can be considered as a partially ordered organizational and technological system of objects, the main and secondary functions of which are involved in the preparation and implementation of processing steps, the movement of parts, and tooling. Therefore, when performing individual processing methods, it is necessary to consider the group of objects of attention corresponding to them. Their characteristics are input parameters for the above-mentioned functions. As the logical operators are executed, and their remaining number decreases, and with them, the considered characteristics of the objects of attention, the certainty of the MC state regarding the purpose of the processing task increases. When all processed parts are in the container for parts and, according to the process control documents, all actions are performed, the MC state can be characterized as a state of maximum certainty.

Tie-in Microelements of Working Motions to the Operating Space To tie-in the microelements of working motions to the operating space, their implementation conditions are formalized in the form of a fuzzy model of operating space fragments, in which the joins between the elements are regulated by fuzzy relations using linguistic variables. For example, for the microelement "ПОО" - to rotate the object around its axis, it is necessary to provide such implementation conditions as:

- hand on the object's surface (after the microelement "ВЗ" - take);
- the object has a rotation freedom degree around its axis at an angle from 300 to 1800;
- the initial object angular coordinates.

Therefore, before implementing such a microelement, the performer needs to process a set of parameters that characterize the mutual position of the hand and the object, evaluate the object freedom degree relative to the system, estimate the space dimensions in the direction of rotation, and then make a turn, controlling the angle between the characteristic surfaces of the object (hand) and the surfaces other objects.

That is, the MC system should receive sufficient information certainty for the implementation of the microelement. The entropy of a system characterized by fuzzy relations can be expressed by a formula (1):

$$HR\left(me_j\right) = -\sum_{nr}^{i=1} \mu_i\left(r_{i,id}\right) \log_2 \mu_i\left(r_{i,id}\right), \tag{1}$$

where me_j - j-th microelement; $j = 1..n_{Me}$, $\mu_i(r_{i,id})$ – fuzzy membership function of the i-th relation of the id-th interval of the linguistic variable; $i = 1..nr$ – set of relationships considered in the context of a given microelement.

Some of the relationships, for example, such as DF_{id} INC_{id}, SOA_{id}, may require preliminary preparation, which means the use of preparatory complexes of movements and the appearance of branching of the machine-manual process algorithm. For example, relation SOA_{id} (accessibility of an object for rotation) requires a preliminary corresponding check $p(SOA_{id}+)$. In the absence of the possibility of turning ($p(SOA_{id}+) = false$) it is necessary to use a set of measures $\{me_l\}$ to obtain the value $p(SOA_{id}+) = true$ as access to the implementation of "ПОО".

Model for Correcting the Implementation Time of a Machine-Manual Production Step The speed of information processing when overcoming uncertainty during the execution of the reception of work [27] can be determined from the expression (2):

$$V\left(Me_{jp}\right) = \frac{1}{t_{jp}}\left[\sum_{n_{ja}}^{ja} HR\left(Me_{ja}\right) + \sum_{n_{ja}}^{ja} HR\left(u\left(r_{ir,ja}\right), Me_{ja}\right) + \sum_{n_j}^{j} HR\left(Me_j\right)\right] \tag{2}$$

where Me_{jp} – working practices; $jp = 1..n_{jp}$ – working practices set as part of the processing step; t_{jp} – time to complete a working practice; $u(r_{ir,ja})$ - microelement fulfillment condition; $r_{ir,ja}$ - relationship to be tested.

The indicator $V(Me_{jp})$ can be interpreted as the intensity of information processing by the human controller during the production step. Micronutrients to correct the time of implementation of a machine-manual production step can be used as the indicator of the relative complexity of the secondary process, which consists of the product of the algorithmic complexity SA_{jp} [27] and the ratio of the intensity levels of the process under study ($\delta_{jp} \cong V(Me_{jp})$) with a similar one, with typical characteristics, taken as the basic process (δ^{b}_{jp}) (3):

$$\Delta_{jp} = SA_{jp} \frac{\delta_{jp}}{\delta^{b}_{jp}} \approx SA_{jp} \frac{V\left(Me_{jp}\right)}{\delta^{b}_{jp}}, \tag{3}$$

In the system of BSM-1 standards, a psycho-physiologically optimal pace of work in the absence of conditions is determined, which allows us to accept it as a basic one. Therefore, if: $\delta^{b}_{jp} = n^{b}_{jp} / t^{b}_{jp}$, where n^{b}_{jp} – the number of microelements in the basic process of working practice, t^{b}_{jp} – execution time of the basic process of working practice (min), then

$$\Delta_{jp} = SA_{jp} \frac{t^{b}_{jp} \cdot V\left(Me_{jp}\right)}{t_{jp} \cdot n^{b}_{jp}}. \tag{4}$$

As a result, the coefficient of information tension, taking into account the coefficient of mental stress of the employee Ks [22], may look like (5):

$$Kit_{jp} = \Delta_{jp} \cdot Ks. \tag{5}$$

The coefficient of mental stress of the employee depends on the ratio of the volume of the remaining process $\sum_{mE} \overline{t^{E}}$ and time to complete $T - \sum t_{jp}$, where $\sum t_{jp}$ - elapsed time so far; T - time for production step. If we assume that the employee emotionally perceives the remaining time in accordance with the measure of achieving the goal, which corresponds to the level of informational entropy, then the indicator of mental stress $s(\sum HR)$ can be expressed as: $s(\sum HR) = \sum HR - HR_{min}$, where $HR_{min} = 0{,}115$ – state of maximum certainty for a human-machine system [23].

The coefficient of mental stress can be calculated (6):

$$Ks = \{ \begin{matrix} -1.8287i^{3} + 3.4718i^{2} - 2.3468i + 1 & if & s\left(\sum HR\right) < Pst \\ s\left(\sum HR\right) + 1 - Pst & if & Pst \le s\left(\sum HR\right) \le Pst + 1 \end{matrix}, \tag{6}$$

where Pst - stress threshold value for the average human ($Pst = 2, 3$ [9]).

As a result, the time spent on performing a machine-manual operation by an employee can be calculated by the formula (7):

$$t_{o} = \sum_{nj}^{j=1} t^{bCM}_{j} \left(Ff, Fk\right) \bullet KF_{j} \bullet Kit_{j} \bullet KP_{j}, \tag{7}$$

where $t_j^{bCM}(Ff,Fk)$ – time for microelement action (calculated according to BSM-1) taking into account quantitative (*Ff*) and quality (*Fk*) factors; KF_j – coefficient, taking into account physical stress; KP_j – coefficient that takes into account the typicality and repeatability of the secondary process performed.

A general view of the manufacturing process analysis algorithm based on information tension is shown in Fig.1.

The Content of a Computer Experiment on Modeling a Secondary Process In this paper, an analysis of the secondary process of manufacturing preparing on a CNC vertical milling machine ГФ2171Ф3 was performed (Fig.2.).

The $ip = \overline{1,mp}$ from Fig. 1 is the batch of details size, and *MAE* is a tooling array.

This operations type is of interest for study due to the presence of a typical range of supporting tools, basing and fixing elements located on the working surface of the machine table, inherent in micro- and small enterprises.

The production step itself consists in milling a keyway on a ∅25.3 mm part of the "Shaft-pinion" part weighing 1.8 kg, with the largest diameter of ∅45 mm, a total length of 173 mm, and a gripping surface length of 100 mm. The workpiece is based on prisms with an emphasis on the end ∅27 without additional orientation in the plane of its cross section and is fixed from above with a clamp with an emphasis on rotating the clamping element with a wrench. In the computer experiment, the process of repeated repetition of the implementation of the microelement structure of the processing step (Fig. 3) to install the part for 2 hours was simulated.

The elements of Fig. 3:

- A1 – tool search process;
- A2 – the process of changing the profile of work to organize events or changing the production task;
- p1 – p8 – logical conditions for the implementation of microelements, which are mainly of a binary nature of implementation – according to the principle of scanning,
- i-th transition in the algorithm,
- m – number of transitions.

Microelement composition was based on the technology data selected by the MES – system (block "Process data"), the choice of auxiliary equipment is made in the block "Supporting tools". In this case, data on the characteristics of the machine tool, fixture, the main tool used, and the production object from the corresponding databases are used.

If a suitable auxiliary tool is not found in the tooling database ("Cash ST" block), then the MES system notificates. Further, for all technological means, the initial state is initiated – in the places of storage at the workplace and the maximum entropy value is assigned relative to the processing goal (block "Initial state TC, $\sum$HR = HR_{max}"), the operation is divided into settings and transitions according to the processing map. The cycles include its limits set by the MES system (the size of the batch (ip = $\overline{1,mp}$), for each transition (jp = $\overline{1,np}$) the synthesis of complexes of

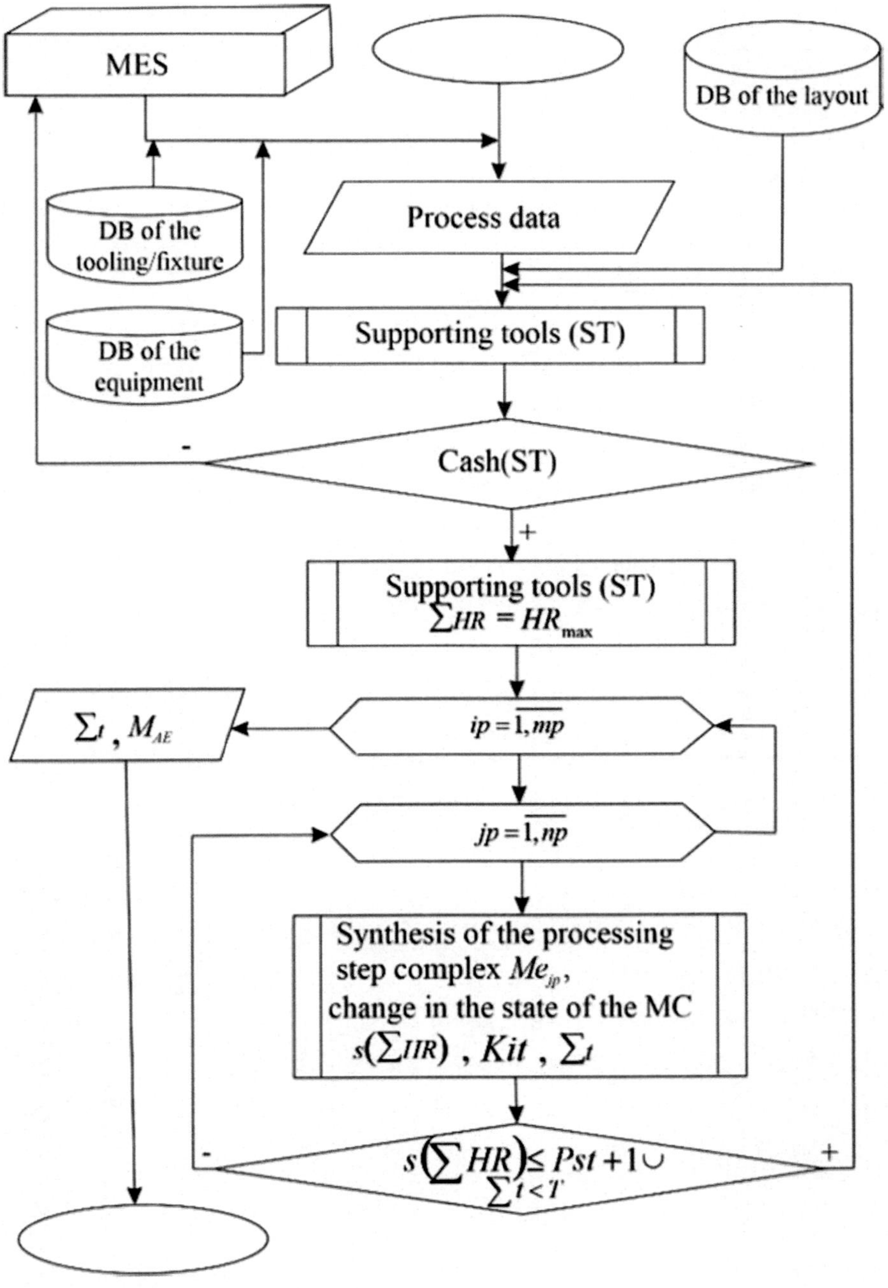

Fig. 1 Algorithm for the analysis of the manufacturing process based on information intensity

No	Micronutrient content	Micronutrient
1	Take a look at the fixture	$\text{ПВ}_1\left(_{S\varphi}\right)$
	Is there a detail in the fixture?	p_1
2	Take a look at the storage location of the auxiliary tool (wrench)	$\text{ПВ}_2\left(_{S\varphi}\right)$
	Is there a necessary auxiliary tool (M24 wrench)?	p_2
3	Extend the hand to the auxiliary tool (to the wrench).	$\text{ПР}_3\binom{\text{К2,ОС1}}{S}$
4	Take an auxiliary tool (wrench)	$\text{В}_4\binom{\text{К2,ОС1,У1}}{M.L}$
5	Move auxiliary tool (wrench) to fixture.	$\text{П}_5\binom{\text{К3,ОС1}}{S.M.L}$
6	Turn the wrench clockwise without force.	$\text{П}_6\binom{\text{ОС2}}{\varphi.}$
7	Install the wrench on the M24 nut.	$\text{УО}_7\binom{\text{П2,ОС1,ОР2,У1}}{M.Lp}$
	Is the wrench set correctly?	p_3
8	Intercept the wrench to transfer force.	$\text{ВП}_8(\ )$
9	Turn the spanner firmly counterclockwise.	$\text{ПОО}_9\binom{\text{ОС2}}{\varphi.F}$
10	Remove wrench from nut.	$\text{Р}_{10}\binom{\text{П2,ОС1,У1}}{M.Lp}$
	Is the nut loose enough? If "Yes" - go to step 11 If "No" - go to step 6	p_4
11	Move the wrench to its storage location.	$\text{П}_{11}\binom{\text{К2,ОС1}}{S.M.L}$
12	Place the wrench in the storage slot.	$\text{ОТ}_{12}\binom{\text{К3,ОС1}}{S.M.L}$

Fig. 2 Trace the element composition of the secondary process for unfastening the part

$$\text{ПВ}_1\left(_{S\varphi}\right)\ p_1\overset{1}{\uparrow}\ \text{ПВ}_2\left(_{S\varphi}\right)\ p_2\overset{41}{\uparrow}\ \text{В}_3\binom{\text{К2,ОС1,У1}}{M,Lp}\ \text{ПР}_4\binom{\text{К2,ОС1}}{S}\ \text{П}_5\binom{\text{К3,ОС1}}{S,M,L}\ \overset{3}{\downarrow}\overset{4}{\downarrow}\ \text{ПОП}_6\binom{\text{ОС2}}{\varphi.}$$
$$\text{УО}_7\binom{\text{П2,ОС1,ОР2,У1}}{M,Lp}\ p_3\overset{3}{\uparrow}\ \text{ВП}_8(\ )\ \text{ПОО}_9\binom{\text{ОС2}}{\varphi,F}\ \text{Р}_{10}\binom{\text{П2,ОС1,У1}}{M,Lp}\ p_4\overset{6}{\uparrow}\ \text{П}_{11}\binom{\text{К2,ОС1}}{S,M,L}\ \text{ОТ}_{12}\binom{\text{К2,ОС1}}{S,M,L}$$
$$\text{ПВ}_{13}\left(_{S\varphi}\right)\ \text{ПР}_{14}\binom{\text{К2,ОС1}}{S}\ \text{ПОО}_{15}\binom{\text{ОС1}}{\varphi}\ \text{ПР}_{16}\binom{\text{К2,ОС1}}{S}\ \text{В}_{17}\binom{\text{К2,ОС1,У1}}{M,Lp}\ \text{П}_{18}\binom{\text{К2,ОС1}}{S,M,Lp}\ \text{УП}_{19}\binom{\text{К2,ОС1}}{M,Lp}$$
$$\text{ОТ}_{20}\binom{\text{К1,ОС1}}{S,M,L}\ \text{ПВ}_{21}\left(_{S\varphi}\right)\ p_5\overset{42}{\uparrow}\ \overset{1}{\downarrow}\ \text{ПР}_{22}\binom{\text{К1,ОС1}}{S}\ \text{В}_{23}\binom{\text{К2,ОС1,У1}}{M,Lp}\ \text{П}_{24}\binom{\text{К3,ОС1}}{S,M,Lp}\ \text{УП}_{25}\binom{\text{К3,ОС1}}{M,Lp}$$
$$\text{ПР}_{26}\binom{\text{К2,ОС1}}{S}\ \text{ПОО}_{27}\binom{\text{ОС2}}{\varphi,F}\ \text{ПВ}_{28}\left(_{S\varphi}\right)\ p_6\overset{41}{\uparrow}\ \text{В}_{29}\binom{\text{К2,ОС1,У1}}{M,Lp}\ \text{ПР}_{30}\binom{\text{К2,ОС1}}{S}\ \text{П}_{31}\binom{\text{К3,ОС1}}{S,M,L}$$
$$\overset{7}{\downarrow}\overset{8}{\downarrow}\ \text{ПОП}_{32}\binom{\text{ОС2}}{\varphi.}\ \text{УО}_{33}\binom{\text{П2,ОС1,ОР2,У1}}{M,Lp}\ p_7\overset{7}{\uparrow}\ \text{ВП}_{34}(\ )\ \text{ПОО}_{35}\binom{\text{ОС2}}{\varphi,F}\ \text{Р}_{36}\binom{\text{П2,ОС1,У1}}{M,Lp}\ p_8\overset{8}{\uparrow}$$
$$\text{П}_{37}\binom{\text{К2,ОС1}}{S,M,L}\ \text{ОТ}_{38}\binom{\text{К2,ОС1}}{S,M,L}$$

Fig. 3 Microelement composition of the processing step "Remove, install, fix part"

microelements – actions is performed, the following are calculated: the time for the complex and for the operation as a whole, the coefficient of information intensity (Kit) and the indicator mental stress s($\sum$HR). A comprehensive assessment of the consistency of the synthesized structure is carried out in terms of the maximum allowable mental stress and the batch processing time limit (comparison block "$(s(\sum HR) \leq P_{st} + 1) \cup (\sum t < T)$". In case of lack of time or exceeding the tension indicator, a new selection of an auxiliary tool is made, taking into account the duration of each of its names in previous attempts. At the same time, the indicator of the urgency of performing a series of processing steps changed through the plannedexecution time T and through the time left to solve essential tasks $\sum_{mE} \overline{t^E}$. As a resultof the synthesis of a group of techniques of a secondary process, consisting of installing and fixing a detail in a fixture during a production step, the dependences of the speed of their implementation on the duration of the manufacturing process were obtained (Fig. 4).

4 Results and Discussion

The graphs (Fig. 4) reflect the changes in the productivity and load of the process of performing installation techniques – removing the detail from the fixture. Performance means the number of microelement actions performed per minute.

In Fig. 3a, we can distinguish the phases of development, peak productivity, and its stable decline due to the growth of information tension. Figure 4b shows a more stable phase of peak performance. This can be explained by the increase in mental stress due to lack of time, but after 71 minutes, there is a transition through the boundary value of stress, which provokes a sharper increase in the drop in productivity than in Fig. 4a.

Figure 4 represents how ($\Pi(t)$) and mental stress ($Ks(t)$) depending on the total time of the working cycle: a – at $\sum_{mE} \overline{t^E} << T - \sum t$ – nonurgent work situation; b – at $\sum_{mE} \overline{t^E} > T - \sum t$ – extremely urgent work situation.

With $K_S = 1$, the process of a machine-manual production step has a priori indicators of the labor intensity of a single process, because from the very beginning the maximum level of information, entropy has been incorporated into the calculation method, and the uncertainty of the situation can increase sharply with the appearance of conditions for branching the structure of microelements and from the conditions for the implementation of a microelement, which require additional work. Therefore, the coefficient of typicality of the work performed is taken to be 0.75.

Thus, the nomenclature-unstable operating conditions of the manufacturing complex provoke an increase in the information intensity of the secondary process through the generation of its algorithmic branching. This greatly enhances the intensity of the information processing process by the human controller. It is also necessary to admit that the object heterogeneity corresponding to different production environments will be quite significant. Therefore, when implementing microelement action complexes, at the stage of determining the appropriate

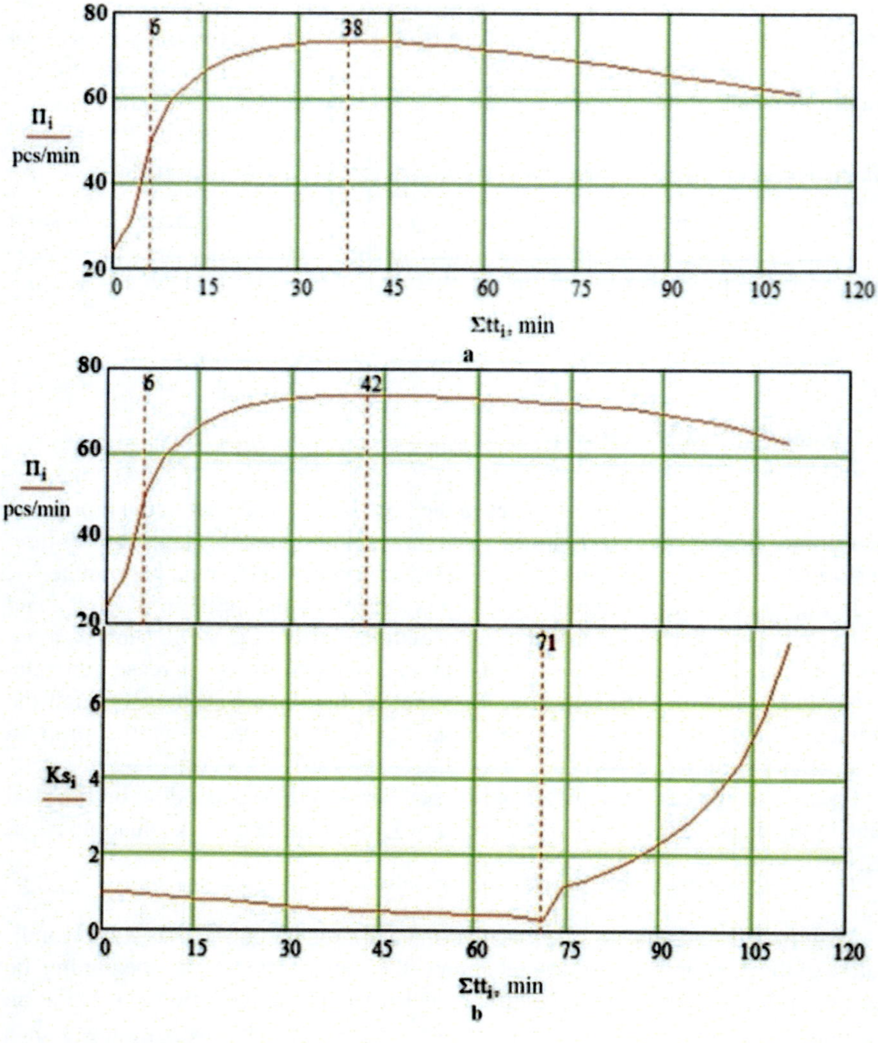

Fig. 4 Curves of performance indicators

relationships between objects, it will be necessary to use specialized technologies for recognizing the states of these objects. As a result, we can conclude that the information tension in conditions when the range of machined parts is constantly changing and expanding, which is typical for SMEs, grows throughout the entire shift. This is logically connected with the appearance of a significant number of uncertainty elements that provoke the fulfillment of new logical conditions, which increases the information load on the human controller.

5 Conclusion

A technique for determining the information tension and an algorithm for analyzing the manufacturing process based on the information tension of a human controller, using the apparatus of fuzzy sets and a model for changing the information entropy of the states of the manufacturing system, is presented. The model allows providing the process of optimizing the choice of organizational and technological conditions for the implementation of the production step to increase the stability of the performance indicator and to approach the issue of planning the work of the manufacturing unit in a more differentiated way. The system of automated determination of the structure of the secondary process obtained dependencies, the nature of which confirms the fact that the growth of the indicator of the urgency of the work leads to an increase in productivity to a limited extent.

In the future, it is planned to replace the repeatability coefficient (KP) with a group of models for calculating the qualitative indicators of the implementation conditions, which, as the number of repetitions of the process increases, would eliminate the need to take into account the logical conditions before the executed fragments of the microelement structure.

To reduce information tension in the conditions of SMEs, in addition to the choice of equipment according to the criteria (suggested by the algorithm), augmented reality tools can be used. In interactive mode, on a 3D map of the organization of the workplace, the placement of priority and alternative equipment can be marked according to the selected operation plan from the automatically generated list of possibilities for the current conditions and scale of work.

In the conditions of large enterprises, the in-line nature of manufacture is very important, and hence the guarantee of the timeliness of the operation. The use in these conditions of the system of analysis of the manufacturing process through the synthesis of its microelement structure in terms of the information uncertainty of the state of the manufacturing complex would allow for a slight expand the range of products.

Acknowledgments This work was supported by the Slovak Research and Development Agency under contract No. APVV-19-0590, and also by the projects VEGA 1/0700/20, KEGA 022TUKE-4/2023 granted by the Ministry of Education, Science, Research, and Sport of the Slovak Republic.

References

1. Strelbitska, N.E., Redko, T.V.: Small and medium enterprises as a factor of social and economic growth in the european union. Scientific bulletin of the Uzhhorod National University 7, 118–122 (2016)
2. Dobrotvorskiy, S., Basova, Y., Dobrovolska, L., Sokol, Y., Kazantsev, N.: Big challenges of small manufacturing Enterprises in Industry 4.0. In: Ivanov, V., Trojanowska, J., Pavlenko, I., Zajac, J., Peraković, D. (eds.) Advances in design, simulation and Manufac-turing III. DSMIE

2020 Lecture notes in mechanical engineering 118–127. Springer, Cham (2020). https://doi.org/10.1007/978-3-030-50794-7_12
3. Wijanarka, B.S., Nuchron, R.D., Habanabakize, T.: The task of machine tool oper-ators in small and medium enterprises in Indonesia. Jurnal Pendidikan Teknologi dan Keju-ruan. **24**(1(5)), 39–45 (2018)
4. Evaluation of Cycle Times in MES Systems, (2019), Electronic resource. - Access mode: https://trex.com.tr/evaluation-of-cycle-times-in-mes-systems
5. Finsterbusch T., Wartig M., Kuhlang P.: Human-oriented design of human work with MTM-HWD - an incipient change in industrial engineering. GfA, Dortmund (ed.) Work in complex systems. Digital, networked, human?! – B.1.9., (2016), Electronic resource. - Access mode: https://gfa2016.gesellschaft-fuer-arbeitswissenschaft.de/inhalt/B.1.9.pdf
6. Morlock, F., Kreggenfeld, N., Louw, L., Kreimeier, D., Kuhlenkoetter, B.: Teaching methods-time measurement (MTM) for workplace design in learning factories. Procedia Manuf., CrossRef. **9**, 369–375 (2017)
7. Alkan, B., Vera, D., Ahmad, M., Ahmad, B., Harrison, R.: A model for complexity assessment in manual assembly operations through predetermined motion time systems. Procedia CIRP. **44**, 429 (2016)
8. Zaeh, M.F., Wiesbeck, M., Stork, S., Schubö, A.: A multi-dimensional measure for determining the complexity of manual assembly operations. Prod. Eng. **3**, 489–496 (2009)
9. Kern, C., Refflinghaus, R.: Assembly-specific database for predicting human reliability in assembly operations. Total Qual. Manag. Bus. Excell. **26**, 1056–1070 (2015)
10. Correia, D.: Improving manual assembly lines devoted to complex electronic devices by applying lean tools. Procedia Manuf. **17**, 663–671 (2018)
11. Judt, D., Lawson, C., Lockett, H.: Experimental investigation into aircraft system manual assembly performance under varying structural component orientations. Proc. Inst. Mech. Eng. **234**, 840–855 (2020)
12. Yang, L., Su, Q., Shen, L.: A novel method of analyzing quality defects due to human errors in engine assembly line. In Proceedings of the 2012 International conference on information management, innovation management and industrial engineering, Sanya, China, 20–21 (2012).
13. Falck, A.C., Örtengren, R., Högberg, D.: The impact of poor assembly ergonomics onprod-uct quality: A cost–benefit analysis in car manufacturing. Hum. Factors Ergon. Manuf. Serv. Ind. **20**, 24–41 (2010)
14. Landau,K., Brauchler, R.,Rohmert, W.:The AET method of job evaluation. In Occupa-tional ergonomics principles of work design, p. 20, CRC Press: Boca Raton, FL, USA,(2003);
15. Falck, A.C., Örtengren, R., Rosenqvist, M., Söderberg, R.: Criteria for assessment of basic manual assembly complexity. Procedia CIRP. **44**, 424–428 (2016)
16. Dobrotvorskiy, S., Balog, M., Basova, Y., Dobrovolska, L., Zinchenko, A.: Conceptof the Software for Materials Selection Using. NET Technologies. In: AdvancedManufacturing Processes. InterPartner 2019. Lecture Notes in Mechanical Engineering, pp. 32–43. Springer, Champions (2020). https://doi.org/10.1007/978-3-030-40724-7_4
17. Dobrotvorskiy, S., Kononenko, S., Basova, Y., Dobrovolska, L., Edl, M.: Develop-ment of optimum thin-walled parts milling parameters calculation technique. In: Ivanov, V., Trojanowska, J., Pavlenko, I., Zajac, J., Peraković, D. (eds.) Advances in design, Simu-lation and manufacturing IV. DSMIE 2021. Lecture notes in mechanical engineering, pp. 343–352. Springer, Cham (2021). https://doi.org/10.1007/978-3-030-77719-7_34
18. Li, D., Fast-Berglund, Å., Salunkhe, O., Fast-Berglund, Å., Skoogh, A., Broberg, J.: Effectsof Information Content in WorkInstructions for Operator Performance. Procedia. Manuf. **25**, 628–635 (2018)
19. Torres, Y., Nadeau, S., Landau, K.: Application ofHuman errors Analysisin Manufactur-ing: A proposed intervention framework and techniques selection. Kongress der Ge-sellschaft für Arbeitswissenschaft; GfA-Press, Dresden, Germany (2019)

20. Stanton, N.A., Stevenage, S.V.: Learning to predicthuman error: Issues of acceptability, reliability andvalidity. Ergonomics. **41**, 1737–1756 (1998). https://doi.org/10.1080/001401398186162. http://www.ncbi.nlm.nih.gov/pub-med/9819584PubMed
21. Bubb, H.: Human reliability: A key to improved quality in manufacturing. Hum. Factors Ergon. Manuf. Serv. Ind. **15**, 353–368 (2005). https://doi.org/10.1002/hfm.20032
22. Ruzmetov, A.R.: Metodika viznachennya funkcij prinalezhnostiv sistemiproektuvannyamikrostrukturitehnologichnogo perehodu.VisnikNacionalnogo tehnichnogo universitetu «Harkivskijpolitehnichnij institut». Harkiv: NTU«HPI». **42**(1085), 139–147 (2014)
23. Ruzmetov, A.R., Shelkovoj, A.N., Shron, L.B., Ishenko, G.I., Semchenko, M.S.: Kvoprosu o imitacionnom modelirovaniimashinno-ruchnyh tehnologicheskih operacij v sistemahobrabotki metallov rezaniem.Suchasni tehnologiyi v mashinobuduvanni, vol. 10, pp. 177–191. NTU«HPI», Harkiv (2015)
24. Ruzmetov, A.R.: Optimizaciya dopomizhnogo procesu mehanichnoyi obrobki rizannyam zurahuvannyam organizacijno tehnologichnihvitrat. Tehnologichnij audit ta rezervi virobnictva4/1 (virobnic-tva4/1). **(24)**, 77–83 (2015)
25. Yurovskii, S. A. M: Ekonomika: Bazovaya sistema mikroelementnyh normativov vremeni (BSM-1). In: Normativno proizvod-stvennoe izdanie, pp. 122 (1989).
26. Trofimov Yu, L.: Inzhenernapsihologiya. Pidruchnik, Libid, Kiyiv. **264** (2002)
27. Romero, D.: Towards an Operator 4.0 Typology: A Human-Centric Perspective on the Fourth Industrial Revolution Technologies. In: Proceedings of the CIE46, pp. 29–31, Tianjin, China (2016)
28. Munir, S., Stankovic, J.A., Liang, C.J.M., Lin, S.: Cyber-physical system challenges for Human-in-the-loop control. Feedback Comput. (2013)

A View on the Digitization of Small and Medium-Sized Enterprises

Romana Hricová

1 Introduction

Business has gone through significant changes, especially in recent years, when many businesses were forced to move to the online space in order to survive. Digitization thus became an essential part of every business that wanted to survive. As a result, the competition has become much closer than in classic business. Sometimes a few clicks on the computer are enough for the customer to change the store where he decides to shop. It is literally a matter of minutes, if not seconds, before a business loses a potential customer to a competitor. In addition, the differences between the products and services themselves are evened out. Therefore, a dissatisfied customer does not hesitate for a long time and finds another seller. However, on the other hand, it also appears that SMEs use digitization to speed up entry into the foreign market and search for new customers there, which increases competition.

Addressing the right customers at the right time, especially with the right offer, has become a challenge for SMEs. The customer is becoming more and more demanding and, in addition, has faster access to a larger volume of information than in the recent past. The customer chooses and compares, and the price is not always the main factor. The product, which is always paramount, was gradually joined by indicators such as the customer's overall impression of the company's approach, the communication process, the speed of providing feedback when necessary, and the perception of the brand as such. Especially the events of the last few years associated with the global COVID-19 pandemic, when the governments of individual countries

R. Hricová (✉)
Faculty of Manufacturing Technologies with the seat in Prešov, Technical University of Košice, Prešov, Slovakia
e-mail: romana.hricova@tuke.sk

M. Balog et al. (eds.), *EAI International Conference on Automation and Control in Theory and Practice*, EAI/Springer Innovations in Communication and Computing, https://doi.org/10.1007/978-3-031-31967-9_17

restricted the movement of their citizens, and they worked from home, changed the behavior of consumers. People not only worked from home but also shopped, which significantly impacted sales SMEs, which were forced to move to the online space if they wanted to maintain their market position. The above-mentioned events created an exceptionally suitable environment for accelerating the digitization of SMEs.

The article presents the individual steps in the digitization and digitalization of SMEs and explains the impact the pandemic had on these processes. The section dedicated to the problems associated with digitalization opens the issue of cyber security, which goes hand in hand with digitalization. SMEs tend to underestimate this area, which may prove to be a problem in the near future.

2 Entrepreneurship and Digitization

The European Union has set digitization as one of its top priorities. Digitization plays an important role in all EU policies. Digital solutions bring with them great opportunities and are very important in restoring the economy after the corona crisis and strengthening the position in the world economy. Therefore, the European Parliament participates in the formation of new legislation that will strengthen European capacities in the field of digital technologies and bring new opportunities for citizens and businesses. Digitization should also help in the transition to a greener economy and in achieving climate neutrality by 2050. The EU aims to improve the digital skills of its population and move to digitization in public services through the gradual training of workers. The European economic recovery plan asks member states to spend at least 20% of the rescue package worth 672.5 billion Euros on digitization. EU investment programs, such as the Horizon Europe research program or the Connecting Europe facility focused on infrastructure, have allocated significant amounts of their budget specifically for digitization [1]. However, digitization in the business environment seems even more interesting, as it has a great impact on the creation of company value [2]. The European Parliament approved the Digital Europe program in April 2021. It is the first-ever financial instrument of the European Union, the aim of which is to bring technology closer to citizens and companies. The program should finance digital infrastructure and strategic technologies that should improve Europe's competitiveness, help in the transition to a greener economy and ensure technological sovereignty. In total, nearly 7.6 billion Euros should be invested in five areas: supercomputers (2.2 billion Euros), artificial intelligence (2.1 billion Euros), cyber security (1.6 billion Euros), advanced digital skills (577 million Euros) and the introduction of digital technologies in all areas (1.1 billion Euros) [1].

According to Lupton (2017), digitization refers to the trend of the spread of digital technologies and their increasing penetration into areas of life [3]. Digitization also brings with it new challenges for businesses, which must realize that if they want to respond correctly to the demands and needs of their customers, they must take into account that business today is different than it was in the recent past. Digitization affects not only organizations, but on the other hand, organizations (as social systems) also affect digitization. Organizations influence digitization in different ways and shape digitization in structurally specific ways [4].

2.1 Digitization of SMEs

Businesses gradually collect information from their activities, starting with information about customers, suppliers, advertising campaigns, employees, sales volumes, etc. Usually, part of the information is in paper form, part in various Word or excels files, part already on social networks, and part only in the notes of a specific employee. Getting the needed information quickly can sometimes be a problem. In addition, the constant search for the necessary information simultaneously slows down the company's communication with the customer.

Thus, digitization in SMEs takes several subsequent steps that cannot be skipped and for which it is necessary to prepare thoroughly. As shown in Fig. 1, the individual stages contain various activities that must be followed.

Digitization is the first stage when the company digitizes information and data. At this level, digitization will not affect business information, the main task is the collection and provision of information. Simplification can be said that it is a transition from analog to digital form. In the context of SMEs, this means that it is

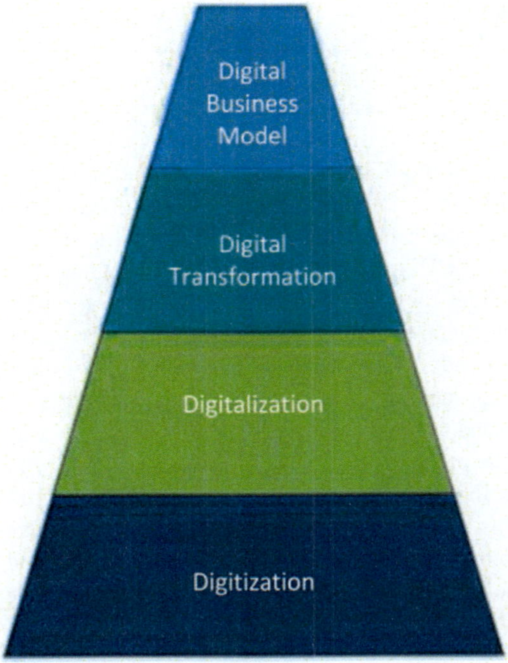

Fig. 1 Digitization procedure [5]

usually the transfer of analog information in the form of ordinary scanning of paper documents, but also, for example, production premises, where records of energy consumption are created, which are further processed. It is about making the company's activities more efficient, thanks to which information is obtained from paper documents. Digitization of documents thus becomes one of the newly emerging digitization trends [6]. It is important to realize that in this step, the company is digitizing information but not digitizing processes.

Digitalization is the second stage, when new digital technologies of SMEs are introduced. Its task is to implement digital technologies into business processes. For a company, this means the digitization of a specific area or process. At this stage, it is important that employees also begin to identify with digitization and that workplaces gradually expand with new technologies, either in the form of mobile devices or through various communication and management platforms. At the time of the company's digitalization, digitalization is used to achieve smaller goals. It is being determined what can be digitized. Data are collected from the machines and evaluated. Furthermore, processes are monitored, various business processes are simulated, multiple systems and technologies are integrated. Digitization is taking the direction of digital transformation. This step already requires changes at the corporate management level.

Digital transformation is understood as a complex and extensive process in which existing business processes are digitally modified, and new ones are subsequently created. The effort is for the company to be efficient in rapidly changing market and business requirements. Digital transformation thus requires a significant redesign of processes so that they are aligned with the digital environment. Simply put, the goal of digital transformation is to improve the business of SMEs, create higher profits, replace and transform business processes, competencies, and change business management and production models. In this step, the information, data, processes and systems that affect the very functioning of the company are digitized, and the company is digitally transformed, and even its strategy is modified. Digital transformation must be seen as a never-ending process because it is constantly necessary to react to environmental changes, re-evaluate business activities, and subsequently transform them.

The Digital business model is the highest, fourth level of the digitization process. Values for customers are brought to the company, which are built on the application of digital technologies. The result of introducing such a business model is a significant advantage, feature, or functionality of the product or service, which brings the customer such a benefit that the customer is willing to pay for. These models are based on a radical rethinking of the way the company cooperates with its customers. The processes are thus gradually linked. They become complex from the beginning to the final outputs. Moreover, all this is digitized. For the company to succeed at this stage, it must not underestimate two basic conditions. The first is that he must know the end customer perfectly, and the second is that he must also know his business range – i.e., its entire length, width, and depth. These dimensions combine to give multiple possibilities for different digital business models that must be focused on creating value for the customer.

Since digitization is currently becoming a supporting innovation tool and affects the increase of the company's performance, it is important to know that before investing money in digitalization tools, the company's management should create a business culture that will emphasize the strategic importance of digitization among employees, carefully plan activities, and progress of digitization itself and, last but not least, also to design an adequate system of digitization analysis with suitable feedback. This will enable the company to continuously optimize its digitization activities and increase its efficiency [7]. Behind the concept of digitization are hidden a number of technologies, strategies, approaches, and processes that are dynamically developing, which is why it is important to constantly follow current trends.

Of course, managers will appreciate the endless flow of fresh customer information that they can use to improve their own products and services. Continuous process improvement in business automation is essential [8].

2.2 *Changes in the Behavior of SME Customers*

The customer and his behavior in the market are changing. The latest surveys [9], show that already today the share of consumers in the EU, more than 60% who buy goods via the Internet. Back in 2004, this ratio was only 21%, while 10 years later, the share increased almost 2.5 times to 50%. This trend continued to grow in the following years. Between 2017 and 2018, almost 60% of consumers made online purchases [10].

The fact that the share of individuals who last shopped online in the past three months is gradually increasing is also proving to be important. This increase is faster than those customers who last shopped online in the past 3 to 12 months. This leads to the clear conclusion that the frequency of e-commerce transactions is increasing. Up to 19.5% of all SMEs (with 10 or more employees) provided online sales in 2017. E-commerce accounted for 17.4% of the total turnover of enterprises. These levels had been stable for several years prior to the outbreak of the pandemic.

However, individual EU countries are involved differently, while Denmark, Ireland and Sweden are at the forefront in this area, and three out of ten companies offer online sales. In Bulgaria and Romania, it is only one out of ten. There are also fairly large differences by industry. For example, the share of companies as well as the share of online turnover, is the highest in the area of accommodation services (the share of companies is 68.1%, and the share of online turnover is 31.6%). In second place is retail trade (28.0% and 10.7%). In contrast, the construction sector is involved the least (4.3% and 1.8%) [9].

2.3 The Impact of the Pandemic on Digitization

During the COVID-19 pandemic, people were forced to change their shopping behavior. There has been a definite move to the online space. And businesses, as a result, under the pressure of an increase in the number of customers and their increased orders, had to react immediately in order to be able to quickly and efficiently cover this increased demand and remain friendly and accommodating. In this case, the companies encountered a new phenomenon and the fact that the competition suddenly ceased to be local and, on the contrary, began to cross geographical boundaries. The data confirm that some Slovaks continue to shop online even after the end of the strict measures because the payment volumes of online stores far exceed the level before the pandemic, but also the levels that were reached in the pandemic 2020. Payment data show that the payment volume in 2020 grew by 40% year-on-year in e-shops [11]. This trend continued in 2021, when online purchases reached an even higher level than in 2020. The main reason was the fact that at the beginning of 2021, measures limiting the movement of residents were again adopted in Slovakia. The highest volume of purchases was recorded in March 2021. At that time, it was up to three-quarters higher than in March 2020 and more than double compared to March 2019 [11]. Although the easing of measures also brought a decrease in online purchases, the volumes are still higher than in 2020. Customer loyalty is also linked to this indicator, which is one of the main factors for the future of the company today. Businesses should be able to continue to function in the set trend and use the possibilities of digitization when creating a strategy focusing on increasing competitiveness and market share. Improving their digital skills can help them do this. Nowadays, there are many digital media available to businesses so that they can reach and retain customers in a more targeted manner, build their brand, and promote products more effectively. Businesses that do not invest in digital media will have a much harder time surviving on the market than those that can quickly adapt to this revolutionary trend [12].

2.4 Reasons for the Creation of a Single Digital Market

The COVID-19 pandemic, which was associated with restrictions on personal contacts, had a significant impact on digitization. Many Slovaks were forced to try online shopping for the first time, and a relatively large part of them use this option even after the measures were eased. The most significant increase was recorded in food delivery, but higher purchases were also recorded in the categories of clothing, shoes, and household goods.

Slovaks not only shopped more but also discovered new stores and brands, as they spent much more time online than before the pandemic. Today, Slovaks buy almost everything online, even though the first quarter of 2022 saw a year-on-year decrease of 28%. The reason is precisely the extremely high increase in online

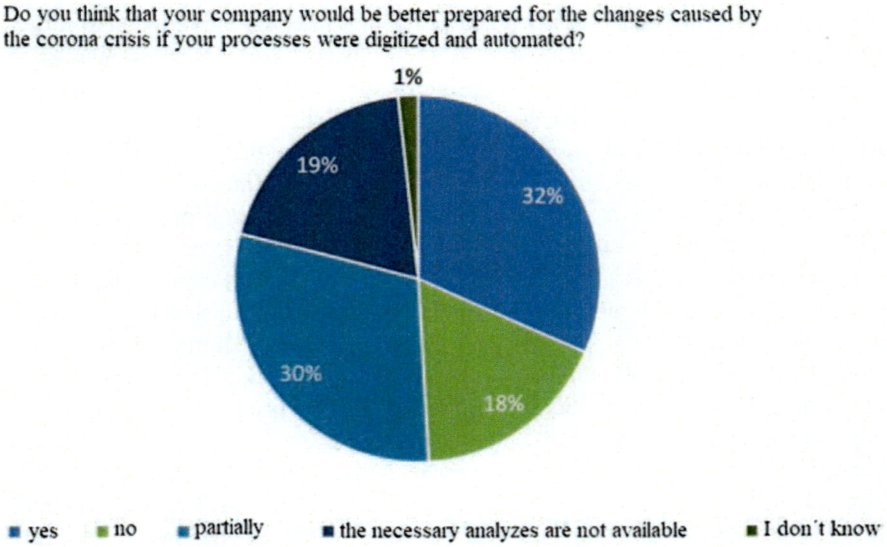

Fig. 2 Survey results [14]

shopping during the pandemic and, at the same time, a partial return to brick-and-mortar stores after the end of the restrictions. But this trend is worldwide, e.g., according to the US Department of the Economy, e-commerce increased during the eight weeks of quarantine in 2020, the same as in the last ten years [13].

The reasons for reduced online shopping by Slovaks are also the current situation associated with the growth of inflation, when customers think more about what they buy. Furthermore, fear of the future is the main reason for more cautious online purchases, and the question remains how long this trend will last. On the other hand, those sellers who expected further growth of online shopping and invested funds in expansion can significantly harm such consumer behavior.

All the above-mentioned changes record the need for SMEs to digitize their information and thus simplify their business. When Industry4UM – the Association of Intelligent Industry conducted its fourth annual survey on digital transformation in industrial enterprises at the turn of August/September 2020, the results showed stagnation compared to 2019. At the same time, up to two-thirds of the 57 surveyed enterprises believed that they would have handled the corona crisis better if would be digitized [14] (Fig. 2).

3 Problems Associated with Digitization

It is an undeniable fact that digital transformation can bring a whole range of competitive advantages, not only for large companies, but also for small and medium ones, which form a key pillar of the national economy of the Slovak Republic.

However, SMEs are facing an important problem – the problem of cyber security. Information is intensively exchanged between business partners, their number is growing, and this creates many opportunities for cyber-attacks. It is SMEs that become an interesting target because cyber attackers expect a lower level of security, which is confirmed in many cases. It is, therefore, essential that SMEs not only deal with cyber security but also try to reduce the risks of cyber threats.

The Slovak Business Agency conducted a survey regarding cyber security in 2022 [15]. 1009 SMEs took part in the survey, while 25.2% were natural persons and 74.8% were legal entities, dominated by limited liability companies (89.8% representation) and the rest by joint-stock companies (2.9%) and companies of another legal form (7.3%), which included public companies, limited partnerships, cooperatives, and private joint-stock companies. According to size, the most represented (69.7%) were micro-enterprises (i.e. enterprises with less than 10 employees), 20.2% were small enterprises (10–49 employees). Medium-sized enterprises (with 50 to 249 employees) were represented by 10.1%. From a territorial point of view, the largest number of subjects came from the Bratislava region, 29.4%, followed by the Žilina Region (11.9%), the Nitra Region (11.0%), the Prešovský Region (10.4%) and the Košice Region (10.2%). The other three regions were relatively balanced – their representation was between 8.5% and 9.4%. The main business activities of the enterprises were services (financial and insurance, real estate services, professional, scientific and technical activities, administrative and support services) – 28.7% and trade – 24.6%. Also represented were companies from the field of construction – 11.9%, industry – 10.6%, transport and information – 9.2%, other services (education, healthcare, and social assistance, art, entertainment, and recreation) – 7.5% and finally the accommodation and catering industry – 3.9% and agriculture – 3.6%. The last monitored information related to the ownership structure, according to which domestic companies were represented to the greatest extent - 93.7% (of which 87.1% were exclusively domestically owned) and 6.6% were predominantly domestically owned. Foreign companies had a share of 6.3% (of which exclusive foreign ownership was 2.7% and predominant foreign ownership 3.6%).

The aforementioned survey showed several interesting facts that indicate that Slovak SMEs are currently not sufficiently prepared for digitization and significantly underestimate the risks associated with it because:

- cyber security is dealt with by almost one-third (31.3%) of SMEs (despite the fact that the legislation does not require them to do so). But two-thirds (66.3%) of SMEs do not deal with cyber security at all, and 2.4% of SMEs do not even have a clear idea of what cyber security is;
- the majority of micro-enterprises do not deal with cyber security - up to 70.8% in the construction sector (85.8%) and accommodation and catering (84.6%) from Trnava (72.8%) and Prešov region (72.4%); mainly domestic entrepreneurs (67.4%) and more natural persons (83.1%);
- on the contrary, medium-sized enterprises (53.9%) deal mostly with cyber security, in the sector of other services (47.4%), transport, information (41.9%),

mainly from the Bratislava region (40.1%), SMEs, legal entities (36.7%), in foreign ownership (46.9%).

The given survey showed great reserves not only in the digitization of businesses, but especially in preparedness for the risks associated with it. Although, on the one hand, more than half of the interviewed companies are aware of the risk (55.6%), on the other hand, many companies do so only because customers (29.4%) or legislation (28.8%) require it, or it is a request from investors (18.2%), or it is the result of an audit (12.1%).

The fact that businesses underestimate this area may also be due to the fact that up to 54.1% of SMEs have not recorded any cyber incidents, and only 19.7% have recorded malicious (intentional) actions.

Another important insight is that up to 51.8% of SMEs do not have cyber security staff, which creates space for hiring or training employees with a focus on cyber security. However, almost a third of those SMEs that deal with cyber security does not provide any opportunity to increase knowledge in the field of cyber security. On a voluntary basis, almost 30.0% of small and medium-sized entrepreneurs provide their employees with the opportunity to increase their knowledge and skills in the field of cyber security. Only 14.1% of SMEs require the increase of knowledge and skills in this area as mandatory. More than a fifth of the surveyed SMEs had no employees.

Another alarming fact is that the majority of companies dealing with cyber security do not have a risk analysis (58.5%). Of the remaining 41.5%, only 8.2% of SMEs have an analysis developed and updated within the last month, and 21.2% have developed and updated it during the year. Other SMEs either have a risk analysis and updated it more than 2 years ago or never updated it.

The approach to risk management can also be seen in how businesses behave when it comes to cyber security funding. In just over half of SMEs dealing with cyber security, this budget is part of the IT budget (52.9%), and 47.1% do not have a budget for cyber security at all or it is not part of the IT budget. At the same time, the largest part of cyber security costs is made up of hardware and software costs (41.7%), followed by backup (12.2%), and only 8.3% consists of costs intended exclusively for maintaining cyber security. However, the costs themselves are very undersized, as more than half of SMEs invest less than 10% of their IT budget in cyber security costs, and 21.7% invest between 10% and 29% of their IT budget. This fact also shows that SMEs do not devote or consider this area so important that they invest larger sums in it. At the same time, if we look at the amount of cyber security costs from the total SME budget, we will find that for most SMEs (62.5%), it is less than 1%. At 9.4% of SMEs, it is 2% to 4%. Only in the case of 3.1% of SMEs were cost greater than 5% and less than 9%. And finally, less than 2% of SMEs invest more than 10% of the total budget.

SMEs that deal with cyber security do not have a specially designated budget for exceptional situations (86.5%), and even only 16.2% of SMEs have insurance in relation to their cyber security.

In this area, the state could and should play an important role, which should encourage businesses and contribute with their activities to businesses paying more attention to cyber security. However, up to 21.4% evaluate the state's information activities in this area as insufficient, and 62.8% of SMEs do not even know the state's information activities in the field of cyber security. Only 7.9% of SMEs consider them sufficient, and 0.9% as very good. 6.9% of SMEs could not answer.

Businesses expect measures from the state, e.g., in the form of grants and vouchers for the purchase of technology and for the provision of educational courses and training (37.5%). They would also welcome information services (34.6%) and, last but not least advice, mentoring, and training (32.1%).

As it follows from the above, digitization is, on the one hand, very important for businesses and brings many competitive advantages, but on the other hand, SMEs encounter problems related to cyber security while not realizing its importance. They do not invest enough funds or time to create backup plans and rely on the fact that it does not affect them because most of them have not encountered any problems yet.

3.1 *Recommendations for SMEs to Ensure Cyber Security*

More and more SMEs in Slovakia are dealing with cyber security, even though there are still many SMEs that do not have a clear idea of what cyber security is. Awareness of the risks associated with digitization is the first step for the subsequent increase in cybernetic security.

Correct and sufficient protection can avert the threat of an incident that would have negative consequences for the company, not only in the financial area but also in the area of loss of reputation among customers.

This area needs to be addressed where the business partner requires it, for example, in customer-supplier chains and where sensitive information may be leaked.

Even though in the case of SMEs, cyber incidents are rare so far and often rather connected to human error or system failure, their increase in the near future is not excluded. Therefore, companies should start paying more attention to this issue. First of all, increase the technological knowledge of employees. Subsequently, it would be advisable for SMEs to develop a risk analysis and also update it at least once a year.

Last but not least, it is important to set aside an adequate part of the finances from the corporate budget, which will serve cyber protection and backup, which should occur regularly. This budget should also include a financial amount designated for insurance in relation to corporate cyber security.

If an SME uses external services, e.g., in the area of Internet connection or due to accounting or for other reasons, it is advisable to know how the cyber security area of the external partner is secured.

Finally, it can be concluded that if the state has listed the digitization of the business environment as one of its priorities, it should especially help SMEs. Firstly, by providing informational activities, advice, and mentoring, but also direct financial support in the form of grants, educational courses, or vouchers for the purchase of necessary technologies.

4 Perspectives of Digitization of SMEs

It was the restrictions associated with the COVID-19 pandemic that showed how important digital skills are for work and communication, not only within work relationships but also in private ones. At the same time, however, shortcomings and, thus the need to increase digital education came to the fore.

It is digital transformation that is a challenge for SMEs, and not only in connection with the single digital market. It is mainly related to increasing competitiveness but also the use of tools that would be accepted not only by employees and business partners, whether in Slovakia or abroad, but also by control authorities or, in case of disputes, by courts.

Already today, SMEs are faced with the requirement to work with various digital means that they must use in contact with the state. This could be using electronic mailboxes to communicate with the public sector or fulfilling legal obligations through a central portal. Or also electronic signing or guaranteed conversion of documents. Activities that entrepreneurs could not imagine ten years ago are now commonplace in the Slovak business environment. Therefore, it is necessary to analyze the individual components of digitization in order to more easily identify the barriers that make it difficult for SMEs to do business on the one hand, but also for public administration authorities on the other hand.

Against the efforts of SMEs to digitalize their activities, there are problems in the field of e-Government, which are connected, for example, to different technical prerequisites for the functioning of individual systems, differences in the setting of access to information systems (e.g. in some places the entrepreneur logs in using a name and password, in other places he also needs GRID card and in some places only an electronic mailbox is enough). Frequent maintenance and unavailability of systems, impossibility to pay administrative and court fees online and, last but not least, problems with ensuring cyber security are also a problem.

Conclusion Micro-enterprises and small and medium-sized enterprises are a pillar not only of Slovak society but of the entire European Union. This is not only because together, they make up 99% of all businesses in the EU, but mainly because they are a decisive employer, as they provide up to two-thirds of jobs in the private sector. In addition, they contribute more than half to the total added value created within the EU. All this creates pressure to increase the competitiveness of SMEs in the market. Digitization creates a unique environment for this because it not only opens up possibilities for companies that were not technically possible just a few years ago, but thanks to it, companies can work and operate more efficiently on the market and thus successfully fight the competition.

Digitization will create conditions for SMEs to place all available and necessary information about customers in one place, where it is possible to work with them rationally and use them.

On the other hand, SMEs are much more vulnerable than large enterprises, which was clearly seen during the COVID-19 pandemic as a result of the restrictive measures taken. Here, it was clearly shown how the company could deal with a crisis situation and how quickly it can react. This ultimately decided which businesses remained on the market and which ones could not handle the situation and had to leave the market. A big role in this was also played by the change in the behavior of consumers who, as a result of the lockdowns, began to use online shopping to a much greater extent and moved to the online world. This turned out to be a great opportunity for those SMEs that were already operating in an online environment or were able to respond very quickly.

However, it is the digitization of SMEs, as shown by various surveys, that presents these companies with new challenges and tasks that they must take into account and for which they are often not prepared. Ensuring the cyber security of SMEs appears to be a very important area, which turns out to be one of the biggest problems in digitization. Part of SMEs not only do not count on security in this area, but are not even willing to invest financially because they do not consider this threat to be too real. Here, it would be appropriate for the state to get more involved and help SMEs pay due attention to this area.

Acknowledgments This work was supported by the Slovak Research and Development Agency under contract No. APVV-19-0590, and also by the projects VEGA 1/0268/22, KEGA 038TUKE-4/2022 granted by the Ministry of Education, Science, Research and Sport of the Slovak Republic.

References

1. Digitization: importance, benefits and EU legislation, https://www.europarl.europa.eu/news/sk/headlines/society/20210414STO02010/digitalizacia-dolezitost-benefity-a-legislativa-eu; last accessed October 21, 2022.
2. Yu, FF., Jiang, D., Zhang, Y., Du, HY.: Enterprise digitalisation and financial performance: the moderating role of dynamic capability. Technol. Analy. Strateg. Manag., ISSN 0953-7325, https://doi.org/10.1080/09537325.2021.1980211 (2021).
3. Lupton, D.: Digital Bodies. Routledge Handbook of Physical Cultural Studies. (2017). https://doi.org/10.2139/ssrn.2606467
4. Buechner, S.: On the Relationship of Digitization and Organization. Zeitschrift fur Soziologie, ISSN:0340-1804. https://doi.org/10.1515/zfsoz-2018-0121 (2019).
5. Industry 4um, https://industry4um.sk/co-je-digitalizacia-a-digitalna-transformacia-urobme-si-poriadok-v-pojmoch-a-vyznamoch/, last accessed October 21, 2022.
6. Patil, R., Karandikar, R.G.: Digitization of documented signals using vertical scanning. In: International Conference on Microwave, Optical and Communication Engineering (ICMOCE), pp. 239–242 (2015)
7. Mahmutovic, K.: The impact of digital marketing orientation on hotel marketing performance. Interdisciplinary Management Research XVII, 1211–1232 (2021) ISSN 1847-0408

8. Jeknic, J., Kraut, B.: Cloud services and Marketing. In: 8th International Convention on Information and Communication Technology, Electronics and Microelectronics (MIPRO), pp. 1492–1498 (2015)
9. European Commision.: https://ec.europa.eu/info/sites/default/files/ccs2015scoreboard_sk.pdf; last accessed October 21, 2022.
10. EÚ Európska komisia, Generálne riaditeľstvo pre spravodlivosť a spotrebiteľov, Hodnotiaca tabuľka spotrebiteľských podmienok : spotrebitelia na jednotnom trhu ako doma : rocnik 2019, Publications Office, https://doi.org/10.2838/745984 (2019).
11. Slovenská sporiteľňa: https://www.slsp.sk/sk/aktuality/2021/10/25/e-shopom-sa-dari-slovaci-si-na-online-nakupy-zvykli,. last accessed October 23, 2022.
12. Mohammad, A.: The impact of digital marketing success on customer loyalty. Marketing and Management of Innovations. **3**, 103–113 (2022). https://doi.org/10.21272/mmi.2022.3-09
13. Online Retail Growth and Expansion: COVID-19 and How It Will Go Next: https://www.ipsos.com/sk-sk/rast-expanzia-online-maloobchodu-covid-19-ako-pojde-dalej; last accessed October 20, 2022.
14. Industry 4um, https://industry4um.sk, zverejnené: 16.11.2020.
15. Survey of the state of cyber security in the SME sector: http://monitoringmsp.sk/wp-content/uploads/2022/08/Prieskum-stavu-kybernetickej-bezpe%C4%8Dnosti-v-sektore-MSP-sprava-z-prieskumu.pdf; last accessed October 21, 2022.

Process Modularity Impact on Manufacturing Lead Time in Mass Customization

Julia Nazarejova, Vladimir Modrak, and Zuzana Soltysova

1 Introduction

Although mass customization sounds quite paradoxically to traditional manufacturing practices, since it combines customization and mass production, this manufacturing strategy presents an important challenge for companies by providing custom products for individual customers at rather low-cost. For manufacturers, it means taking individual requirements from the customers and passing them directly to the assembly floor without an additional engineering effort. The such ability can be achieved only by systematic implementation of well-known instruments of Industry 4.0 [1–3]. In general, it is possible to state that manufacturing lead times (MLT) in terms of mass-customized production are longer than in the case of standard production. However, manufacturers try to minimize MLT of mass-customized products in order to shorten delivery times and thereby increase competitiveness. Principally, this objective can be achieved in two ways. The first of them is based on using common methods which are used in the production of standard products. The second possible way is to apply the so-called tailor-made approach to the given problem. It is known that modularity enables to produce many product variants by limiting the costs of variety in mass-customized environment. In this context, our intent here is to present a new approach which is aimed to increase performance of assembly processes in terms of mass customization. This approach is based on the assumption that process modularity can improve manufacturing process performance, especially in terms of mass customization [4, 5]. Moreover, motivation for this research stems from the reality that modularity as a design principle is

J. Nazarejova (✉) · V. Modrak · Z. Soltysova
Faculty of Manufacturing Technologies with the seat in Prešov, Technical University of Košice, Prešov, Slovakia
e-mail: julia.nazarejova@tuke.sk

M. Balog et al. (eds.), *EAI International Conference on Automation and Control in Theory and Practice*, EAI/Springer Innovations in Communication and Computing, https://doi.org/10.1007/978-3-031-31967-9_18

considered as an important criterion in designing assembly process structures [6, 7], and in addition, we also took into account the idea that modular design is a useful tool for managing complexity in manufacturing [8, 9].

2 Methodological Framework

As mentioned in the "Introduction" section, it seems to be reasonable to anticipate that process modularity could positively influence MLT in the production of mass-customized products. In line with this assumption, the proposed methodical approach to solve the given problem is based on the combination of several methods/procedures which are arranged into the compact framework presented in Fig. 1.

Further, the individual steps of this framework are described.

Step 1. Determination of input data.
At first, all needed process parameters are extracted from the process planning phase. Let us say that production task requires to produce the three product types

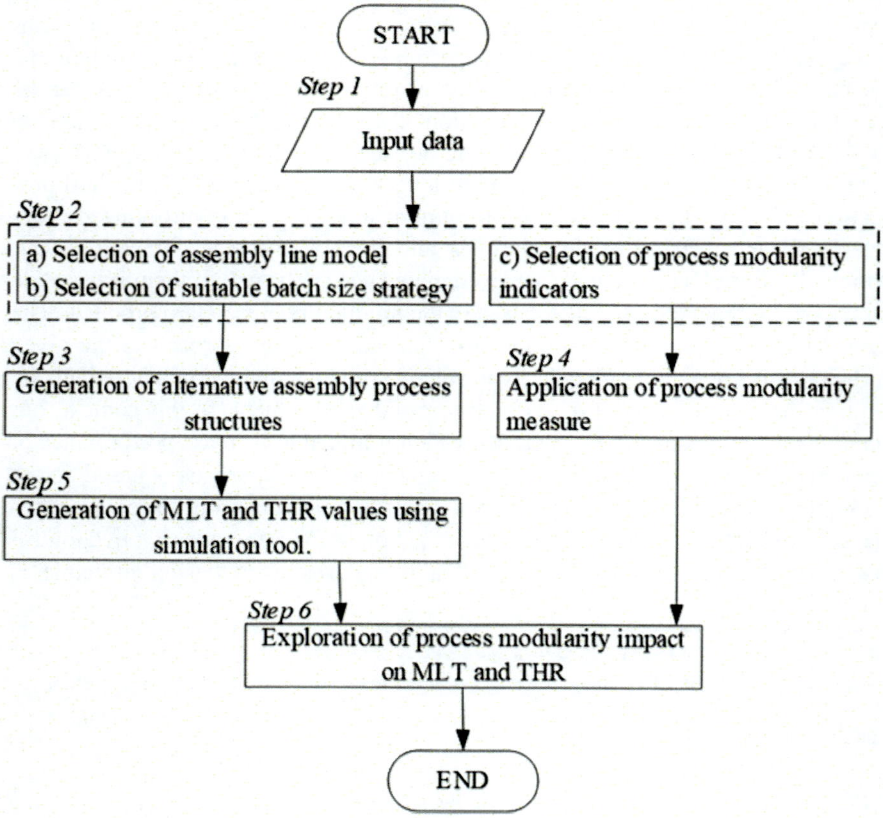

Fig. 1 A methodological framework

(P1, P2 and P3) while each of them is consisting of three components (A, B and C) using three working stations (W_1, W_2 and W_3). Further, production batches (PBs) are as follows product type P1–3 parts, product P2–6 parts and product P3–15 parts, while all the processing times are identical. Moreover, it has been found that the optimal sequence for releasing parts into the manufacturing system is P1 → P2 → P3. Then, the task is to solve very simple flow shop problem with m-machines and n-products (m x n), which is presented in Table 1.

Step 2a. Selection of assembly line model

As it is known, assembly lines are divided into three types of assembly line models, namely single model assembly line, mixed model assembly line, and multi-model assembly line [10]. In this paper, mixed-model and multi-model assembly lines are considered, and they will be compared later, in the case study section.

Step 2b. Selection of suitable batch size strategy

For this purpose, the two possible scenarios will be applied, namely Scenario 1-uniform number of transport batches (L), and Scenario 2 – uniform size of transport batches. Then, the most appropriate transport batch size (TBS) method will be selected for MLT minimization. Input data for these two scenarios are shown in Table 2.

By using Gantt charts (see Fig. 2.), one can compare the two batch size strategies. As it is visible, the second Scenario with MLT = 30 minutes brought better results against the first one, where MLT = 34 minutes.

A posteriori Scenario 2 will be prioritized when choosing one of the two possible batch size strategies.

Table 1 Flow shop $m \times n$ problem, where processing times are in minutes.

$m \times n$	P1	P2	P3
W_1	1	1	1
W_2	1	1	1
W_3	1	1	1

Table 2 The two possible batch size strategies/scenarios

Scenario 1	P_1	P_2	P_3	Scenario 2	P_1	P_2	P_3
PB	3	6	15	PB	3	6	15
L	3	3	3	L	1	2	5
TBS	1	2	5	TBS	3	3	3

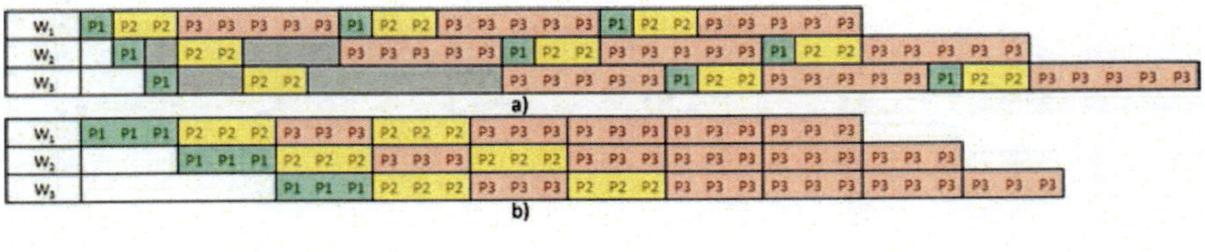

Fig. 2 (**a**) Scenario 1, (**b**) Scenario 2

Step 2c. Selection of process modularity indicators

In reality, there are only few known methods to measure modularity of process structures, from and two of them seems to be almost equally effective. They are Cross-module independence (CMI), Relative modularity indicator (RNM).

CMI is defined through the ratio of the sum of relations inside all modules to the sum of all relations and may be formally expressed as follows [11]:

$$CMI = 1 - \sum_{\text{int}=1}^{n} \frac{R_{int}}{T}, \quad (1)$$

where:

int denotes a specific module,
n stands for the number of modules,
R is the sum of the relations within a module,
T is the sum of all relations in a given assembly process structures (APS) model. Modules of APSs are represented in the trees by internal nodes [11].

Indicator RNM is expressed through the following formula [12]:

$$RNM = \frac{n}{\sum_{i=1}^{v} \deg(v)_i \bullet \log_2 \deg(v)_i}, \quad (2)$$

where:

$\deg(v)_i$ is the degree of vertex of $(v)_i$ in graph *G*, while *G* consists of a set of *V* vertices $\{V\} = \{v_1, v_2, \ldots, v_v\}$,
n is the number of modules in the network [12].

Step 3. Generation of alternative assembly process structures

In this step, alternative assembly process structures (APSs) are selected. Theoretically, it is possible to generate all the possible APSs, but the total numbers of APSs for higher number of input components can be very huge. For this sake, it is more practicable to select some representatives from the overall portfolio of APSs. For the given flow shop problem described in Table 1, the two process structures are realistic, which are shown in Fig. 3.

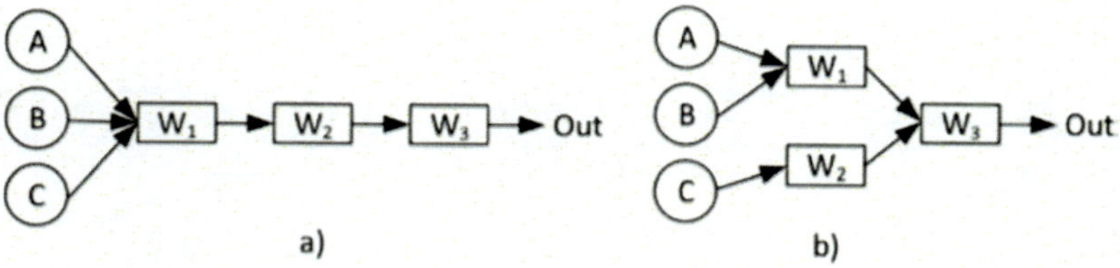

Fig. 3 The two process structure alternatives

Step 4. Application of process modularity measures

Subsequently, above described modularity measures – Eqs. (1) and (2), are applied for the two process structure alternatives, as follows:

$$CMI_{\text{Alternative1}} = 1 - \frac{4}{6} = 0.33;$$

$$CMI_{\text{Alternative2}} = 1 - \frac{4}{6} = 0.33;$$

$$RNM_{\text{Alternative1}} = \frac{3}{4 \cdot (1 \cdot \log_2 1) + 4 \cdot \log_2 4 + 2 \cdot \log_2 2 + 2 \cdot \log_2 2} = 0.25;$$

$$RNM_{\text{Alternative2}} = \frac{3}{4 \cdot (1 \cdot \log_2 1) + 3 \cdot \log_2 3 + 3 \cdot \log_2 3 + 2 \cdot \log_2 2} = 0.26.$$

Step 5. In this step, MLT and THR values using simulation tool are generated

For this purpose, Tecnomatix Plant Simulation is employed. Examples of the two APS models in Tecnomatix environment are shown in Fig. 4.

Manufacturing lead time (MLT), which is the total time required to process a product through the plant, is usually calculated by the following formula [13]:

$$MLT_j = \sum_{i=1}^{no_j} \left(Tsu_{ij} + Q_j * Tc_{ij} + Tno_{ij} \right), \tag{3}$$

where:

MLT_j means MLT for a batch of part or product j;
Tsu_{ij} is setup time for operation i on part or product j;
Q_j is quantity of part or product j in the batch being processed;
Tc_{ij} is cycle time for operation i on part or product j;
Tno_{ij} is nonoperation time associated with operation i, where i indicates the operation sequence in the processing, $i = 1, 2, \ldots, no_j$ [13].

For our purpose, the previous formula can be slightly modified:

$$MLT_j = \sum_{i=1}^{no_j} \left(Tsu_{ij} + Q_j * Tc_{ij} + Tt_{ij} \right), \tag{4}$$

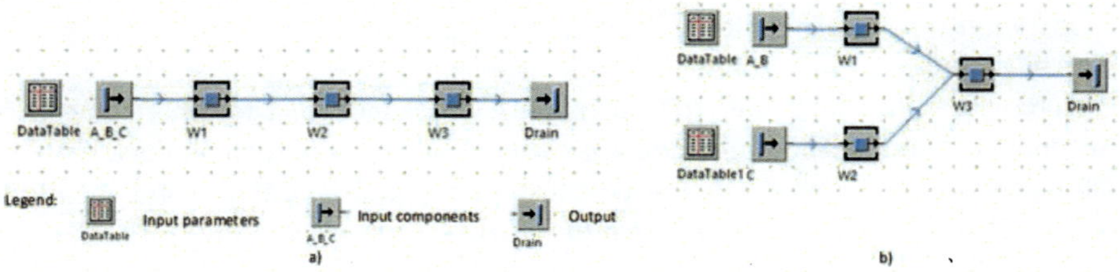

Fig. 4 The two process structure alternatives modeled using a simulation tool

where:

Tt_{ij} is transport time associated with operation i, where i indicates the operation sequence in the processing, $i = 1, 2, \ldots, no_j$.

The second selected performance indicator – Throughput rate (THR) calculates the rate at which units going through the production process from the beginning to the end of production process (e.g., how many units were produced, e.g., per minute, hour or day).

By applying simulation models within the framework, the obtained results for MLT and THR are as follows:

$$MLT_{\text{Alternative1}} = 30\,\text{minutes}\,(0.5\,\text{hour}) \qquad THR_{\text{Alternative1}} = 16\,\text{parts / hour}$$

$$MLT_{\text{Alternative2}} = 27\,\text{minutes}\,(0.45\,\text{hour}) \qquad THR_{\text{Alternative2}} = 17.78\,\text{parts / hour}$$

Step 6. Exploration of process modularity impact on MLT and THR
Finally, above mentioned methodological steps are applied to the theoretical case study of assembly process structures with the aim of describing the relation between process modularity and MLT and THR in the next section.

3 Theoretical Case Study

Let us consider the following case of the assembly process in terms of mass customization environment completing 44 variants of one product type. The assembly process consists of five assembly operations, which can be processed on individual workstations, or on integrated workstations. Each assembly process produces four types of products (P1, P2, P3, and P4). Products consist of six mandatory components (A, B, C, D, E, F) and two optional components (V1, V2). The compositions of the products are as follows P1 = {A, B, C, D, E, F}; P2 = {A, B, C, D, E, F, V1}; P3 = {A, B, C, D, E, F, V2}; P4 = {A, B, C, D, E, F, V1, V2}. As this assembly process consists of eight input components, theoretically it would be possible to generate 261 APS alternatives. For the purpose of this study, ten of them have been selected., which are depicted in Fig. 5.

In order to explore the influence of process modularity on assembly process efficiency, the two types of simulation experiments will be conducted. For the first of them, Mixed Model Assembly Line model (MiMAL) will be employed, while for the second one Multi-Model Assembly Line model (MuMAL) will be used.

Operation times for individual alternatives of products and structures are provided in Table 3.

Moreover, set up time for each operation is 5 seconds, and transport time of 8 seconds is assumed. The simulation was done for 25 pieces of each product, so the total batch size is 100 products. For demonstration purpose, a model of assembly processes No. 1 from Fig. 5. is depicted in Fig. 6.

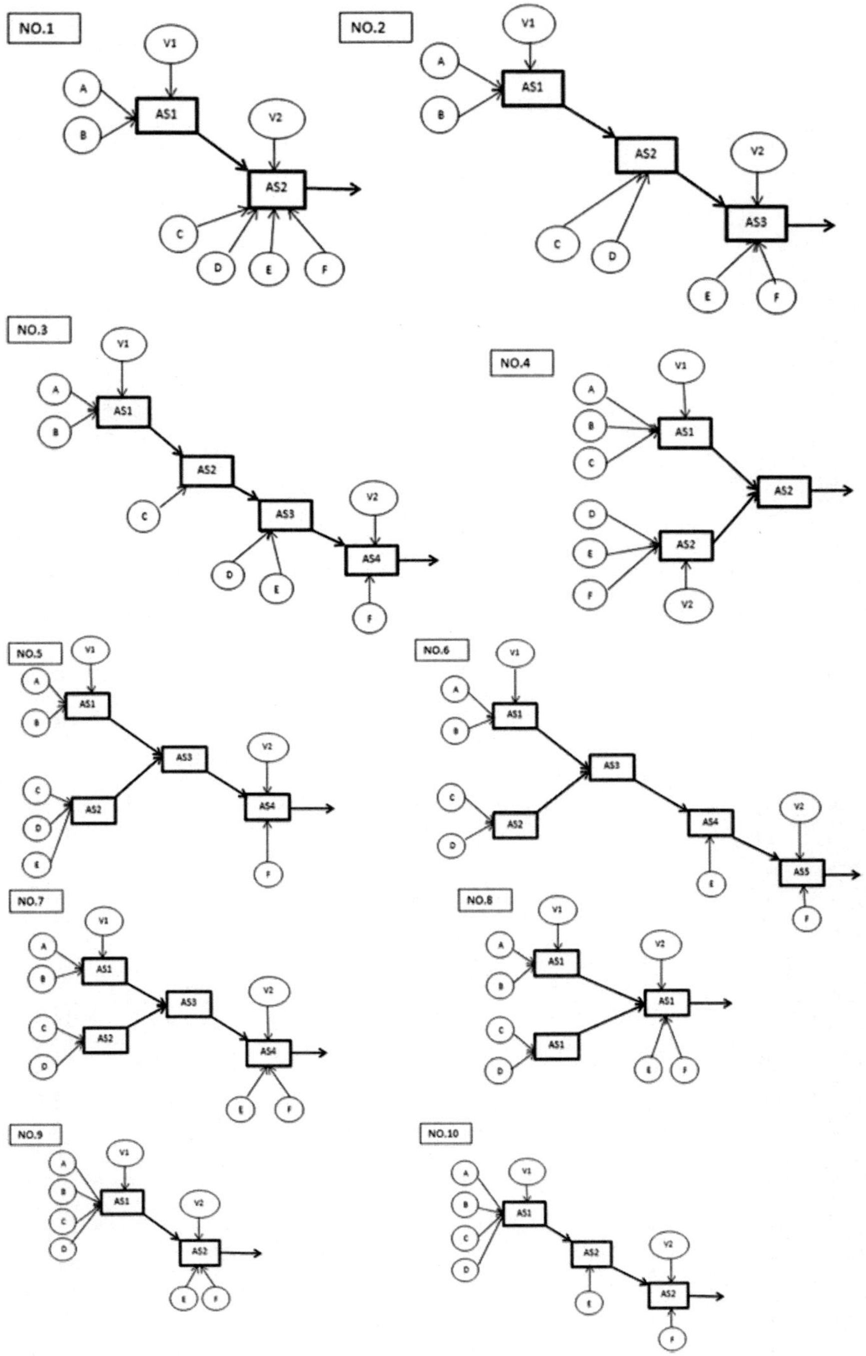

Fig. 5 All APSs considered for the case study

Table 3 Assembly time for individual alternatives of products for all structures in seconds

Parts assembly	No.1	No.2	No.3	No.4	No.5	No.6	No.7	No.8	No.9	No.10
A + B	**20**	**20**	**20**	–	**20**	**20**	**20**	**20**	–	–
A + B + V1	**30**	**30**	**30**	–	**30**	**30**	**30**	**30**	–	–
AB+C + D + E + F	**50**	–	–	–	–	–	–	–	–	–
AB+C + D + E + F + V2	**60**	–	–	–	–	–	–	–	–	–
ABV1 + C + D + E + F	**50**	–	–	–	–	–	–	–	–	–
ABV1 + C + D + E + F + V2	**60**	–	–	–	–	–	–	–	–	–
AB+C + D	–	**30**	–	–	–	–	–	–	–	–
ABV1 + C + D	–	**30**	–	–	–	–	–	–	–	–
ABCD+E + F	–	**30**	–	–	–	–	**30**	–	**30**	–
ABCDV1 + E + F	–	**30**	–	–	–	–	**30**	–	**40**	–
ABCD+E + F + V2	–	**40**	–	–	–	–	**40**	–	**30**	–
ABCDV1 + E + F + V2	–	**40**	–	–	–	–	**40**	–	**40**	–
AB+C	–	–	**20**	–	–	–	–	–	–	–
ABV1 + C	–	–	**20**	–	–	–	–	–	–	–
ABC + D + E	–	–	**30**	–	–	–	–	–	–	–
ABCV1 + D + E	–	–	**30**	–	–	--	–	–	–	–
ABCDE+F	–	–	**20**	–	**20**	**20**	–	–	–	**20**
ABCDE+V1 + F	–	–	**30**	–	–	–	–	–	–	–
ABCDE+F + V2	–	–	**30**	–	**30**	**30**	–	–	–	**30**
ABCDE+V1 + F + V2	–	–	**40**	–	–	–	–	–	–	–
A + B + C	–	–	–	**30**	–	–	–	–	–	–
A + B + C + V1	–	–	–	**40**	–	–	–	–	–	–
D + E + F	–	–	–	**30**	–	–	–	–	–	–
D + E + F + V2	–	–	–	**40**	–	–	–	–	–	–
ABC + DEF	–	–	–	**20**	–	–	–	–	–	–
ABCV1 + DEF	–	–	–	**20**	–	–	–	–	–	–
ABC + DEFV2	–	–	–	**20**	–	–	–	–	–	–
ABCV1 + DEFV2	–	–	–	**20**	–	–	–	–	–	–
C + D + E	–	–	–	–	**30**	–	–	–	–	–
AB+CDE	–	–	–	–	**20**	–	–	–	–	–
ABV1 + CDE	–	–	–	–	**20**	–	–	–	–	–
ABCDEV1 + F	–	–	–	–	**20**	**20**	–	–	–	**20**
ABCDEV1 + F + V2	–	–	–	–	**30**	**30**	–	–	–	**30**
C + D	–	–	–	–	–	**20**	**20**	**20**	–	–
AB+CD	–	–	–	–	–	**20**	**20**	–	–	–
ABV1 + CD	–	–	–	–	–	**20**	**20**	–	–	–
ABCD+E	–	–	–	–	–	**20**	–	–	–	**20**
ABCDV1 + E	–	–	–	–	–	**20**	–	–	–	**20**
AB+CD + E + F	–	–	–	–	–	–	–	**40**	–	–
ABV1 + CD + E + F	–	–	–	–	–	–	–	**40**	–	–
AB+CD + E + F + V2	–	–	–	–	–	–	–	**50**	–	–
ABV1 + CD + E + F + V2	–	–	–	–	–	–	–	**50**	–	–
A + B + C + D	–	–	–	–	–	–	–	–	**40**	**40**
A + B + C + D + V1	–	–	–	–	–	–	–	–	**50**	**50**

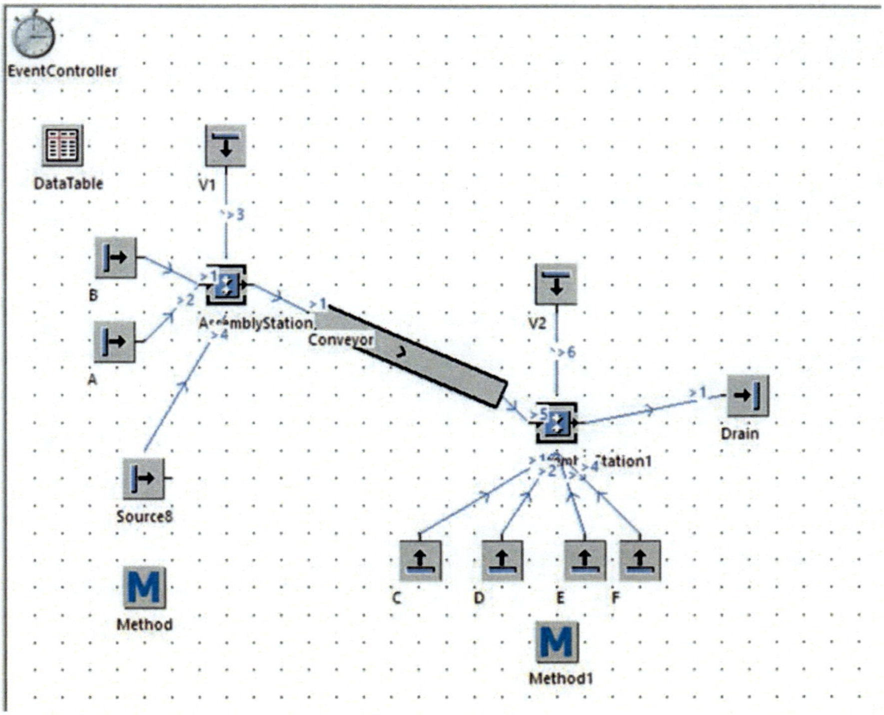

Fig. 6 The print screen from the Simulation software showing assembly line structure with input components

Table 4 Assembly time for individual alternatives of products for all structures in seconds

APSs	CMI	RNM	MLT for MiMAL (minutes)	MLT for MuMAL (minutes)	THR for MiMAL (parts/hour)	THR for MuMAL (parts/hour)	Number of nodes/ modules
No. 1	0.1	0.224	100.6	92.3	56,0.67	65.01	2
No. 9	0.1	0.228	84.22	75.88	71.25	79.07	2
No. 8	0.18	0.302	92.13	75.63	65.12	79.33	3
No. 10	0.18	0.302	84.6	76.18	70.92	78.76	3
No. 4	0.18	0.305	75.5	67.3	79.5	89.49	3
No. 2	0.18	0.308	67.9	60.1	88.32	99.83	3
No. 7	0.25	0.358	67.7	59.68	88.65	100.53	4
No. 3	0.25	0.362	60.15	51.82	99.75	115.79	4
No. 5	0.25	0.362	59.52	51.02	100.81	117.61	4
No. 6	0.31	0.397	51.73	43.24	115.8	138.25	5

In accordance with the above outlined methodological framework, the influence of modularity on MLT and THR will be explored using the two modularity indicators, namely RNM and CMI. The results of the given indicators for individual structures for the both types of experiments are shown in Table 4.

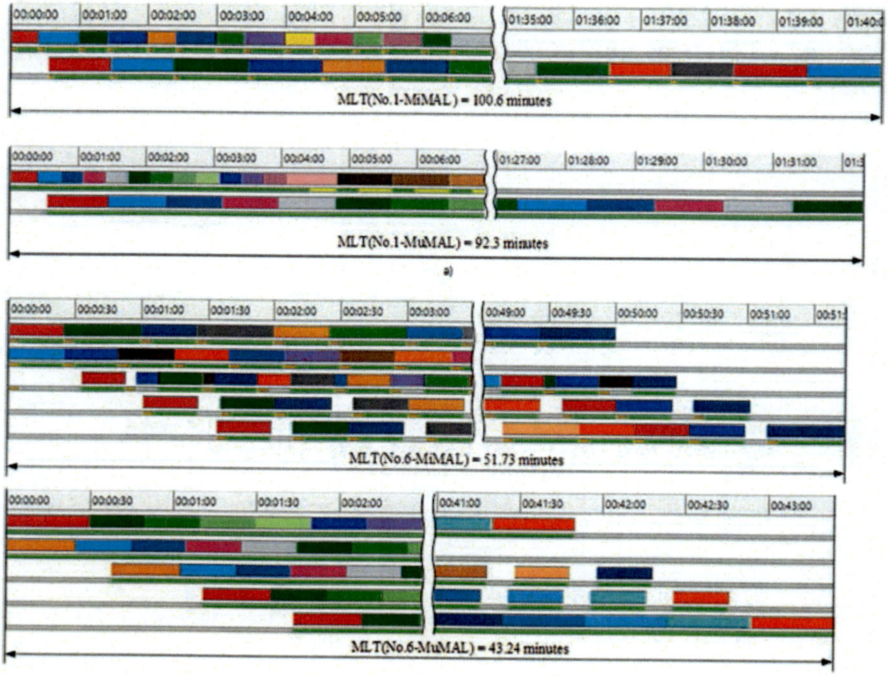

Fig. 7 Gantt charts using MiMAL and MuMAL for APS (**a**) No. 1 and (**b**) No. 6

The obtained results can be also demonstrated via selected Gantt charts generated through the simulation software as shown in Fig. 7.

When analyzing the results from Table 4, one can state that the best MLT and THR values were obtained using APS No. 6, which is the optimal modular APS from the ten considered alternatives according to RNM and CMI indicators.

4 Conclusions

Based on the obtained results from the two independent experiments, it is possible to state, that the three steps/substeps of the proposed approach influenced the targeted parameters, i.e., MLT and/or THR. They are:

- Selection of assembly line model (Step 2a). When comparing results from the two experiments, one can see, that MLT and THR values of experiment no. 2 (using MuMAL) are better than the values of experiment no. 1 (using MiMAL).
- Selection of suitable batch size strategy (Step 2b). As it was shown in Fig. 2, batch size strategy applying the uniform transport batch size for given product types provides shorten MLT than the strategy based on the uniform number of transport batches. In our case (when the production due date is not specified), the one-piece flow concept was applied in the case when the uniform transport batch

size equals 1, which is the optimal batch size strategy if its negative effects are omitted.

- Topology of assembly process structures (Step 3). When comparing MLT and THR values from Table 4, it is clear that those values also depend on the structural properties of assembly processes.

However, the main finding of this research is that APSs of a higher level of relative modularity are more suitable than those with a lower level of relative modularity from the viewpoint of MLT and THR minimization. It has been found that MLT values vary from 43.24 minutes to 100.6 minutes (56.5%) between modular APS (No. 6) and integral APS (No. 1). In addition, when comparing the same two APSs (No. 6 and No. 1) THR values differ from 132.25 parts/hour to 56.67 parts/hour. These significant differences justify us to confirm the positive impact of process modularity on the efficiency of assembly processes in a mass customization environment.

Acknowledgments This research was funded by the KEGA Project No. 025TUKE-4/2020 granted by the Ministry of Education of the Slovak Republic.

References

1. Matt, D.T., Modrak, V., Zsifkovits, H.: Industry 4.0 for SMEs: Challenges, Opportunities and Requirements. Palgrave Macmillan, London, UK (2020)
2. Müller, J.M., Kiel, D., Voigt, K.: I.: what drives the implementation of industry 4.0? The role of opportunities and challenges in the context of sustainability. Sustainability. **10**(1), 247 (2018)
3. Modrak, V.: Mass Customized Manufacturing: Theoretical Concepts and Practical Approaches. CRC Press (2017)
4. Jacobs, M., Droge, C., Vickery, S.K., Calantone, R.: Product and process modularity's effects on manufacturing agility and firm growth performance. J. Prod. Innov. Manag. **28**(1), 123–137 (2011)
5. Da Cunha, C., Agard, B., Kusiak, A.: Design for cost: module-based mass customization. IEEE Trans. Autom. Sci. Eng. **4**(3), 350–359 (2007)
6. Suh, N.P., Suh, N.P.: Axiomatic Design: Advances and Applications (Vol. 4). Oxford University Press, New York (2001)
7. Modrak, V., Soltysova, Z.: Process modularity of mass customized manufacturing systems: principles, measures and assessment. Procedia CIRP, 67, 36–40. LNCS Homepage, http://www.springer.com/lncs. Last accessed 2016/11/21 (2018)
8. Ethiraj, S.K., Levinthal, D.: Modularity and innovation in complex systems. Manage. Sci. **50**(2), 159–173 (2004). https://doi.org/10.1287/mnsc.1030.0145
9. Modrak, V., Soltysova, Z.: Novel complexity indicator of manufacturing process chains and its relations to indirect complexity indicators. Complexity. **2017**, 1 (2017)
10. Niaz Morshed, M.D., Kazi Saifujjaman, P.: Assembly line balancing to improve productivity using work sharing method in apparel industry. Global J. Res. Eng. **14**, 1–3 (2014)
11. Blackenfelt, M.: Managing Complexity by Product Modularisation., Doctoral dissertation,. Royal Institute of Technology, Sweden (2001)
12. Modrak, V., Soltysova, Z.: Development of the modularity measure for assembly process structures. Math. Probl. Eng. **1-9**, 1 (2021)
13. Groover, M.P.: Automation, Production Systems, and Computer-Integrated Manufacturing, 4th edn, pp. 1–816. Pearson, London (2016)

Part VI
Simulation, Modelling, Human-robot Collaboration, Customization, Digital Twin, Assembly

Testing and Verification of the Proposed Method for the Assembly Process Based on the Human–Robot Collaboration

Marek Vagas, Alena Galajdova, Patrik Sarga, Robert Rakay, and Jaroslav Romancik

1 Introduction

The system approach to achieving the optimal design of collaborative assembly activity is derived from the initial planning of cooperative assembly as a whole through the determination of individual sequential handling operations, the selection of peripheral devices, spatial distribution, the breakdown of the process into programmable sections, the installation of collaborative workplace equipment, operating conditions, and operator safety up to after training workers for operation and maintenance [1].

The central and most prominent element in these solutions is the collaborative robot. Today, we can characterize them as true collaborators or colleagues of people, which opens new possibilities for technical and nontechnical applications. Their size, skill, and sensitivity predestinate them for small jobs "of all kinds." They can move in limited spaces or directly on the working table together with a person. They are no longer just mechanical arms with fixed and repetitive movements but thinking and learning intelligent devices equipped with new sensory technologies and artificial intelligence [2]. Their connectivity and networking make it possible to store big data in cloud storage, which is then used, for example, to analyze or monitor the progress of a given process in automated production. By applying the approaches of the concept of intelligent industry 4.0 to these existing solutions, we arrive at modern requirements oriented to business conditions, technologies, and innovative environments. Deploying a collaborative robot, however, can also be

M. Vagas (✉) · A. Galajdova · P. Sarga · R. Rakay · J. Romancik
Department of Industrial Automation and Mechatronics of the Faculty of Mechanical Engineering, Technical University of Košice, Košice, Slovakia
e-mail: marek.vagas@tuke.sk; alena.galajdova@tuke.sk; patrik.sarga@tuke.sk; robert.rakay@tuke.sk; jaroslav.romancik@tuke.sk

M. Balog et al. (eds.), *EAI International Conference on Automation and Control in Theory and Practice*, EAI/Springer Innovations in Communication and Computing, https://doi.org/10.1007/978-3-031-31967-9_19

perceived as counterproductive. Its synonym and characteristic are low speed (250 mm/s) or limitation of peak current (protection against impacts, overload). However, this increases the cycle time (e.g., the time of component assembly, production, etc.) and simultaneously requires that the payload be as small as possible [3]. However, the advantage is its high safety and positive environmental impact. On the other hand, we should also consider that a classic end effector has a severe risk of injury to the human operator. Therefore, we do not meet the condition of safety and collaboration for the workplace imposed on such types of workplaces.

Therefore, the overall and complete design of an automated workplace with the participation of a collaborative arm can be called a kind of "collaborative solution" only if all elements or devices are of a collaborative nature. The advantage of deploying such complete collaborative solutions is their versatility and easy adaptation to other applications [4]. The critical term is, therefore, application because some tasks are not even suitable for collaboration alongside a human, even if a collaborative robot will perform them. Collaborative robots can perform both collaborative tasks and the tasks of a standard industrial robot [5]. We can determine the difference between these by the application's deployment method, the form and difficulty of programming, or the exact chronology of the implementation of the given task [6]. Even though these are collaborative applications, they must be subjected to honest assessment and risk assessment just like standard automated and robotic workplaces.

2 Method for the Experiment

We can assume the compilation of the methodological design for collaborative assembly activity based on our previous experience and knowledge from the design and operation of similar collaborative applications, as well as concerning other realized applications of a similar nature or technology or according to Fig. 1.

One of the determining factors for the design is the concentration of individual elements of the assembly object in the workspace of the collaborative robot and operator [8]. The construction of such a scenario depends on general methodologies

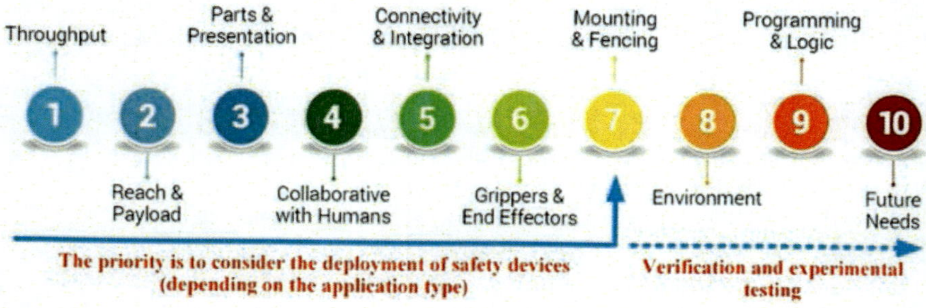

Fig. 1 The example of steps for automated workplace proposal [according to 7]

for assembly technology using an arm (or two arms, in the case of a two-handed type) of a robot and a person, considering the specification of assembly activities to the assembly object [9]. For a correct and problem-free course of the collaborative assembly activity, we consider it justified to emphasize and give due importance to the basic setup of the collaborative application of an assembly nature before it is launched in the "sharp" operating mode.

Furthermore, practical experience has shown that noncompliance with instructions and recommendations can significantly impact downtime caused by human service, e.g., by finding the right parts of the assembly object on the workbench with hand tools. Therefore, considering these requirements, we recommend that cooperative operation is carried out only by staff trained for this purpose and adequately instructed about the safety and risks of this collaborative workplace [10]. Protection is always critical in implementing collaborative assembly activities. It means whenever a robot and a person must work on one joint action within one workspace. Traditional barriers are mostly not necessary. Instead, technological protective devices are more often used, which must be subjected to careful analysis to minimize the risk at the workplace. Collaborative application security planning is essential to security strategy and identification of potentially dangerous situations, although it is sometimes confused with reactive control techniques. Even a person in such a workplace must gain some sense of security in every state of collaborative operation [11].

We must find a correlation between the robot's behavior and possible human damage to prevent injuries. Therefore, it is usually necessary to ensure that the robot does not exceed the maximum safe zone and, thus, does not meet a person. Furthermore, a responsibly implemented selection of the optimal assembly structure of the assembly object presupposes the use of already-known knowledge and methodologies to design assembly activities. Therefore, the proposal is also based on learning from system approaches from the design and operation of similar collaborative applications [12]. Based on the mentioned facts, we can define the following criteria:

- The number of changes in position and orientation of individual parts of the assembly object.
- Automated assembly time.
- The number of tools and equipment.
- The complexity of the assembly activity.

In the initial phase of the creation and selection of suitable assessed criteria, it was necessary to develop a philosophy that optimizes the conditions of this assembly process concerning changes in the position and orientation of individual parts of the assembly object during the process, Fig. 2.

The stages of assembling a suitable object assembly structure include the requirement of the feasibility of the assembly sequence as evidence that the object can be "assembled" together with criteria that support this idea (such as energy consumption, total assembly time, etc.). The last stage in designing a suitable structure is the possibilities and methods of using the simulation of automated projection systems

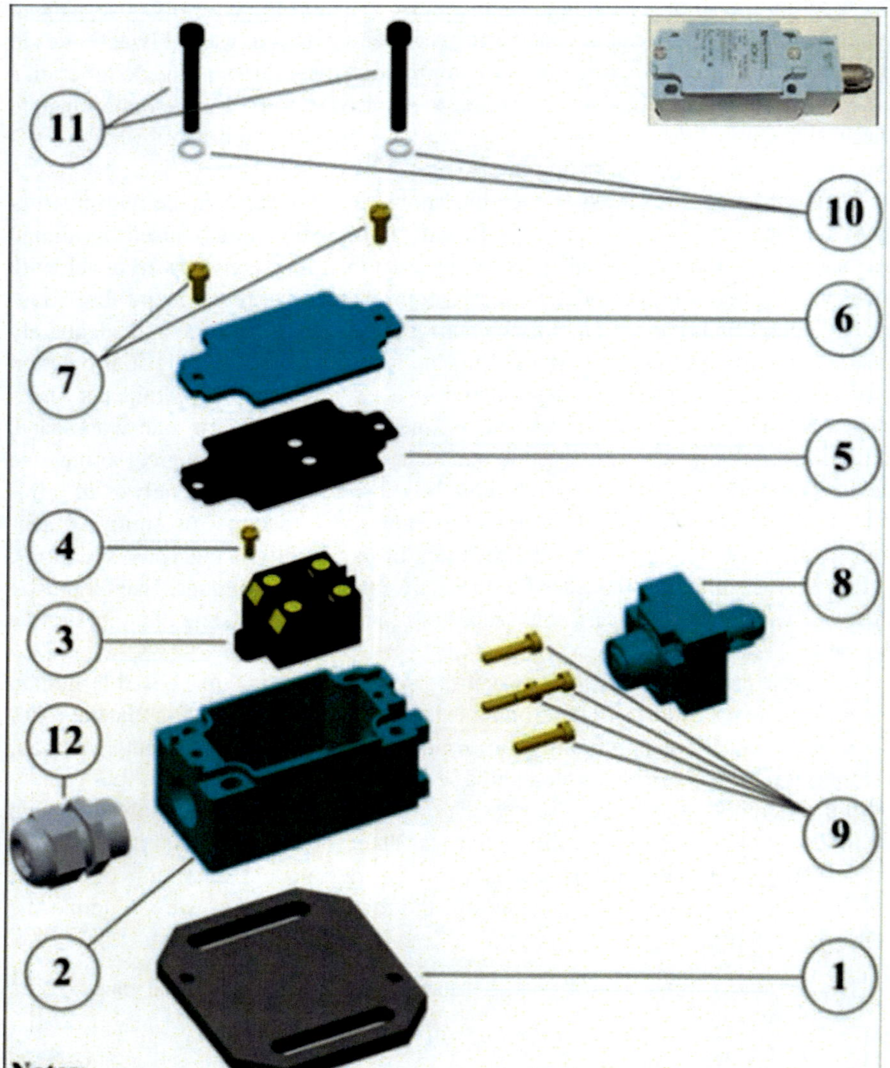

Fig. 2 The assembly of the industrial limit switch in several steps

in the form of CAD/CAM techniques. It enables the creation of an assembly scenario in such a way that it comes as close as possible to the optimal assembly structure of the assembly object, considering its complexity or the number of tools in the assembly workplace [13].

3 Case Study

When designing a collaborative application of an assembly nature, the assumption of the most negligible risk of mutual collision was considered, which was verified using offline programming systems with an emphasis on collision detection. In the first phase of the solution, we mainly worked with the CAD system, and for feasibility, individual movement simulations were tested while the actual application was gradually built.

3.1 Layout

The collaborative application is supposed to substitute for single-purpose applications requiring greater flexibility. However, the vast possibilities of this application require mastery in the form of a design solution based on the modular principle, which enables the implementation of several variant dispositions, where the basis of the whole thing is intelligent control [14]. The collaborative application (Fig. 3) consists of a two-armed robot, which includes a working table and an operator

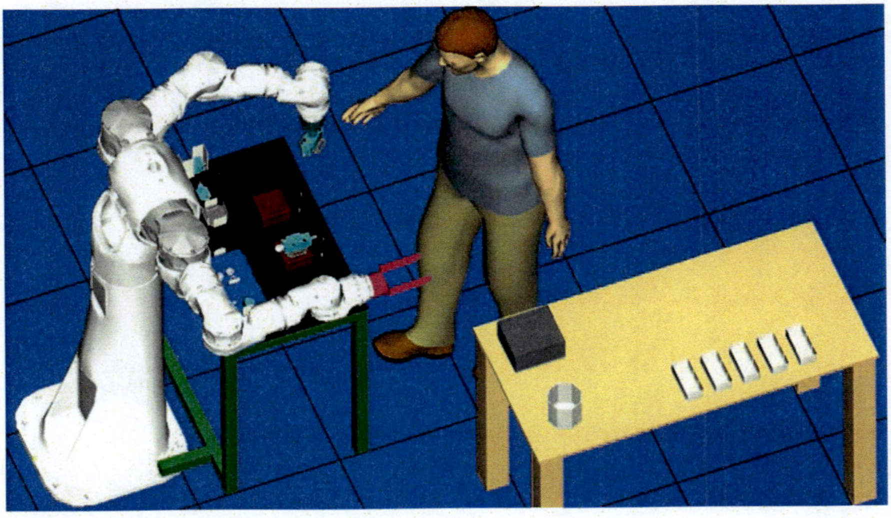

Fig. 3 Layout of the proposed collaborative workplace

(human). This person controls activities using the control panel on the adjacent working table, where selected parts of the assembly object are located.

3.2 Participated Devices

Dual arm robot. The primary device of the collaborative assembly application is a two-armed robot from MOTOMAN, type SDA10F. Its main area of application is assembly, packaging, and handling processes. The workspace and dimensions of the robot are shown in Fig. 4.

Deploying a collaborative robot in an assembly application requires perfect preparation both in the field of design and in the field of software equipment. Therefore, the role of the collaborative robot will primarily be feeding and preparation activities, which will facilitate the operator's work and speed up the overall time of the assembly cycle. End effectors. The two-armed robot is equipped with appropriate effectors.

The choice of effector depends on various factors, such as the mounting object's size, shape, and weight. Since in the collaborative workplace, the arm of the R1

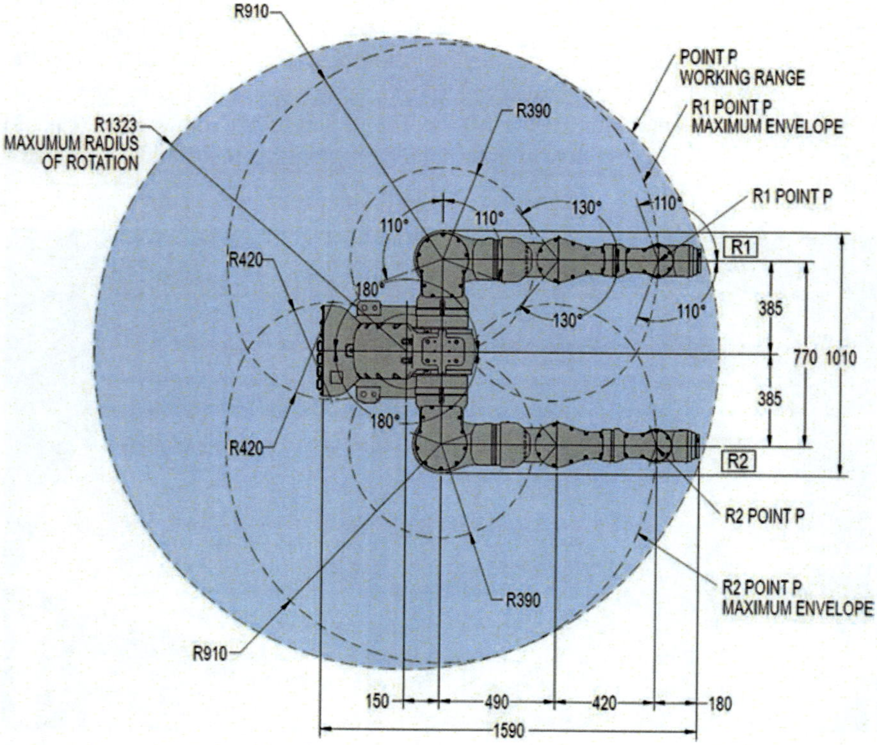

Fig. 4 Dual arm operating and working workspace [15]

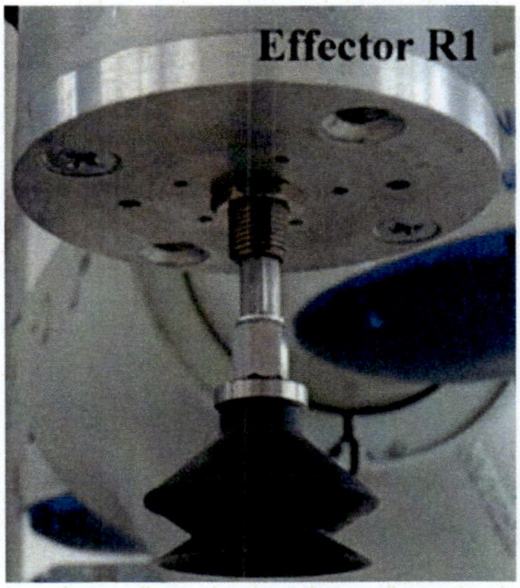

Fig. 5 The end effector for robotic arm R1

Table 1 Conditions for calculating a sufficient diameter of the effector suction cup R1

Weight of the assembly object	$m = 0.75\text{Kg}$
Safety factor	$s = 4$
Achieved ejector vacuum	$P = 52.3$ kPa
Number of suction cups	$n = 1$
Gravitational acceleration	$g = 9.81$ m/s^2

robot will grasp parts of the assembly object of different shapes and sizes with a relatively flat surface, a vacuum effector with a vacuum suction cup was chosen, Fig. 5.

If we need to determine the sufficient diameter of the suction cup, it is necessary to perform the calculation according to Table 1, which we implement for the most unfortunate case. It is the case where the robot lifts finished objects and gives them to the operator, while we consider that the assembly object weighs 0.75 kg.

$$D = \sqrt{\frac{4 * m * g * s * 1000}{\pi * p * n}} \tag{1}$$

$$D = \sqrt{\frac{4 * 0.75 * 9.81 * 4 * 1000}{\pi * 52.3 * 1}} = 26.8\,\text{mm} \tag{2}$$

Control of the effector is provided by a universal output on the robot's control system. The rule for activation and deactivation is defined by ON and OFF commands. Activation means deriving negative pressure and grasping the components of the assembly object. Since the robot's primary task of the R2 arm is manipulating and feeding magazines with its parts, a simple structural adjustment was proposed, enabling a sufficiently rigid attachment of the magazines, Fig. 6.

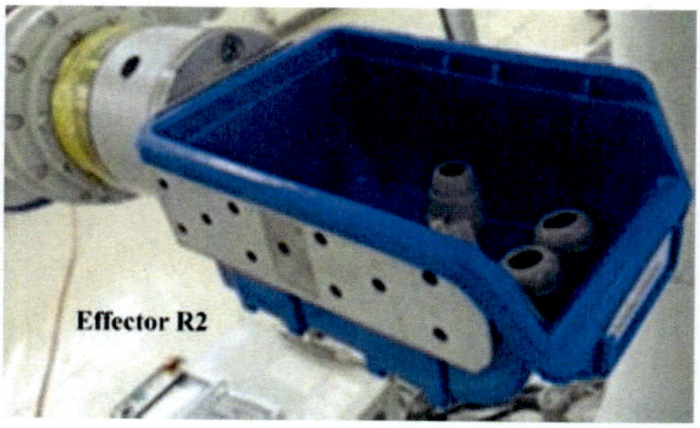

Fig. 6 The end effector for robotic arm R2

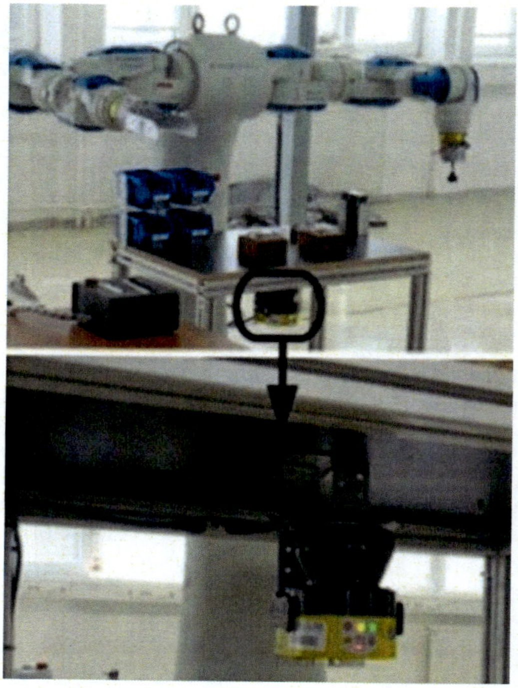

Fig. 7 2D safety laser scanner

2D laser scanner. Since the two-armed robot does not have directly integrated safety sensors, we approached the possibility of using an external safety device as a 2D laser scanner, Fig. 7.

With the small 2D laser scanner S300 mini, we can monitor up to three (one protection and two warning) security zones contactless, which can be programmed using the CDS software by connecting to a computer. The advantage is setting individual parameters, such as the zone's shape, the range of scanned zones, the response, and many other parameters.

Working tables. Directly in front of the robot is a working table with containers for individual parts of the assembly object, which have a precisely defined position during the assembly process. In its lower part, there is a 2D laser scanner. The table dimensions are 750 × 500 × 800 mm, Figs. 8 and 9.

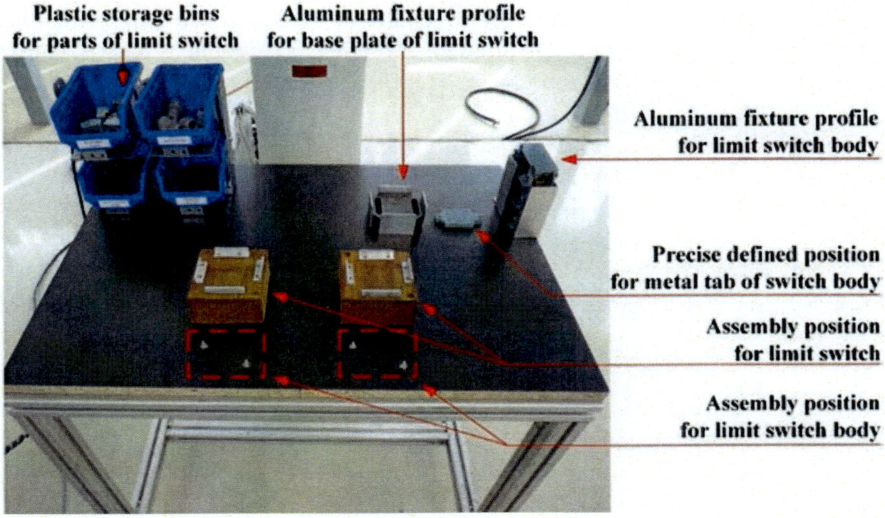

Fig. 8 Working table for the robot

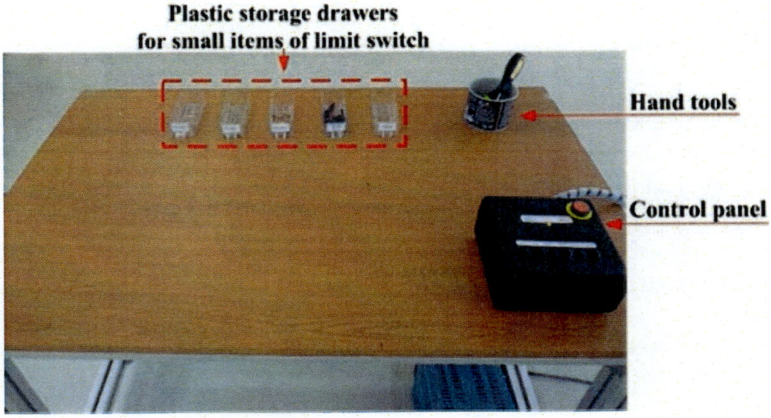

Fig. 9 Working table for the operator

4 Discussion and Conclusion

Based on the cooperative workplace concept, we construct the actual workplace. The requirements at the workplace were considered within the mechanical, electrical, and software structure and the implementation of the human–robot cooperation interface. In the context of mechanical design, mechanical interfaces were constructed and produced. Furthermore, we manually make the nonpurchasable components like the jaws of grippers and other devices. The electrical structure includes connections for the 2D laser scanner of the robot control system, the same as for the peripheral control panel. Regarding the software, we use a programmable unit for information processing and program synchronization at the whole assembly cooperative workplace. To ensure the safety of humans cooperating with robots, a 2D laser scanner that provides information on the state of the assembly collaborative workplace will be implemented to slow and, if necessary, stop the processes. Then, we experimentally tested three possible statuses of the 2D laser scanner:

4.1 The First Status Consists of Decreasing the Robotic Arm's Speed

When the marked zone is interrupted (Fig. 10), the robot's control system activates the "Speed Limit" function and mode. We can characterize this mode by limiting the robot's speed to a constant value of 250 mm/s. An orange LED indicates the mode on the control panel (left part of the picture). It also ensures that the operator is sufficiently informed about the state of the collaborative application.

4.2 The Second Status Consists of the Robot Stops

Suppose the operator approaches the 2D laser scanner to a value smaller than 500 mm. In that case, the robot stops its operation completely, Fig. 11. Stopping is necessary so that the operator cannot enter the robot's workspace during the automatic process and thus prevent possible unwanted contact or injury. The relevant zone is drawn in red. The control panel detects its crossing. The stop is realized thanks to the "External Hold" function.

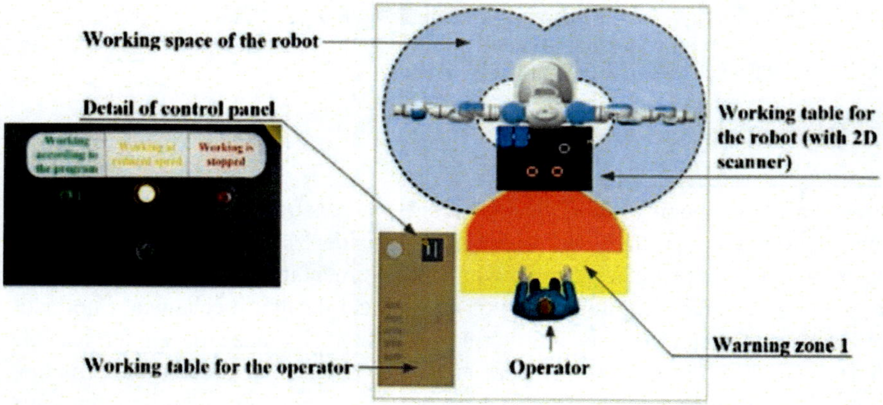

Fig. 10 Activating the "Speed Limit "function

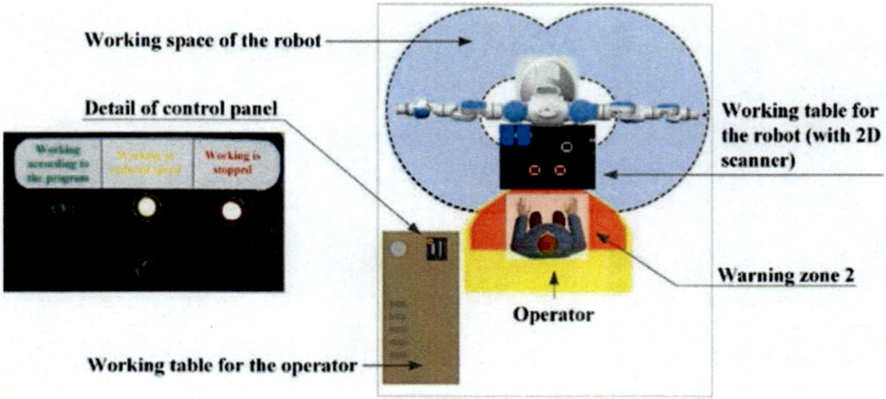

Fig. 11 Activating the "External Hold" function

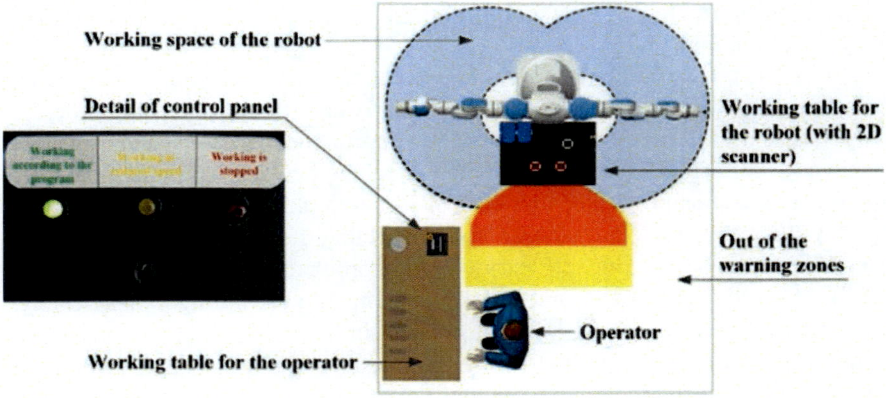

Fig. 12 Activating the "External Start" function

4.3 The Third Status Consists of Working a Full Speed Again, According to the Program

To restart the robot according to the program activity, we must leave the marked zones in Fig. 12. At the same time, the logical "1" value responsible for the "External Start" function is set on the external input DIN 01, which again allows the start (continuation) of the robot program.

Restarting the collaborative application is conditioned by manually turning on the servo motors on the robot's programming unit. The control of the assembly process and the operating status of the two-armed robot is also indicated by the corresponding light signal of the 2D laser scanner. At the same time, orange (and one red strip, which means entering the warning zone 1) means the speed limitation of the two-armed robot. The orange color (and two red stripes, which indicate the

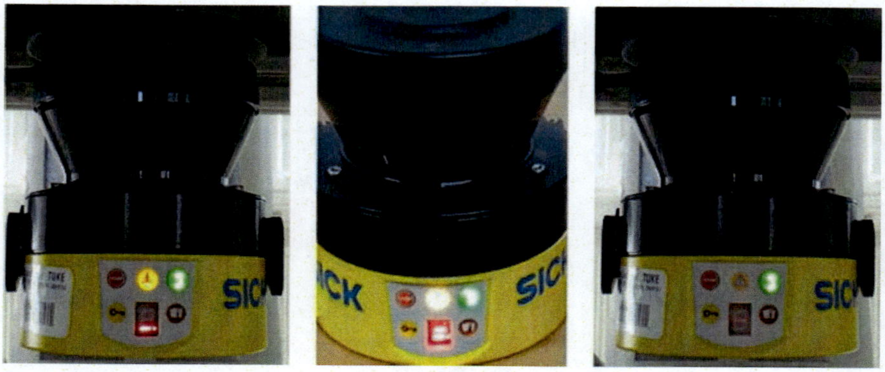

Fig. 13 Three possible statuses of the 2D laser scanner

entry into warning zone 2) show a complete stop of the robot. A green color (without an indication of red lines) signals the normal movement of the robot according to the program, Fig. 13.

A novel approach to human–robot collaboration was developed. A two-armed industrial robot was used in this case for the first time to classify possibilities of cooperation near a person. Furthermore, co-development between human–robot teamwork and their safety was studied and matching between assembly processes and synchronization proved the existence of co-development between them.

This study was limited to several factors affecting the cooperative workplace's overall assembly time, but at least two are problematic. First, the two-arm MOTOMAN SDA10F robot cannot work at maximum speed during joint assembly. Since he works with a person, it is necessary to limit his speed and working space from the point of view of safety. The second factor is the task-oriented assembly process, i.e., the robot continues assembly only when the human operator gives information about the completion of the assembly subtask and an instruction to continue.

We want to implement a complex 3D camera system in the future in the cooperative workplace to make a closer connection with the robot's control system. This connection will better ensure purposeful testing and implementation of assembly applications requiring image recognition of objects. On the other hand, the fact that slows down the assembly process is the manual supply of all parts of the limit switch to the cooperative workplace by human operators. Therefore, it is advisable to eliminate this disadvantage by automating it using conveyors and containers. Considering all these notes, the presented project provides an efficient solution to face the initially mentioned and realized human–robot cooperation projects.

Acknowledgments This work has been supported by the Slovak Grant KEGA 044TUKE-4/2021-Remote access to laboratory exercises for industrial automation, KEGA 031TUKE-4/2022 Innovation of methods and instruments in the education of automation in the context of Industry 4.0, and Slovak Grant VEGA 1/0169/22 New methodics approaches to data from automated and robotized workplaces.

References

1. Virgala, I., et al.: Power and force limiting technique at collaborative workplace. MMS Journal. **2021**, 4424–4427 (2021)
2. Vargas, R.A., et al.: Introduction and configuration of a collaborative robot in an assembly task as a means to decrease occupational risks and increase efficiency in a manufacturing company. Robot. Comput. Integrat. Manufact. **57**, 315–328 (2019)
3. Bozek, P., et al.: Methods, models, algorithms for diagnostics of mechatronic systems. Stud. Syst. Decis. Contr. **345**, 17–26 (2021)
4. Grushko, S., et al.: Intuitive spatial tactile feedback for better awareness about robot trajectory during human-robot collaboration. Sensors. **21**(17), 5748–5774 (2021)
5. Vagas, M., et al.: Application of speed and separation monitoring technique at automated assembly process. MM Sci. J. **2021**, 4420–4423 (2021)
6. Cini, F., et al.: The relevance of signal timing in human-robot collaborative manipulation. Sci. Robot. **6**(58), eabg1308 (2021)
7. Universal robots' eBook, downloaded and available at: https://www.universal-robots.com/sg/ebook-get-started-with-cobots/
8. Holubek, R., Ruzarovsky, R.: The methods for increasing of the efficiency in the intelligent assembly cell. Appl. Mech. Mater. **470**, 729–732 (2013)
9. Jimenez, P.: Survey on assembly sequencing: a combinatorial and geometrical perspective. J. Intell. Manufact. **24**(2), 235–250 (2013)
10. Dado, M., et al.: Internet of things as advanced technology to support mobility and intelligent transport. In: International Internet of Things Summit, pp. 99–106. Springer, Cham (2016)
11. Halenar, I., et al.: Virtualization of Production Using Digital Twin Technology, pp. 1–5. 20th International Carpathian Control Conference (ICCC) (2019)
12. Kulic, D., Croft, E.: Safe planning for human-robot interaction. In: IEEE International Conference on Robotics and Automation, pp. 1882–1887 (2004)
13. Beniak, J., et al.: Research on parameters optimization for the Additive Manufacturing process. In: 13th International scientific conference on sustainable, modern and safe transport. Transcom 2019. **40**, 144–149 (2019)
14. Saukkoriipi, J., et al.: Programming and control for skill-based robots. Open Eng. **10**, 368–376 (2020)
15. Yaskawa Europe GmbH. Robot Assistants Collaborative and Easy-to-Use (2021). Available on:https://www.yaskawa.it/yaskawa.it/dowloads/YEUR_RobotAssistants_EN_06.2021.pdf

Using Mathematical Modelling for Experimental Obtaining of Initial Data when Creating Digital Twins of Complex Technical Products

Dmitriy Volontsevich, Ievgenii Veretennikov, Vadym Karpov, and Vratislav Hladký

1 Introduction

Already in the mid-1980s, technologies for Continuous Acquisition and Lifecycle Support (CALS) of complex technical products began to develop, now widely spread in many countries of the world.

Like many other achievements of scientific and technological progress, the CALS concept was primarily developed for the design, production, operation and disposal of military products. In previous years, CALS concepts have become widespread in the defense industry and military-technical infrastructure of NATO countries.

The result of the implementation of the CALS concept is the minimization of costs during the entire life cycle of an innovative engineering product and an increase in its quality and competitiveness.

The issue of implementing the CALS concept for enterprises producing high-tech innovative engineering products is no longer a matter of competitiveness but of survival. In developed countries, it has been long enough, and soon throughout the world, products that are not accompanied by integrative electronic documentation and do not have the means of integrated logistics support for the postproduction part of the life cycle will not be in demand among consumers [1].

One of the most important achievements in the implementation of the CALS concept and the related Industry 4.0 is the use of digital twins.

D. Volontsevich · I. Veretennikov · V. Karpov
National Technical University "Kharkiv Polytechnic Institute", Kharkiv, Ukraine

V. Hladký (✉)
Department of Industrial Engineering and Informatics of the Faculty of Manufacturing Technologies with a seat in Prešov, Technical University of Košice, Prešov, Slovakia
e-mail: vratislav.hladky@tuke.sk

M. Balog et al. (eds.), *EAI International Conference on Automation and Control in Theory and Practice*, EAI/Springer Innovations in Communication and Computing, https://doi.org/10.1007/978-3-031-31967-9_20

A study [2] shows that digital twins reflect real objects with their data, functions, and communication capabilities in the digital world. They are the nodes of the Internet of Things and, in the process of providing networking, also provide automation of complex value chains. The use of modeling tools enables the use of digital twins for experimentation, so basic digital twins become experimental digital twins. In the first stage, experimental digital twins interact with each other in the virtual world. The networks resulting from interacting experimental digital twins run various scenarios on virtual test benches, thereby providing information for simulation-based systemic, integrated design. As a result, complete digital analogs of the corresponding real objects and their behavior are consistently created. The network of experimental digital twins, together with real objects, allows you to create hybrid systems in which experimental digital twins are used on an equal footing and in combination with real equipment [1].

The issues of creating experimental twins of high-tech engineering products need further development, as this is a very new and promising area of research.

Mathematical modeling in the development of complex science-intensive technical products, especially military equipment, has been widely used in Ukraine since the advent of computer technology. This, in particular, in the Kharkiv Polytechnic Institute is well traced by the publication of monographs [3], textbooks [4], and other literature. Modern trends in the development of the CALS concept and the related Industry 4.0 only strengthen and structure the use of mathematical models and digital twins in the life cycle of complex science-intensive technical products in a certain way.

In some approaches, it is proposed to create a digital twin of an object or process at the stage of its development [5, 6]; in others, based on filling the profile with data from sensors at the operation stage [7].

Within the framework of this study, an approach is proposed in which a digital twin is created at the very first stage of the life cycle of a high-tech engineering product – the definition and analysis of needs. Further, it is filled with data at all stages of the life cycle.

Previously, with the classical approach, as a rule, mathematical models began to be used in technical design for structural and parametric optimization of the designed product. However, for the correct operation of a complex mathematical model, a large array of adequate initial data is needed that correctly describes the behavior of certain elements or materials under given conditions, which can be very difficult to find in the reference literature and requires additional physical experiments. Therefore, the purpose of this work is to substantiate the feasibility of using mathematical modeling together with a physical experiment to simplify and reduce the cost of the process of obtaining initial data when creating adequate digital twins. The substantiation is carried out on the example of experimental determination of the damping coefficient of the track roller tire for a tracked vehicle.

2 Literature Review

There are a number of methods for determining the damping properties of the system. So, for example, in the author's certificate [8] a method is described in which the elastic suspension of a mechanical object acts as an object, which is installed on a vibrating table and the corresponding level of resonant vibrations is set. With the help of strain gauges, the movements of the object and the table are measured, and the damping coefficient is calculated from the ratios of the oscillation amplitudes. The disadvantages of this method are the complexity of measurements, for which it is necessary to have special equipment, and the high qualification of the experiment performer.

Another author's certificate [9] describes a method for determining the damping coefficient using a vibration machine consisting of a chute mounted on an elastic suspension and having a vibration exciter of any type. Above the chute is a load with the possibility of dropping it from different heights. When dropping, the damping time of free oscillations is fixed. When calculating the damping coefficient, the ratio of the drop heights and the difference in the damping time of the oscillations are taken into account. The disadvantages of this method are the complexity of the stand and the inaccuracy of determining the damping time of oscillations, since it is quite difficult to determine the moment of complete damping of oscillations during each measurement.

The work [10] describes a method for determining the damping properties of a pneumatic tire using a lever-type stand. The stand is equipped with a dynamometric platform, a light-beam oscilloscope, and a displacement sensor. The wheel is subjected to stage-by-stage loading with fixation, followed by the recording of the elasticity characteristics, as well as the recording of the damped oscillations of the wheel axle with an oscilloscope. Like the previous methods, this method requires precise expensive equipment and highly qualified personnel to perform the experiment.

The papers [11, 12] describe the research method based on the EUSAMA methodology. It makes it possible to obtain dissipative characteristics of the wheel or suspension as a whole but also requires relatively expensive specialized experimental equipment.

The paper [13] describes an approach to a fairly accurate determination of the dissipative characteristics of an elastomeric suspension of a car, but it also requires expensive specialized experimental equipment and qualified personnel performing experiments.

The closest to method proposed in this article for determining the damping properties of a rubber tire is the method described in [14]. The method includes the operation of a free fall of a calibrated load with a wheel onto the supporting surface from the undeformed state of the pneumatic tire to the state of maximum normal deformation, as well as the operation of measuring the amount of normal deformation when a wheel falls and the operation of measuring the time of the wheel falling. In addition, an additional operation is carried out to determine the average speed of the fall of the wheel on the supporting surface. At the same time, to calculate the damping coefficient (viscous resistance), a theorem is used about the change in kinetic energy for a wheel that falls on the supporting surface from the position of normal zero deformation to the position of maximum dynamic deformation recorded by a high-speed video camera. The disadvantage of this method is the difficulty of fixing the deformation of the wheel, which is associated with the use of a high-speed video camera. In addition, to calculate the damping coefficient according to the kinetic energy theorem, it is necessary to determine the rate of complete deformation of the wheel, which leads to an increase in the requirements for this video recording.

Thus, it can be said that the vast majority of studies on the smoothness of the ride and vibration loading of a car, as well as the assessment of the absorbing capacity of a tire, it is carried out using a viscous model of the inelastic resistance force, which assumes a linear relationship between this force and the rate of radial tire deformation [10]. In addition, all the analyzed methods require precise expensive equipment and highly qualified personnel. And this is especially hard on small manufacturing enterprises that cannot afford large expenses for these purposes [15].

3 Research Methodology

The proposed technique for using mathematical modeling for experimental obtaining of initial data is that a previously created and verified mathematical model is found that describes a more technologically simple experiment and which includes the unknown characteristic of the material or product we need as a whole. Further, this more technologically simple physical experiment is carried out, and it is also carried out on the found mathematical model. When comparing the results of physical and numerical experiments, we get the opportunity to determine a previously unknown characteristic of our material or product as a whole.

The described technique is implemented on the example of the experimental determination of the damping coefficient of the track roller tire for a tracked vehicle.

Consider a flat mathematical model that simulates a wheel falling from a fixed height onto a rigid flat surface.

The equation for the vertical movement of the wheel is written as its vertical acceleration, which has a positive sign when moving up:

$$\ddot{z} = \frac{P_{sp} + P_d - m_w g}{m_w}, \tag{1}$$

where m_w – the mass of the wheel;

g – acceleration of gravity;

P_{sp} – springy force of deformation of the wheel only in contact with a solid surface, determined by the equation $P_{sp} = -c_w(z - r_w)$ and if $P_{sp} < 0$, then $P_{sp} = 0$;

P_d – damping force of wheel deformation only in contact with a solid surface, determined by the equation $P_d = -k_w\dot{z}$. At the same time, if $P_{sp} + P_d < 0$, then $P_{sp} + P_d = 0$.

Here in the formulas:

c_w – wheel tire stiffness coefficient;
r_w – undeformed wheel radius;
z – the height of the wheel geometric center above the surface;
$\dot{z}$ – wheel vertical speed;
k_w – desired wheel damping coefficient.

This variant is intended for use with air-inflated tires. For them, with a sufficient degree of accuracy in the working range of deformations, it is possible to take a directly proportional dependence of the elastic force on the magnitude of the deformation. For track rollers of tracked vehicles with solid rubber tires, the quadratic dependence of the elastic force on the amount of deformation should be taken. This is due to the significantly changing volume of rubber involved in the formation of the elastic force of resistance to deformation. And unlike the air pressure, which is the same throughout the tire, in solid rubber, there is almost no transfer of the stress state to the undeformed layers. Accordingly springy force of deformation of the wheel only in contact with a solid surface, determined by the equation: $P_{sp} = -a_w(z - r_w)^2 - b_w(z - r_w)$ and if $P_{sp} < 0$, then $P_{sp} = 0$.

We accept as initial conditions. $\dot{z} = 0$ and $z = h$ – the height from which the wheel is thrown. The value of k_w in the first approximation is taken arbitrarily. After double integration using any software, we determine the height of the first jump of the wheel and compare it with the result of a full-scale experiment. Next, we change the value of k_w until the height of the wheel's first jump in the numerical and natural experiments do not match.

In the study, to simulate the fall of a track roller on a flat solid surface, based on the described mathematical model, a program was written based on a standard program for integrating a system of differential equations using the Runge-Kutta method with a constant step.

This program, as well as the standard program for integrating a system of differential equations, was implemented in the Borland Pascal programming language in the Delphi 7 environment.

To experimentally determine the height of the first rebound of the wheel after its free fall on a hard surface, we use a rather simple experimental stand, schematically shown in Fig. 1. The wheel (1) is fixed in a vertical position relative to the surface with a flexible rope (2) and suspended at a fixed height through a block (3). After that, the wheel falls freely on a hard surface. With the help of a video camera (5) and a measuring scale (4), the maximum height of the first rebound of the wheel is recorded. The lens must be mounted at a height close to the height of the first rebound of the wheel to avoid the effect of optical distortion. For greater accuracy, it is necessary to take several measurements with different initial wheel lift heights.

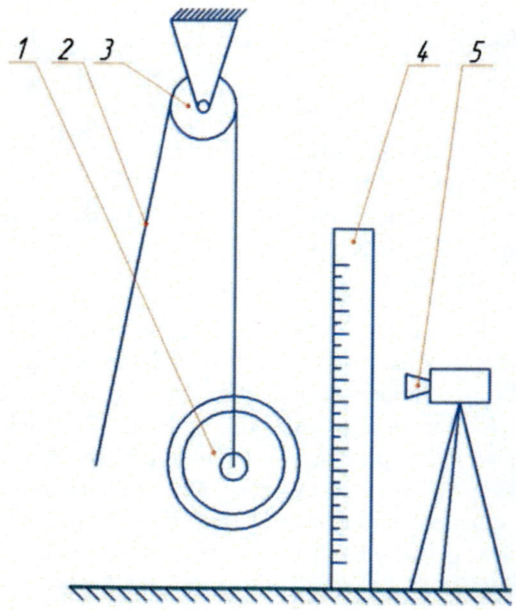

Fig. 1 Scheme of the experimental stand: 1 – wheel, 2 – flexible rope, 3 – block, 4 – measuring scale, 5 – video camera

4 Results

The full-scale experiment was carried out with the track roller of the MT-LB transporter-tractor. When the experiment was carried out according to the above algorithm, an array of data was accumulated, which is presented in Table 1.

In accordance with the data obtained in the mathematical model, the damping coefficient was changed until the maximum height of the first wheel rebound in the numerical experiment did not reach the average rebound value obtained in the full-scale experiment.

The results of mathematical modeling are presented in Table 2.

A graph illustrating an example of the change in the height of the track roller from time to time when falling from a given height and rebounding from a solid, even horizontal surface and an arbitrary value of the damping coefficient is shown in Fig. 2.

Fig. 3 shows the time dependence of the change in the value of the deformation of the rubber bandage, elastic and damping forces arising in the contact zone of the track roller with the surface.

Based on the results obtained, the average value of the wheel damping coefficient k_w is $2433\frac{N \bullet s}{m}$.

The average error between the experiment and the numerical method for the rebound height is 3.2%.

Table 1 Experiment results

Distance from the center of the wheel to the surface, mm	Experiment number	Rebound height, mm
635	1	525
	2	535
	3	515
	Mean	525
935	1	635
	2	665
	3	655
	Mean	651
1235	1	825
	2	845
	3	835
	Mean	835
1535	1	1015
	2	945
	3	975
	Mean	978

Table 2 The results of mathematical modeling

Distance from the center of the wheel to the surface, mm	Rebound height, mm	Wheel damping coefficient
635	490	2370
935	658	2460
1235	838	2390
1535	1020	2510

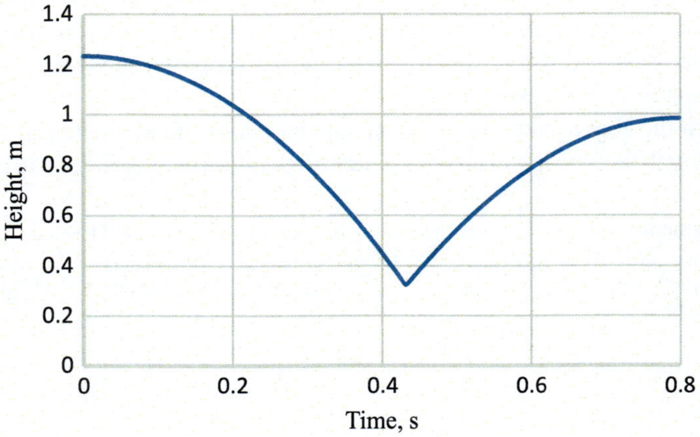

Fig. 2 Change in the height of the track roller from time

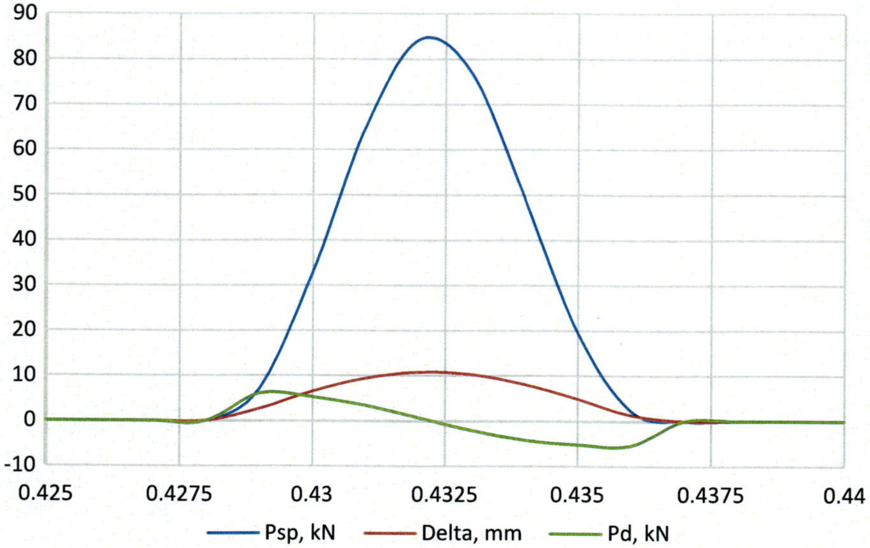

Fig. 3 Time dependency of the deformation of the rubber bandage, elastic and damping forces

5 Conclusion

The proposed method makes it possible to accurately determine the damping coefficient both for wheels with a pneumatic tire or for wheels with a rubber tire without determining their chemical composition and without the need to use expensive experimental equipment. To confirm this conclusion, we used the results of experiments conducted in the 1970s at VNIITransMash (USSR) according to the methods described in [8, 10]. The results of these experiments were not published, however, they were included in the calculation tables for modeling the Kharkov Tractor Plant, as the manufacturer of the specified conveyor-tractor. The discrepancy between these tabular data and the results obtained by the proposed method in the entire range of tests did not exceed 5%.

Based on the above materials, a patent application was filed in Ukraine on behalf of the National Technical University "Kharkov Polytechnic Institute" [16].

Acknowledgments This work was supported by the Slovak Research and Development Agency under contract No. APVV-19-0590, and also by the projects VEGA 1/0268/22, KEGA 038TUKE-4/2022 granted by the Ministry of Education, Science, Research and Sport of the Slovak Republic.

References

1. Volontsevich, D., Skvorchevsky, A.: Restoration and transformation of high-tech machine building industry by implementing the principles of the calsconcept in the context of industry 4.0 development. Eastern-Eur. J. Enterprise Technol. **3**(1-117), 15–24 (2022). https://doi.org/10.15587/1729-4061.2022.260045
2. Schluse, M., Priggemeyer, M., Atorf, L., Rossmann, J.: Experimentable Digital Twins–Streamlining Simulation-Based Systems Engineering for Industry 4.0. IEEE Transact. Industr. Informat. **14**(4), 1722–1731 (2018). https://doi.org/10.1109/tii.2018.2804917
3. Alexandrov, Y., Volontsevich, D., Karpenko, V., Lebedev, A., Peregon, V., Samorodov, V., Turenko, A.: Dynamics of Transport and Traction Wheeled and Tracked sVehicles, 1st edn. Publishing house KhGADTU (KhADI), Kharkiv, Ukraine (2001). (in Russian)
4. Alexandrov, Y., Volontsevich, D., Dushchenko, V., Yepifanov, V., Kokhanovsky, N.: Mathematical Modeling of the Perturbed Movement of Armored Vehicles Aggregates and Systems: Textbook, 1st edn. Publishing house NTU KhPI, Kharkiv, Ukraine (2012). (in Russian)
5. Melesse, T., Di Pasquale, V., Riemma, S.: Digital twin models in industrial operations: a systematic literature review. Procedia Manufact. **42**, 267–272 (2020). https://doi.org/10.1016/j.promfg.2020.02.084
6. Suhail, S., Hussain, R., Jurdak, R., Hong, C.: Trustworthy digital twins in the industrial internet of things with Blockchain. IEEE Internet Comput. **26**(3), 58–67 (2021). https://doi.org/10.1109/MIC.2021.3059320
7. Huang, Z., et al.: A Survey on AI-Driven Digital Twins in Industry 4.0: Smart Manufacturing and Advanced Robotics. Sensors. **21**(19), 63–40 (2021). https://doi.org/10.3390/s21196340
8. Tyulenev, V. A.: Method for Determining Energy Dissipation in Materials During Vibrations. Patent of SU №200259, Int.CL G01H11/16, 1966. Date of publication: 29.07.1966, Bull. 16. https://patents.su/?search=%E2%84%96+200259&type=number
9. Varsanofev, V., Varsanofev, D.: Method for determining the damping coefficient of the elastic suspension of a mechanical object. Patent of SU No 911170, Int.CL G01H11/00, 1982. Date of publication: 07.03.1982, Bull. No 9. https://patenton.ru/patent/SU911170A1
10. Lomakin, V., Cherepanov, A., Vermeuk, V.: Study of the elastic and damping characteristics of passenger car tires at the stand. Automot. Ind. **8**, 25–26 (1976) (in Russian)
11. Mazurek, I., Klapka, M.: method of measuring damping ratio of unsprung mass of half axles of passenger cars using a suspension testing rig without disassembling. Patent EP No 3193152A1, Int.CL G01M17/04, 2017. Date of publication: 19.07.2017, Bull. No 29. https://patents.google.com/patent/EP3193152A1/en
12. Stanislav Koláček, S., Dostál, P.: Evaluation of dampers using a resonance adhesion tester. Acta Universitatis Agriculturae et Silviculturae Mendelianae Brunensis. **61**(6), 1749–1753 (2013). https://doi.org/10.11118/actaun201361061749
13. Stepanov, E., Maleev, S., Musarsky, R., Ogorodnov, S.: Experimental and calculation method for determining the parameters of a car suspension with an elastic elastomeric element. IOP Conference Series. Mater. Sci. Eng. **1086**, 012023 (2021). https://doi.org/10.1088/1757-899X/1086/1/012023
14. Tarasov, V., Bojarkina, I., Degtjar, V., Bojarkin, A.: Method of Determining Coefficient of Viscous Friction of Wheels of Ground Vehicles. Patent of RU No 2578827, Int.CL G01M17/02, 2015. Date of publication: 27.03.2016, Bull. No 9. https://patenton.ru/patent/RU2578827C1.pdf
15. Dobrotvorskiy, S., Basova, Y., Dobrovolska, L., Sokol, Y., Kazantsev, N.: Big challenges of small manufacturing Enterprises in Industry 4.0. In: Ivanov, V., Trojanowska, J., Pavlenko, I., Zajac, J., Peraković, D. (eds.) Advances in Design, Simulation and Manufacturing III. DSMIE 2020. Lecture Notes in Mechanical Engineering 118–127. Springer, Cham (2020). https://doi.org/10.1007/978-3-030-50794-7_12
16. Volontsevich, D., Veretennikov, I., Karpov, V., Kostyanik, I.: Method for determining the damping coefficient of a rubber tire. Application for a patent of Ukraine a202202658, Int.CL G01M 17/02, from 25.07.2022

Calibration of Panasonic TM-2000 Welding Robot Using Simulation Software

Karol Goryl and Martin Polläk

1 Introduction

The ISO 9283 standard forms the basis for testing the performance characteristics of industrial robots. Unidirectional position repeatability is one of the most important parameters to be monitored from both a robot quality and end-customer perspective. Most manufacturers declare repeatability and position accuracy based on tests performed according to ISO 9283. In the case of Panasonic, it could not be clearly established how and according to what the manufacturer tests the industrial robots. The following factors contribute to the deterioration of position repeatability: irregular calibration of the robot axes, overloading of the arm or collisions after which the robot has not been calibrated and inspected.

The basic performance characteristics for the selection of an industrial robot include:

A. Payload: This is the weight that the robot can carry without causing deviations in other specifications, such as accuracy, during the execution of the task. The maximum load on the robot arm may be greater than the specified payload. If the load is greater than the payload, it is very likely that the robot will be less precise, have a problem with the trajectory, or have very large deviations [1, 2].
B. Range: The range can be defined as the maximum distance the robot reaches within its working space. The robot's reach is determined by the function of the joints, the length of the robot arm, and other configurations [1, 2].

K. Goryl (✉) · M. Polläk
Faculty of Manufacturing Technologies with a seat in Prešov, Technical University of Košice, Prešov, Slovakia
e-mail: karol.goryl@tuke.sk

M. Balog et al. (eds.), *EAI International Conference on Automation and Control in Theory and Practice*, EAI/Springer Innovations in Communication and Computing, https://doi.org/10.1007/978-3-031-31967-9_21

C. Accuracy: Accuracy is generally defined as the error between the simulated, i.e., the required position and the actual position of the robot end point. When using robotics, accuracy is not always paramount [1, 3].
D. Repeatability: The robot's repeatability expresses how precisely a robot can achieve the same position if the movement is repeated several times. The repeatability of the position is influenced by the accuracy of the robot. For example, when testing a robot, where the desired position is chosen and repeated 30 times, the robot will not always reach the same position, but deviations will occur, and the robot will move around from the desired point. The radius, which is formed by repeated movements, is called repeatability [1, 3].

1.1 Standard ISO 9283 as a Basis for Testing Industrial Robots

This standard describes the test methods for determining the performance characteristics of industrial robots. The standard contains 14 tests that can be used for the development, testing, or comparison of industrial robots. Each test is characterized by the number of measuring points and the number of cycles to be completed. Subsequently, the tests are described where it is determined which measurement conditions have to be complied with. Each performance test shall be performed at the maximum possible load and speed. The standard itself does not indicate which metrological system should be used to measure the characteristics, which can be seen as an advantage, since more affordable and accurate measurement systems can be used. Most manufacturers of industrial robots test according to the mentioned standard. In the specifications when choosing a robot, only one-way position accuracy is indicated by the manufacturers. Most of the characteristics are tested using an imaginary ISO cube, which is inserted into the robot's workspace so that it is in the place where the robot works most often. It is appropriate to give an example of such a cube. In Fig. 1, it is possible to see a cube on which a diagonal has been selected using points C1 to C8. In this illustrative example, these are points C1-C2-C7-C8. Subsequently, it is necessary to select measuring points, the number of which may vary depending on the measured characteristic. In this example, the measured characteristic is unidirectional position accuracy, which means that five measurement points (P1-P5) were selected [4–7].

2 Calibration Techniques for Industrial Robots

Just like any industrial mechanical device, after a certain time, robotics is subject to various changes that have arisen due to the wear of parts, and this also results in poorer accuracy and repeatability. In general, calibration is defined as a tool to improve positioning accuracy by software without the need to change the robot or control system [8, 9].

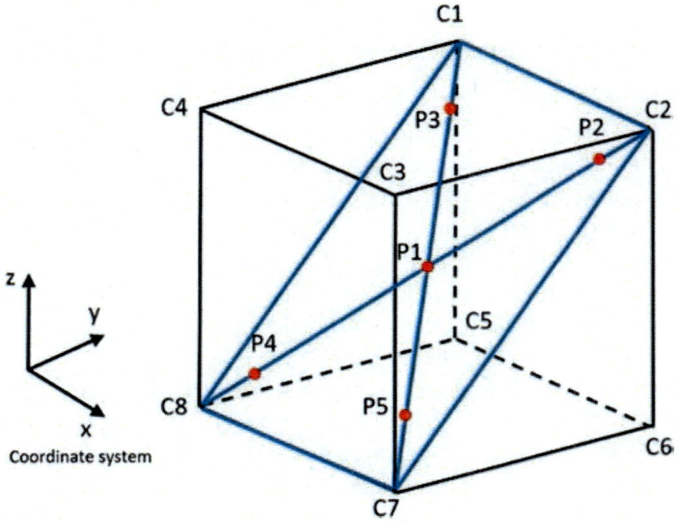

Fig. 1 Example of an ISO cube

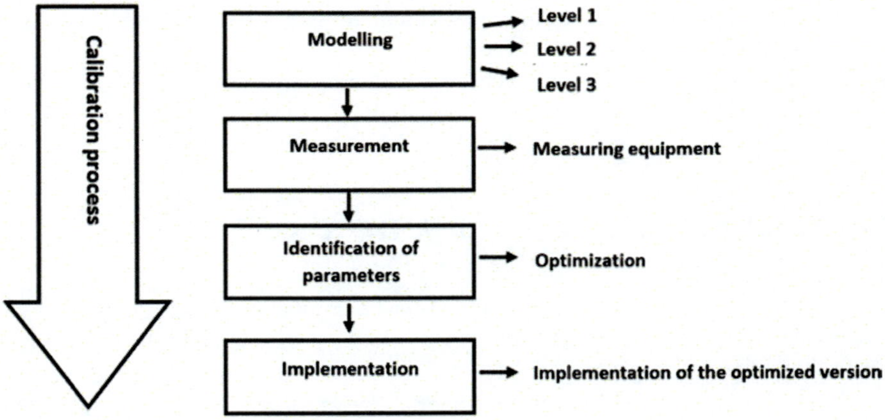

Fig. 2 Schematic of the calibration process [11]

Total calibration can be divided into four parts Fig. 2:

1. *Modeling*:

   Level 1: The objective of this level is to find and determine the relationship between the actual joint shift and the signal generated by the converter.
   Level 2: This level, also referred to as kinematic calibration, contains the overall kinematic diagram of the robot, which contains the length of joints and angular shifts.
   Level 3: The aim of the latter level is to detect nonkinematic errors in the positioning of TCP that may be caused by friction, will, and joint yields.

2. *Measuring devices*: The measuring equipment depends on the parameters of the mathematical model that need to be identified and is divided into two groups:

   A. Internal measuring devices: These are sensors that are part of the robotic system. Among the most used are Encoder and Resolver. An encoder is a device that detects angular or linear displacements. The accuracy of encoder calibration depends on its resolution. A resolver is a device that measures the movement of a rotary joint.

      The output of the resolver is a digital signal that shows the difference between the programmed and the real angle achieved. The accuracy of the device is high, but compared to an encoder that has a comparable resolution, the encoder is more affordable.

   B. External measuring devices: These are devices that are used to measure the position of the effector. These devices are temporarily attached to the robotic arm or located near it. An example of such a device is the LEICA LT/LTD 500 (laser tracker), which consists of a mirror system, two encoders, a laser interferometer, a distance sensor, and a reflector, which is mounted on a robotic arm.

3. *Identification of parameters:* It is a mathematical process in which values that have been measured internally and externally are used to determine the coefficients of the mathematical model.
4. *Implementation:* This is the process of including an updated version of the robotic mathematical model in the control unit [10, 11].

2.1 Simulation Software Panasonic DTPS

It is software that was exclusively designed for Panasonic industrial robots for the purpose of offline and 3D programming Fig. 3. In the program, and it is possible to create the entire welding process with calculations of cycles welding times up to simulation. Furthermore, the software supports several formats for inserting a CAD model (STL, IGES, IGS, and DXF). Panasonic has also integrated robot calibration and calibration of external axes, i.e., positioners, into the software, which saves time and costs.

2.2 Principle of Calibration of Panasonic Industrial Robots

The Panasonic TM series of robots have the so-called integrated calibration, where with the help of the DTPS software, it is possible to calibrate all axes. The robot's calibration consists of fixed stops, more precisely in the calibration screws, which are located on each calibrated axis and can be seen in Fig. 4. During calibration, the robot repeatedly hits the calibration stops, and thus the so-called collisions. To

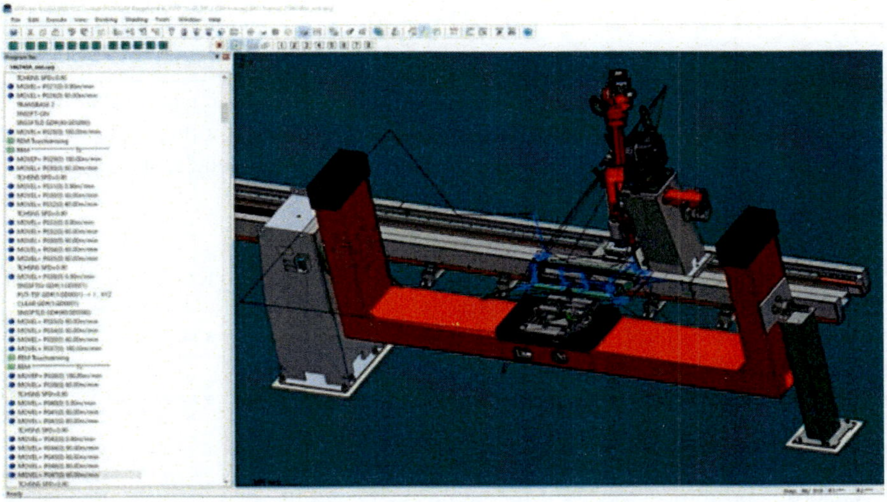

Fig. 3 The working environment of the DTPS software

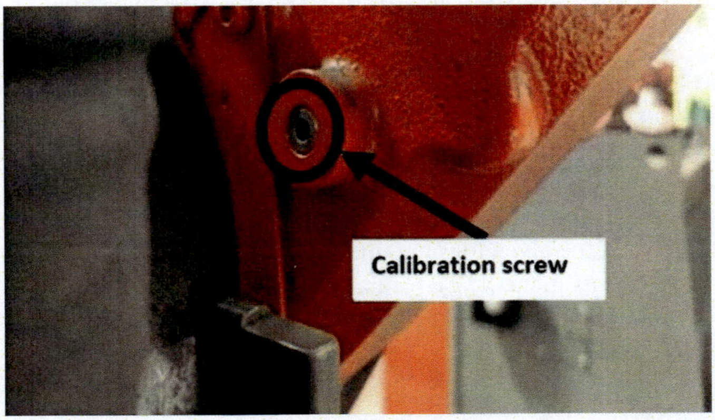

Fig. 4 Calibration screw

prevent deformation of these calibration screws, the touch sensitivity is set to 100% during calibration. As soon as a touch occurs, the robot stops, and the smart encoder writes down the obtained value.

After the calibration, all axes were calibrated, and the values from the smart encoder were set and saved. Next, a prompt was displayed on the teach pendant that the calibration was successful, and it was necessary to screw back the calibration stops and remove the calibration pin. After completing these actions, the robot checked whether all stops were in place.

2.3 Simulation Software RoboDK

It is a simulation software that offers an intuitive environment in which to program various robotic tasks, such as welding, palletizing, packing, milling, 3D printing, etc. The software supports basic formats for inserting CAD models, such as STEP, IGES, and STL.

The advantages of this program compared to other software are:

- an online library containing a large number of robots, positioners, and tools from various manufacturers,
- the possibility of inserting your own designed robot in the CAD program,
- the possibility of testing the accuracy of robots,
- the software includes performance tests according to the ISO 9283 standard, Renishaw ballbar test,
- calibration of robots using laser trackers,
- exporting postprocessors for different types of controllers.
- Another addition from the RoboDK developers is the possibility to use the web interface of the software for robot programming, which works on a similar principle to the classic software interface. RoboDK also provides online technical support, which can be used by end customers in case of problems (Fig. 5).

The basis for the simulation was the creation of CAD components of the welding station according to the model of a real workplace Fig. 6. First, it was necessary to measure the most important parts of this workplace, namely the positioner, the welding table and the robot's carriage.

The next step was the insertion of preprepared CAD components and their layout. In order to insert the given files into the software environment, it was necessary to save the components in the STEP format, which RoboDK supports.

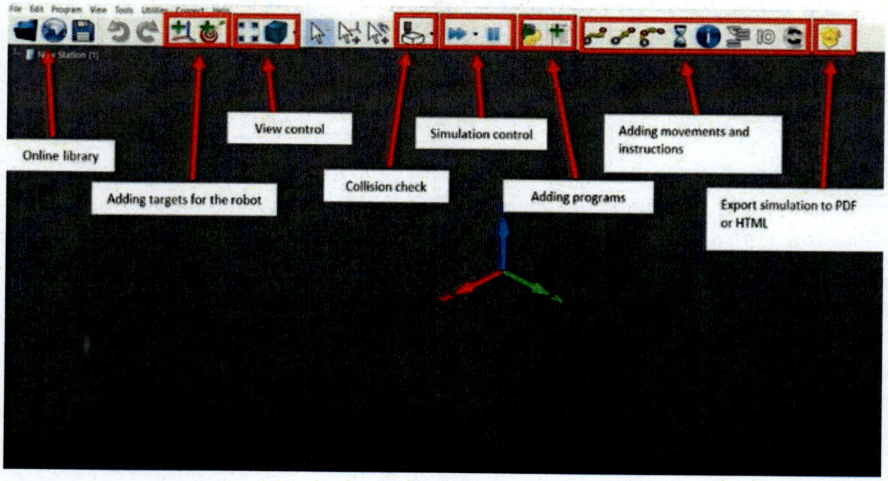

Fig. 5 Basic RoboDK software interface with description

Fig. 6 Robot workplace

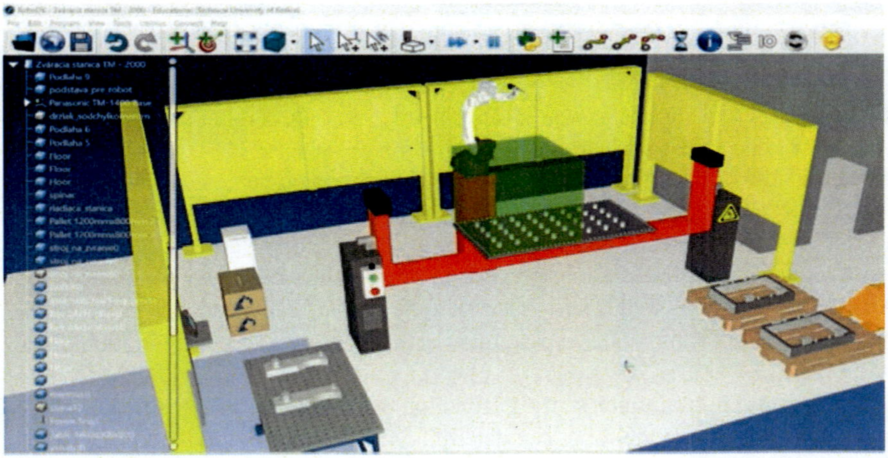

Fig. 7 Complete welding station

The other added components that complete the welding station were: welding table, pallet car, control box, welding wire, positioning switch, teach pendant, warning signs, welded parts, and robot. The ready welding station for simulation can be seen in Fig. 7.

A digital indicator was chosen as the appropriate metrology system because it is an accurate, simple, and affordable measuring device. Specifically, it is an INSIZE deviation meter model 2112-251F, with an accuracy of 0.005 mm, data output, a measuring range of 25.4 mm, and a digital step of 0.001 mm. As repeatability was measured on a welding robot that has a welding wire at the end of the welding nozzle, a flat touch of 20 mm diameter was chosen for the repeatability measurement. This was due to the impossibility of attaching another suitable object, e.g., a precision machined cube, to the end of the robot arm. A suitable fixture was subsequently designed for attachment to the selected indicator on the welding table (Fig. 8).

Once the measurement system was selected, it was necessary to design an ISO cube with the selected measurement plane and five measurement points. The

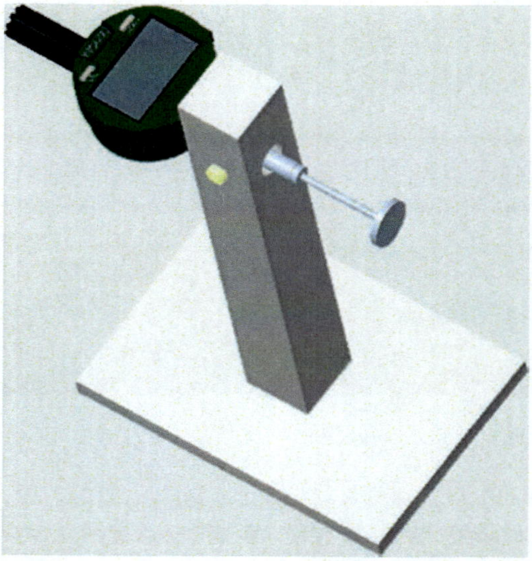

Fig. 8 Fixture with attached digital indicator

dimensions of the ISO cube were determined based on the robot's workspace and where the robot most frequently welds. The length of the edge is 1200 mm. This type of welding robot uses the "touch sensing" function, which means that the robot locates the position of the workpiece to be welded before welding. This type of search is entered directly when programming the welding task of the robot and works on the principle of the tension generated when the welding wire is approached with welded material. In this way, inaccurate welds can be avoided, which may have been caused, for example, by displacement of the workpiece or miscalibration of the positioner.

It is because of the "touch sensing" function that the Z-axis was omitted from the measurement because the robot can determine the distance to the workpiece in the Z-axis when welding. The proposed ISO cube can be seen in Fig. 9, where points C5-C6-C7-C8 represent the selected measurement plane, and points P1 to P5 represent the points to be measured.

Commonly used welding speeds range from 30 to 60 m/min. Therefore, the measurement speed will be set at 90 m/min, which is 50% of the maximum possible speed. The chosen speed is sufficient, and the reason for this is that several studies carried out indicated that testing does not need to be carried out at maximum speeds, as there is no deterioration in performance parameters. The payload is 3 kg, in this case, a welding torch.

The measuring device was inserted into the software environment as a classic CAD component, which was placed in the required position. In Fig. 10, the five measuring devices can be seen, which are placed on the measuring points for the purpose of programming the simulation.

Programming a robotic task in RoboDK, and thus preparing a program, consists in assigning targets that the robot should achieve. In this case, the targets for the robot were added using the "Teach target(s) on surface" function found in the "Program" tab (Figs. 11 and 12).

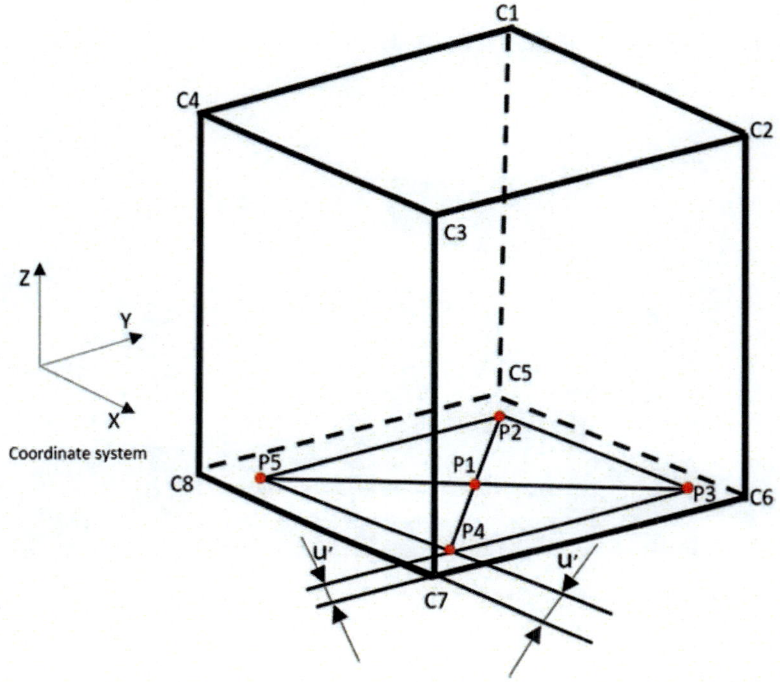

Fig. 9 Designed ISO cube

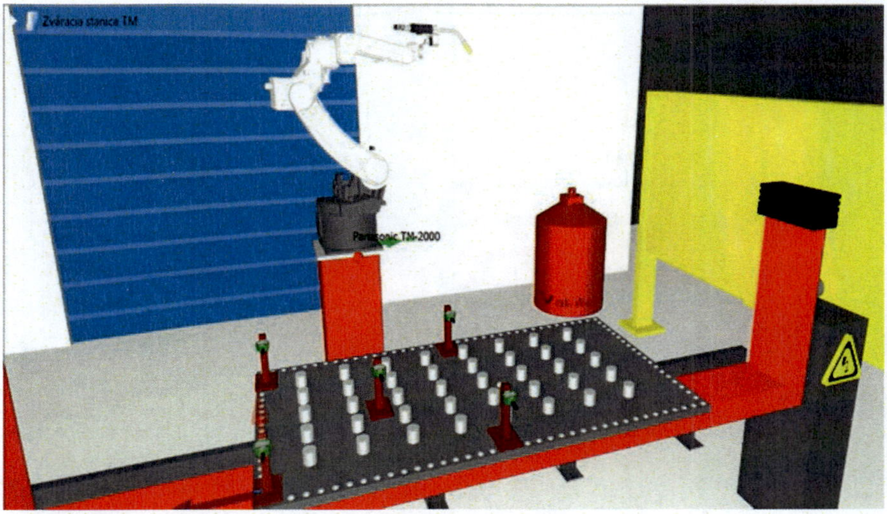

Fig. 10 Inserted measuring devices at measuring points

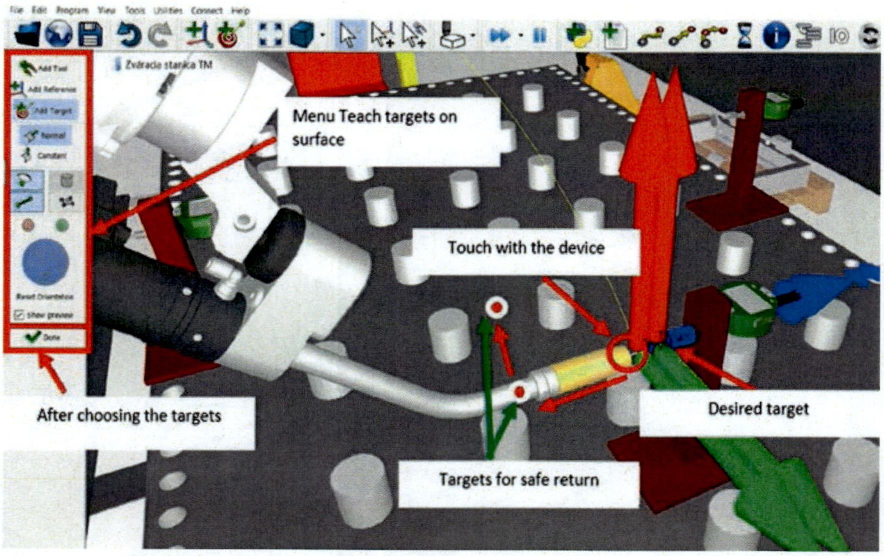

Fig. 11 Addition of targets for a robot

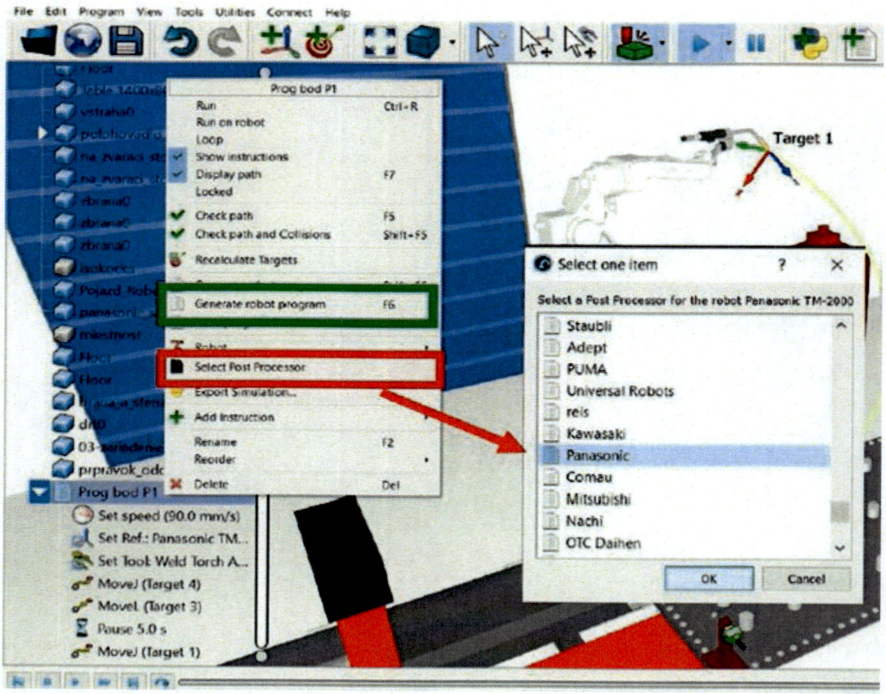

Fig. 12 Choosing the right postprocessor

After programming all targets, a suitable postprocessor was chosen for exporting the program. RoboDK offers almost 70 processors from different manufacturers. The list of offered postprocessors is displayed by clicking on the program and then selecting "Select Postprocessor." In this case, the postprocessor was chosen for the Panasonic controller.

3 Conclusion

The ISO 9283 standard forms the basis for testing the performance characteristics of industrial robots. One-way repeatability of the position represents one of the most important monitored parameters from the point of view of the quality of the robots, as well as from the point of view of the end customer. Most manufacturers declare repeatability and position accuracy based on tests performed according to the ISO 9283 standard. Simulation software is a great help in programming and detecting possible risks associated with robot operation. There are many simulation software on the market, but from the point of view of flexibility, simplicity, and generality, the RoboDK software was the right solution. The aim of this paper was to highlight the simplicity of creating simulations in RoboDK software combined with the preparation for measuring the repeatability of the welding robot.

Acknowledgments The authors thank the Ministry of Education of the Slovak Republic for supporting this research with the grant KEGA no. 038TUKE-4/2021, VEGA no. 1/0080/20 and project APVV-18-0316.

References

1. Niku, S.B.: Introduction to Robotics - Analysis, Control, Applications 3rd Edition: Analysis, Control, Applications. Wiley-Blackwell, Hoboken, NJ (2020)
2. Zhang, D., Peng, Z., Ning, G., Han, X.: Positioning accuracy reliability of industrial robots through probability and evidence theories. J. Mech. Des. N. Y. **143**, 1–30 (2021). https://doi.org/10.1115/1.4047436
3. Zhenhua, W., Hui, X., Guodong, C., Rongchuan, S., Sun, L.: A distance error based industrial robot kinematic calibration method. Ind. Rob. **41**, 439–446 (2014). https://doi.org/10.1108/ir-04-2014-0319
4. Chladek, S., Sveda, J., Sika, Z., Benes, P.: Increase of industrial robot accuracy based on kinematic errors compensation. MM Sci. J. 2022, (2022). https://doi.org/10.17973/mmsj.2022_12_2022149
5. Garcia, L., Bielke, U., Neumann, C., Hageney, T., Banzhaf, K., Wiedenmann, E., Börret, R.: Online position correction approach of an industrial robot by using a new photogrammetric measurement system. Proc. Inst. Mech. Eng. Pt. B: J. Eng. Manuf. 095440542211311 (2022). https://doi.org/10.1177/09544054221131156
6. Hu, M., Wang, H., Pan, X., Tian, Y.: Optimal synthesis of pose repeatability for collaborative robots based on the ISO 9283 standard. Ind. Rob. **46**, 812–818 (2019). https://doi.org/10.1108/ir-03-2019-0056

7. McGarry, L., Butterfield, J., Murphy, A.: Assessment of ISO standardisation to identify an industrial robot's base frame. Robot. Comput. Integr. Manuf. **74**, 102275 (2022). https://doi.org/10.1016/j.rcim.2021.102275
8. Chen, T., Lin, J., Wu, D., Wu, H.: Research of calibration method for industrial robot based on error model of position. Appl. Sci. (Basel). **11**, 1287 (2021). https://doi.org/10.3390/app11031287
9. Asif, S., Webb, P.: Realtime calibration of an industrial robot. Appl. Syst. Innov. **5**, 96 (2022). https://doi.org/10.3390/asi5050096
10. Wang, W., Gao, G., Zhang, H., San, H., Wu, X., Sun, G.: Kinematic calibration for industrial robots using articulated arm coordinate machines. Int. J. Model. Identif. Control. **31**, 16 (2019). https://doi.org/10.1504/ijmic.2019.10017469
11. Nof, S.Y. (ed.): Handbook of Industrial Robotics. John Wiley & Sons, Nashville, TN (2008)

Modelling of Technical Condition Control of Heavy Loaded Gears Teeth

Anatoliy Gaydamaka, Yuriy Muzikin, Volodymyr Klitnoi, Andrew Ruzmetov, and Tomáš Čakurda

1 Introduction

Gear wheels of heavily loaded reducers, for example, roller drives of rolling mills, normally work with significant short-term overloads [1] in an environment of increased vibration and temperature [2], often with insufficient lubrication and a low level of air and hydraulic filtration [3]. This causes the premature appearance of fatigue cracking of the working surfaces of the gear wheels teeth and their fatigue destruction. In the absence of information about the actual technical condition of the gear wheels of the specified gearboxes, in order to exclude emergency breakdowns, the replacement of the gear wheels is performed after the end of the specified service life. Summarizing the very limited information on determining the technical condition of individual gear wheels of heavily loaded machines, especially with large modules (m > 20 mm), allows us to outline the prospects for development and implementation in real conditions of control of the technical condition of gear wheels directly during operation. Therefore, there is a need for scientifically based modeling of the control of the technical condition and forecasting the accident-free guaranteed operation of the gear wheels of heavily loaded machines. The specified control modeling problem can facilitate the selection of materials [4] and, in general, is an integral part of the development of Industry 4.0. [5].

The article provides an overview of modern approaches to modeling the control of the technical condition of gears during their operation. An analysis of the damage

A. Gaydamaka · Y. Muzikin · V. Klitnoi (✉) · A. Ruzmetov
National Technical University "Kharkiv Polytechnic Institute", Kharkiv, Ukraine
e-mail: Volodymyr.Klitnoi@khpi.edu.ua

T. Čakurda
Faculty of Manufacturing Technologies with a seat in Prešov, Technical University of Košice, Prešov, Slovakia

M. Balog et al. (eds.), *EAI International Conference on Automation and Control in Theory and Practice*, EAI/Springer Innovations in Communication and Computing, https://doi.org/10.1007/978-3-031-31967-9_22

detected on the teeth of the gear wheels of the vertical roller reducer of the "Slabbing 1150" crimping mill and the main lift reducer of the pincer crane after 1 year of their operation in production is presented. The most expedient method of technical condition control based on the level of accumulated fatigue damage has been established, and an algorithm for modeling the technical condition control and forecasting the accident-free guaranteed operation of gear wheels of heavily loaded machines has been developed.

2 Literature Review

The technical condition of machines in operation is determined on the basis of knowledge of the informative parameter of the technical condition, the limit state criterion of the object, and the results of periodic or continuous control [6]. When assessing the technical condition of gears, parameters such as the intensity of temperature change [7], kinematic transmission error [8], and vibroacoustic signals [9] are widely used.

Registering the change in temperature of the gear as an integral method of assessing the technical level indicates only the deterioration of the working conditions without establishing a specific cause. Therefore, this method of control is not very effective.

The method of assessing the technical condition of gears, based on the analysis of the nature of the change in the kinematic error of the transmission, allows you to identify single defects of the teeth, to assess the uneven wear of both the pair and the multistage gear.

The vibroacoustic method is the most developed and informative. In the research, four main parameters of gearbox vibration are distinguished, each of which corresponds to a certain type of damage - pitting, cracks, chipping, breakage.

The general drawback of the last two gear control methods is the dependence of the accuracy of the technical condition assessment accuracy on the qualification of the diagnostician, his knowledge of the features of construction and operation, as well as the nature of vibration processes. At the same time, the analysis and interpretation of vibroacoustic signals have a great deal of subjectivity in deciphering the type of defect and its quantitative assessment. In addition, each of the above methods gives an integral assessment of a group of wheels and does not allow for detection of the technical condition of an individual gear wheel, to assess the wear or pitting of its teeth.

In connection with the above, directions for the control of individual technical objects, such as civil aviation objects [10] or offshore wind turbines [11], according to the level of accumulated fatigue damage in operation, have recently appeared. The most relevant in the framework of the mentioned problems are the directions of control of metal hardness [12], parameters of deformation [13], and magnetic hysteresis loops [14]. A change in the properties of the metal surface layers of gear wheels during operation can be a generalized control criterion. By measuring the process of planned and preventive works, for example, the hardness of the metal of the teeth in the zones of possible destruction and comparing the obtained values with the maximally achieved hardness of the defective wheel, its technical condition is evaluated [15]. However, in the application of metal hardness control, it must be taken into account that the fatigue phenomenon is known to be sensitive to dents or other breaks in the material, which cause stress concentrations and can be formed during hardness measurement [12]. The use of modern measuring devices eliminates the mentioned shortcomings of the method; thus, it can be stated that the method can be considered nondestructive [16]. In addition, work [17] clarifies that since polishing of a sample damaged by fatigue partially, if not completely, removes the surface grains hardened by deformation, therefore, if possible, the samples should be polished by an electrochemical method. This should be done before carrying out cyclic processing. Taking into account the specified requirements, when conducting hardness measurements, it should be noted that the results of the measurements are very well correlated with the process of fatigue failure [12]. The technical condition of the part can also be determined by the coefficient of internal energy dissipation of the test object by measuring the coercive force – an informative characteristic of the hysteresis properties of the metal [18]. The coefficient of internal energy dissipation of the part is compared with the data of the experimental curve previously obtained for a similar part with 100% resource, after which the failure-free guaranteed operation of the part is determined [19].

The method of determining the residual magnetization of the metal part is based on a close correlation between structurally sensitive magnetic parameters of the metal and the amount of mechanical stress. The residual resource of metal structures, according to the results of magnetic control, is determined by the previously established experimental dependence of the value of the magnetic parameter on the number of load cycles. A widely used magnetic parameter – the coercive force is highly sensitive to changes in the metal structure under load: the increase in the value of the coercive force from the initial position to the state of destruction, depending on the grade of the material, reaches 100–400%. When using the specified method, it should be taken into account that it works only for ferromagnetic materials. In [20], they concluded that the hysteresis parameters usually do not show significant changes until the last 10–20% of the service life during fatigue failure; that is, the use of the method beyond the yield point requires sensitive equipment. The residual resource of many metal structures was determined by the method of magnetic structuroscopy: tower cranes, gantry cranes, bridge cranes, and rolling mills [21].

3 Research Methodology

The aim of this study is to develop an algorithm for modeling the technical condition control and prognostication of the accident-free guaranteed operation of toothed wheels of heavily loaded machines.

To achieve this goal, it is necessary to solve the following tasks:

- based on the literature analysis and the analysis of the damage detected on the teeth of the gear wheels of the vertical roller reducer of the "Slabbing 1150" crimping mill and the reducer of the main lift of the tong crane, establish the most appropriate way of controlling the technical condition according to the level of accumulated fatigue damage;
- develop a procedure for modeling technical condition control based on the results of determining the level of accumulated fatigue damage in operation.

3.1 Research of Damage on the Surface of Teeth

The first task was performed by inspecting the appearance and analyzing the damage found on the gear wheels of the vertical roller reducer of the "Slabbing 1150" crimping mill and the reducer of the main lift of the pincer crane, presented by the crimping workshop of PJSC "Zaporizhstal," after 1 year of their exploitation. This task is characterized by a high degree of subjectivity, since the performed inspection gives only an expert assessment of the observer, and in addition, during 1 year of operation of the reducers under consideration, their operating conditions change significantly, which is not reflected in the operation log. Therefore, the analysis of detected damages on the working surfaces of teeth should be perceived as a qualitative picture of real tribotechnique processes. In this case, the reliability indicator of the toothed gear will act as a multi-parameter dependence, the real value of which can only be obtained experimentally.

The inspection results of the gear tooth of the vertical roller reducer of the "Slabbing 1150." From the analysis of the technical characteristics of the toothed gear, it can be assumed that all the errors that occur during installation and operation will significantly affect the nature of tooth damage, taking into account the dimensions of the gears, as well as the ratio of the hardness of the tooth material and the hardness of its surface, it is necessary to condition the step-by-step quality control of each technological operation during the heat treatment of the gears; otherwise, various damages are possible during their operation. Taking into account that the length of the tooth is 450 mm and the nature of surface damage differs significantly in individual areas of the tooth, the inspection was carried out both in local areas and in large areas. In addition, since the gear works in a reversible mode, the inspection took into account the side of the tooth "A" and "B," which conditionally correspond to the forward and reverse movement of the slab.

A more detailed consideration of individual zones of side "A" of the gear tooth is given in Fig. 1. The nature of damage in these local zones is random and cannot be represented as a certain regularity. In addition, these zones are scattered both along the entire length of the tooth and in different areas along its height

In Fig. 1: in zone "a" one deep and several small sea shell-shaped spalls are clearly visible, in zone "b" small sea shell-shaped spalls are observed; in zone "c"

Fig. 1 The results of checking the gear tooth of the vertical roller reducer "Slabbing 1150" ("A" side)

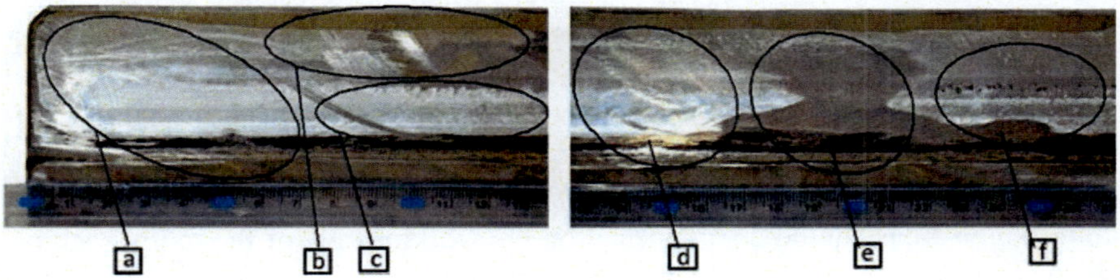

Fig. 2 The results of checking the gear tooth of the vertical roller reducer "Slabbing 1150" ("B" side)

the zone of the tooth begins, which is less loaded as traces of grinding are visible, in zone "d" it is shown - but the part of the tooth that is the most loaded, as traces of scuffing are visible.

Local spalls and dents on the surface of the tooth can be caused both by poor oil filtration and by random particles caught in horizontal cleaning. Sufficiently large areas of the tooth, which partially or completely do not participate in the operation of the gear transmission, or, on the contrary, experience loads that lead to jamming, can be caused either by installation and operating errors, or by significant deformations of the bearing elements of the reducer.

A superficial examination of the surface of tooth "B" shows that it is significantly different from surface "A," both in terms of the nature of the detected damages and their location. Marked pitting, areas receiving maximum loads, and areas that do not transmit torque at all have place on the tooth (see Fig. 2).

Three zones are observed in the area of the tooth from 0 to 140 mm (Fig. 2a, b, c.) In zones "a" and "c" you can see how the lower part of the tooth works. The load on the tooth is unevenly distributed here, and it decreases from left to right. The upper part of the tooth is practically not loaded, so in the "b" zone, traces of the grinding wheel and traces of flanking of the tooth are clearly visible. In zone "d," the loading decreases from left to right. The "e" zone on this part does not work at all, because the grinding marks are slightly worn only in the upper part of the tooth in front of the flanking area. At the section from 220 mm, the zone of the main loading of tooth "f" begins, here, the zone of deep spalls, located along the pitch line almost to the end of the tooth, is clearly visible.

Cracking of the material of the lateral surface of the tooth at the level of a pitch line about 200 mm long has characteristic signs of destruction in the form of limited pitting. In addition, a qualitative check of the hardness of the tooth on the surface in several places showed that H < 350 HB.

Despite the fact that the analyzed toothed gear works in reverse mode, the nature of tooth damage from sides "A" and "B" is significantly different, which indicates a difference in the nature of the load on the teeth during forward and reverse movement of the working elements of the drive. The absence of any damage in the local areas of the tooth, where flanking was performed, shows that there are no edge impacts during the engagement process, and therefore, the flanking effect has been achieved. The presence of almost all types of damage on the surface of the tooth shows that none of them is dominant, and their totality only indicates the extremely complex nature of the interaction between the specific elements of the gearbox and the external force parameters acting on them.

The inspection results of the gear tooth of the main lifting gear of the pincer crane. Three areas can be distinguished on the working side of the tooth, each of which corresponds to a certain type of damage (see Fig. 3).

In zone "a" there are pronounced traces of plastic deformation of the tooth material, directed toward the linear speed of sliding of the mating surfaces in engagement. The characteristic angle of the location of the plastic deformation in relation to the generating cylinder corresponds to the angle of inclination of tooth *8*°. As a result of the plastic deformation, a ridge was formed in section "c," which limits the contact zone of the two gears that form the engagement. In zone "c," which completely covers the entire length of the tooth, a ridge was formed, bent in the direction opposite to the angular speed of rotation. Its size and characteristic shape indicate significant plastic deformation of the tooth under the influence of all factors that

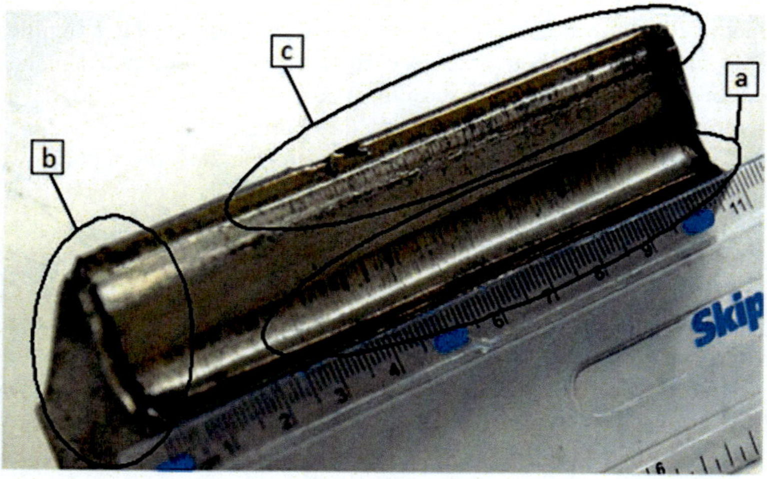

Fig. 3 The results of checking the gear tooth of the main lifting gear of the pincer crane

affect the operation of the gear transmission. This form of damage to the working surface of the tooth is not predicted by any kind of strength calculations.

Critical analysis and assessment of the observations. The nature of damage to the surface of the researched gear tooth of the reversible reducer of the "Slabbing 1150" crimping mill showed a significant difference in damage from surfaces "A" and "B," i.e., lack of symmetry of damage to the sides. This, in turn, enables a simplified calculation of the reverse when calculating the strength of the tooth by introducing a special coefficient that takes into account the bilateral load on the tooth. In addition, on each surface "A" and "B," at least four different types of destruction can be distinguished, which does not allow declaring any type of destruction as dominant. A qualitative comparison between them does not allow judging the probable priority of any of them. Some local areas of the tooth do not participate in the work at all, as traces of mechanical processing are completely preserved on them. Therefore, it is not obvious to determine the local zone on the tooth, where the probability of destruction will be maximum. Based on this, when predicting both the actual state of the tooth in terms of strength and the residual resource of its work, it is necessary to evaluate the strength of all problem areas of the tooth, both on one side and on the other.

The analysis of the appearance of the tooth, taken from the reducer of the main lift of the pincer crane, showed that under the influence of external load, as well as the environment (high temperature and dustiness), the latter received one-sided destruction in the form of plastic deformation of the material of the tooth surface with characteristic two zones of cavities on the leg and head. The shape and dimensions of the cavities are determined by the direction of the linear speed of sliding of the contact surfaces of the connected teeth. In addition, a ridge was formed on the tooth head from the side of the plastic deformation, bent in the direction opposite to the angular speed of rotation. This type of tooth destruction is not provided for by the standards during strength calculations. Considering that this zone is not local, which can be considered only as a correction for the main type of destruction, but basic, which completely covers the entire surface of the tooth, this type of destruction requires a new approach to assessing the strength of the tooth.

The use of such parameters as vibration, noise level, temperature, quality of lubricant according to the chemical and granulometric composition of particles, the magnetic permeability of the material of the gear wheels, as well as their combinations, as criteria for assessing the condition of the gear wheels in the gearboxes of the considered type, cannot be considered satisfactory. This is due to the fact that their magnitude and the dynamics of change are affected by a large number of reasons, many of which are functionally unrelated to the operation of the gear transmission. In addition, when using the specified criteria for assessing the condition of the gear wheel, it is necessary to differentiate each possible type of failure and rank them according to the probability of possible accidental tooth breakage.

According to the results of detected damage to the surface of gear wheels, it is proposed to use the hardness value of the material in the studied local area as a criterion for assessing their technical condition. Based on the energy theory, it follows: where the deformation is greater, there is a greater change in hardness, which in this case is a function of the probability of tooth destruction. By comparing changes in material hardness in local areas of the tooth with the maximum allowable value for a given type of fracture, it becomes possible to use the hardness value as an indicator of determining the residual service life, and therefore, to prevent emergency failures according to the type of fracture under consideration. The dangerous zone of expected destruction should be recognized as the one where the actual hardness of the material is closest to the maximum permissible hardness; that is, it has a minimum reserve factor according to the hardness criterion.

Since tooth damage in different local areas varies significantly, hardness control must be carried out not only in one selected area, but also in other problem areas. This is due to the fact that in a number of cases during the work, the dominant zone of possible destruction can become any other one lying within the limits of the tooth surface. The hardness test in the control zone should be conducted at one or more points depending on the shape and size of the damage under consideration.

3.2 *Modeling of the Technical Condition Control Based on the Results of Hardness Measuring of the Surface Layers*

Based on the analysis, the tooth hardness measurement zones can be located on the working surfaces of the teeth – around the pole zone and near the root of the tooth. The accuracy of constructing the resource depletion function depends on the number of experiments to measure the hardness of the metal over the entire working period up to the limit state of the appearance of fatigue cracks in the vicinity of the pitch diameter or fatigue cracks at the tooth legs. Assuming that gear wheel resource depletion is a nonincreasing function, we can limit ourselves to minimal (five) experiments that determine any second-order curve [11]. If there are uncharacteristic values of the hardness of individual points, it is necessary to carry out additional measurements close to the previous ones, since there may be not only measurement errors but also metal inhomogeneity. Therefore, the number of measurement points in each experiment should be at least three...five. The relative location of the measurement points in each experiment is determined by the purpose of the study, the measurement method, and the type of hardness tester. When measuring hardness, the distance from the edge of the working surface of the toothed wheel to the center of the first impress of the indenter of the hardness tester, as well as between adjacent impress, should be at least three diameters of the impress. An increase in the distance between the measurement points worsens the accuracy of the assessment of the technical condition and makes it difficult to predict the residual resource of the gear wheel.

Forecasting the accident-free guaranteed operation of gears based on the results of diagnostics of their technical condition is carried out according to recommendations [22].

Accident-free guaranteed operation of a technical object is operation during the term, which ensures the preservation of the parameters of regulatory and technical documentation. Prediction of accident-free guaranteed performance is possible if three conditions are present: a known parameter of the technical condition, the possibility of periodic (continuous) control of the parameter of the technical condition, and the presence of a limit state criterion.

Forecasting of the guaranteed operation of the object is usually carried out according to the scheme shown in Fig. 4.

After certain periods of operation *t1, t2.... tn* measure the maximum values of damage *h1, h2....hn,* and extrapolate the dependence to the maximum allowable value of damage *hn.* This method makes it possible to obtain a fairly accurate estimate of the guaranteed working time, if the dependence $h(t)$ is known.

In this work, it is proposed to prognosticate the guaranteed service life of gear wheels, taking into account the stages of running-in and stable accumulation of damage (linear function), since the wheels are removed from service in advance of their possible destruction. Assuming that the damage (control parameter) in the gear wheel accumulates monotonically according to a linear function, and their dispersion does not change, the method of standardized calculation norms is used for prognostication the guaranteed service life. The introduced assumptions are valid in the case when the control parameter obeys the normal distribution law. The latter is true for tooth damage, which depends on many factors. Moreover, it is believed that each of the factors has an equal influence on the occurrence of damage. Therefore, the proposed type of control parameter change function is fully justified.

The linear function of the control parameter change is taken in the form:

$$x(t) = c_1 + c_2 \cdot \tau, \tag{1}$$

where c_1, c_2 – the unknown coefficients; $\tau = t_i - t_0$, $\tau \geq 0$, $t_0 \geq 0$ is the initial value of working time.

The destination parameter (for example, damage) is evaluated by changing the control parameter according to the expression:

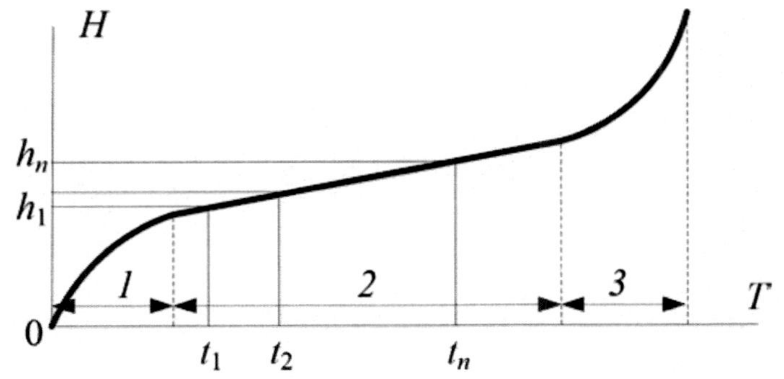

Fig. 4 A typical diagram of the dependence of the value (H) of damage on the term (T) of operation: 1 – practice; 2 – stable accumulation of damage; 3 – intense destruction

$$y_i = x(t_i) + \xi_i, i = 1 \ldots N, \tag{2}$$

where t_i is the work-up value and the i – measurement term, $t_0 \le t_1 \le t_2. \ldots \le t_N, \xi_i$ – an unbounded random variable with variance σ^2, which is symmetrically located relative to the zero mathematical expectation or symmetrically distributed over a limited random variable, for which the condition is fulfilled for all work-up values:

$$-\xi \le \xi(t) \le \xi, 0 \le \xi \le \infty. \tag{3}$$

At the same time, the number of measurements is chosen under the condition $N \ge 2m$, where is the number of unknown coefficients of the parameter change law (recommended $N \ge 11$), and the measurement terms t_i are such that the random variables ξ_i are practically independent.

The following forecasting procedure is recommended for N measurements of a quantity y_i at arbitrary intervals of operation:

Values are calculated.

$$Y_1 = \sum_{N}^{i=1} y_i, Y_2 = \sum_{N}^{i=1} \tau_i \bullet y_i. \tag{4}$$

$$D = N \sum_{N}^{i=1} \tau_i^2 - \left(\sum_{N}^{i=1} \tau_i \right)^2; D_{11} = \frac{\sum_{N}^{i=1} \tau_i^2}{D};$$
$$D_{12} = D_{21} = -\frac{\sum_{N}^{i=1} \tau_i}{D}; D_{22} = \frac{N}{D}. \tag{5}$$

Estimates of coefficients c_1, c_2 are calculated.

$$c_1 = Y_1 \cdot D_{11} + Y_2 \cdot D_{21}, c_2 = Y_1 \cdot D_{12} + Y_2 \cdot D_{22}. \tag{6}$$

Calculate the estimate of the mean square deviation

$$\sigma = \sqrt{\sum_{N}^{i=1} (Y_i - c_1 - c_2 \cdot t_i)^2 / (N-2)}. \tag{7}$$

Calculate the mean square deviations of the coefficients c_1, c_2.

$$\sigma_1 = \sigma \sqrt{D_{11}}, \sigma_2 = \sigma \sqrt{D_{22}}. \tag{8}$$

Guaranteed coefficient c_1, c_2 estimates are calculated.

$$c_{1G} = c_1 \pm K \cdot \sigma_1; \tag{9}$$

$$c_{2G} = c_2 \pm K \cdot \sigma_2 .. \tag{10}$$

where $K = 1.282$ with confidence probability = 0.9; $K = 1.646$ (= 0.95); $K = 2.32$ (= 0.99); (+) is taken when the "damage" assignment parameter increases, (−) when it decreases.

Calculate the average (expected) accident-free operating time

$$T_\gamma = \left(Y_{\lim} - c_1\right) / c_2 - t_k, \tag{11}$$

where $Y_{\lim}$ – the limit value of the destination parameter; t_k– the time of the last measurement.

Guaranteed accident-free operation is calculated

$$T_\gamma = \left(Y_{\lim} - c_{1G}\right) / c_{2G} - t_k. \tag{12}$$

4 Results

The analysis of the damage detected on the teeth of the gear wheels of the vertical roller reducer of the "Slabbing 1150" crimping mill and the reducer of the main lift of the tong crane was carried out. According to the results of detected damage to the surface of gear wheels, the feasibility of using the hardness value of the material as a criterion for assessing their technical condition is indicated. An algorithm for modeling the technical condition control and forecasting the guaranteed accident-free operation of heavily loaded machines gears based on the results of measuring the hardness of the surface layers was developed.

Based on the results of the presented research, an agreement was prepared with PJSC "Zaporizhstal" during which, with the help of the presented algorithm for modeling the control of the technical condition and prognostication the accident-free guaranteed operation of the gears of heavily loaded machines, it is planned to carry out calculations of the residual resource of the gear wheels of the vertical roller reducer of the "Slabbing 1150" crimping mill and the reducer of the main lift of the tong crane. The measurement of the hardness of the surface layers of the teeth is planned for the teeth of a gear wheel that has not yet been operated (a new wheel), for the teeth of a gear wheel in the process of operation (the measurement schedules are determined by the sequence of scheduled maintenance work) and for the teeth of a gear wheel whose operation has been stopped after a certain period of time according to regulatory documents.

The results of the agreement implementation are planned to be presented in the next publication.

5 Conclusions

According to the analysis of the technical literature and the analysis of the damage detected on the teeth of the gear wheels of the vertical roller reducer of the "Slabbing 1150" crimping mill and the main lift reducer of the pincer crane, it was established that the optimal criterion for assessing the technical condition is the value of the hardness of the material in the investigated local area.

Hardness must be controlled in several problem areas. The hardness check in the control zone should be conducted at one or more points depending on the shape and size of the damage under consideration.

The proposed algorithm for modeling the control of the technical condition and prognostication of the accident-free guaranteed operation of the gears of heavily loaded machines is based on the results of measuring the hardness of the surface layers. When determining the accident-free guaranteed service life based on several indicators of the limit state in different areas, the final decision is made based on the indicator that determines the minimum term.

Acknowledgments This work was supported by the Slovak Research and Development Agency under contract No. APVV-19-0590, and also by the projects VEGA 1/0061/23, KEGA 022TUKE-4/2023 granted by the Ministry of Education, Science, Research and Sport of the Slovak Republic.

References

1. Klitnoy, V., Klitnoi, V., Batrak, P.: Optimization of planetary gear of onboard gear-box using the method of differential evolution. Avtomobilnij Transp. **47**(15), 15–20 (2020)
2. Klitnoi, V., Gaydamaka, A.: On the problem of vibration protection of rotor systems with elastic adaptive elements of quasi-zero stiffness. Diagnostyka. **21**(2), 69–75 (2020)
3. Gaydamaka, A., Muzikin, Y., Klitnoi, V., Basova, Y., Dobrotvorskiy, S.: Selecting the method for pre-tightening threaded connections of heavy engineering, pp. 69–77. International Conference on Reliable Systems Engineering (ICoRSE) (2021)
4. Dobrotvorskiy, S., Balog, M., Basova, Y., Dobrovolska, L., Zinchenko, A.: Concept of the software for materials selection using .NET technologies. Advanced manufacturing processes. InterPartner 2019. Lecture notes in mechanical engineering, pp. 32–43. Springer, Champions (2020)
5. Dobrotvorskiy, S., Basova, Y., Dobrovolska, L., Sokol, Y., Kazantsev, N.: Big challenges of small manufacturing Enterprises in Industry 4.0. In: Ivanov, V., Trojanowska, J., Pavlenko, I., Zajac, J., Peraković, D. (eds.) Advances in Design, Simulation and Manufacturing III. DSMIE 2020. Lecture Notes in Mechanical Engineering, pp. 118–127. Springer, Cham (2020)
6. Frini, M., Soualhi, A., El Badaoui, M.: Gear faults diagnosis based on the geometric indicators of electrical signals in three-phase induction motors. Mech. Mach. Theory. **138**, 1–15 (2019)
7. Yuejian, C., Xihui, L., Ming, J.Z.: An improved singular value decomposition-based method for gear tooth crack detection and severity assessment. J. Sound Vib. **468**, 115068 (2020)
8. Omar, D.M., Matti, R.: Gear fault models and dynamics-based modelling for gear fault detection – a review. Eng. Fail. Anal. **117**, 104798 (2020)

9. Anil, K., Gandhi, C.P., Yuqing, Z., Rajesh, K., Jiawei, X.: Latest developments in gear defect diagnosis and prognosis: a review. Measurement. **158**, 107735 (2020)
10. Yucesan, Y.A., Dourado, A., Viana, F.A.C.: A survey of modeling for prognosis and health management of industrial equipment. Adv. Eng. Inform. **50**, 101404 (2021)
11. Worden, K., Cross, E.J., Dervilis, N., Papatheou, E., Antoniadou, I.: Structural health monitoring: from structures to systems-of-systems. IFAC-Papers OnLine. **48**(21), 1–17 (2015)
12. Bjorheim, F., Siriwardane, S.C., Pavlou, D.: A review of fatigue damage detection and measurement techniques. Int. J. Fatigue. **154**, 106556 (2022)
13. Mocko, W., Grzywna, P., Kowalewski, Z.L., Radziejewska, J.: An influence of cyclic loading on the form of constitutive relationship for DP500 steel. Mater. Des. **103**, 183–193 (2016)
14. Serbin, E.D., Kostin, V.N., Vasilenko, O.N., Ksenofontov, D.G., Gerasimov, E.G., Terentev, P.B.: Influence of the two-stage plastic deformation on the complex of the magnetoacoustic characteristics of low-carbon steel and diagnostics of its structural state. NDT & E Int. **116**, 102330 (2020)
15. Drumond, G., Roudet, F., Chicot, D., Pinheiro, B., Pasqualino, I.: A damage criterion to predict the fatigue life of steel pipelines based on indentation measurements. J. Offshore Mech. Arct. Eng. **143**(1), 011701 (2021)
16. Sulko, M., Chmelko, V., Kepka, M.: Possibility of fatigue damage detection by non- destructive measurement of the surface hardness. Procedia Struct. Integr. **7**, 262–267 (2017)
17. Ye, D.Y., Wang, D.J., An, P.: Characteristics of the change in the surface microhardness during high cycle fatigue damage. Mater. Chem. Phys. **44**(2), 179–181 (1996)
18. Gubskij, S., Suhomlin, V., Voloh, V.: Control of the stress state of steels by coercive force. Mashinobuduvannya. **13**, 6–10 (2014)
19. Venkatachalapathi, N., Jameel Basha, S., Raju, G.J., Raghavulu, P.: Characterization of fatigued steel states with metal magnetic memory method. Materials Today: Proceedings. **5**(2) Part 2, 8645–8654 (2018)
20. Chen, Z.J., Strom, A., Jiles, D.C.: Micromagnetic surface measurements for evaluation of surface modifications due to cyclic stress. IEEE Trans. Magn. **29**(6), 3031–3033 (1993)
21. Boiko, H., Boiko, T., Yepikhin, R.: Methods for assessing the technical condition of cranes Visnyk skhidnoukrainskoho natsionalnoho universytetu im. Volodymyra Dalia. **2**(243), 42–49 (2018)
22. Gajdamaka, A., Klitnoi, V., Muzikin, Y.: Diagnostics of the technical condition and predic-tion of accident-free guaranteed operating time of heavily loaded machines gear wheels. Vehicle Electr. Innov. Technol. **20**, 54–61 (2021)

Automated Subsystem for Cutting Modes Calculations

Vitalii Ivanov, Svitlana Vashchenko, Ivan Pavlenko, Michal Hatala, Vitalii Kolos, and Vladyslav Andrusyshyn

1 Introduction

The successful operation of enterprises depends on the ability to accumulate, systematize, and process information [1]. The introduction of computer technologies increases the efficiency and productivity of the work. Computer-aided design (CAD) systems rapidly develop in various production fields, including manufacturing engineering, particularly fixture design. Numerous engineering tasks can be solved using CAD technologies, notably creating drawings, three-dimensional modeling for product design and the assembly process, design of the auxiliary equipment, development of the technological documentation, generating programs for CNC machines, and maintenance of the electronic archive [2].

A significant part of modern CAD consists of several modules: assembly, machining, engineering data management, etc. A common methodology and tools combine them. Combining the design and production planning modules into a single CAD system reduces the cost and time of bringing new products to the market [3, 4]. It allows designing parts considering their technology and the material.

V. Ivanov (✉) · I. Pavlenko · V. Andrusyshyn
Sumy State University, Sumy, Ukraine

Faculty of Manufacturing Technologies with a seat in Prešov, Technical University of Košice, Prešov, Slovakia
e-mail: ivanov@tmvi.sumdu.edu.ua

S. Vashchenko · V. Kolos
Sumy State University, Sumy, Ukraine

M. Hatala
Faculty of Manufacturing Technologies with a seat in Prešov, Technical University of Košice, Prešov, Slovakia

M. Balog et al. (eds.), *EAI International Conference on Automation and Control in Theory and Practice*, EAI/Springer Innovations in Communication and Computing, https://doi.org/10.1007/978-3-031-31967-9_23

In addition, the modular architecture of CAD facilitates the expansion of the system and its adaptation according to the user's requirements, allowing you to acquire only the necessary components [5, 6].

Considering the high labor intensity of design-technological and optimization tasks and a large amount of monotonous work that does not require creativity from the engineer, it can be concluded that designing fixtures is expedient to automate. Currently, there are software tools for automating certain stages of the fixture design, but such partial automation does not give the desired result. Therefore, computer-aided fixture design (CAFD) technology was proposed, which automates the fixture design process [7].

This research aims to develop an information model for determining force characteristics during drilling, milling, and boring operations and to develop a subsystem for cutting modes calculation that is a component of the CAFD system.

2 Literature Review

For the production of parts in manufacturing engineering, the development of the manufacturing process and the design of the necessary tooling using specialized CAD software, particularly CAFD, is of great importance.

The developed CAFD system [8] ensures the selection of the optimal composition of the fixtures for the given production conditions based on the implementation of multicriteria optimization. CAFD provides information support, data storage, and processing during the automated fixture design. A feature of the CAFD system is an optimization module, which is intended to perform multicriteria optimization during fixture design choosing. The system‘s main features include the following: an intuitive interface; operation with the libraries; integration with Autodesk Inventor; creation of a report based on the results of calculations. The program consists of two parts [9]. The first one is designed to work with the database. The second part is represented by five software modules that function sequentially and use information from the database.

Fixture design is a time-consuming process and needs a number of data considerations [10]. Fixture layout selection depends on the machine tool's capabilities [11–13], workpiece material and coatings properties [14, 15], accuracy [16], and quality [17] parameters. It is essential to ensure the dynamic stability of the cutting process based on the working conditions, cutting and auxiliary tools design, and economic indicators [18, 19].

An important stage is determining cutting modes and power characteristics when performing machining operations, which is the most common method of parts machining in manufacturing engineering. The research study [20] focuses on the digitalization of the machining calculations based on the geometrical parameters, material properties, information about the machine tool and cutting tool, removal data, and cutting parameters. The study [21] presents experimental research for determining the optimal machining parameters based on resource consumption, machining quality, and production efficiency when turning, grinding, and drilling. To improve the productivity and quality of the machining process, the research method of drilling nontechnological holes in polymer composite materials was proposed using optimization of the cutting modes [22]. The research study [23] presents a comprehensive experimental investigation of machining parameters for sustainable machining considering surface roughness, tool wear, cutting forces, chip morphology, and chip reduction coefficient.

Worldwide manufacturers of cutting tools provide online and offline calculations of cutting modes based on their developed algorithms. Seco Tools AB [24] developed "Machining Calculator," which is available for desktop and mobile solutions and provides calculation feeds, speed, material removal rates, and search and compare indexable inserts for cutting tools. "Machining Calculator" from Sandvik Coromant [25] is designed to help engineers and operators optimize the performance of their turning, milling, drilling, and tapping machines by calculating optimal cutting settings based on working conditions. Walter AG [26] developed an application to precisely determine cutting parameters for various machining operations in the field of turning, drilling, threading, and milling. It calculates torque, driving power, machining volume, main cutting time, cutting force, and chip thickness. Moreover, it allows easy cost comparison between two tool solutions. "TaeguTec Tool CutZZ" was developed for calculating cutting modes for TaeguTec Tool Manufacturer [27]. One of the main features of this application is that it is linked to the TaeguTec Tool Adviser software, which, after entering essential information, selects the best tool for machining needs. In addition, the application is linked to the TaeguTec e-catalog, which contains various tools and inserts. "Iscar Cutting Info" [28] provides machining data for all of Iscar's cutting tool geometries. "Speed & Feed" [29] allows calculating cutting modes for Kennametal's cutting tools. Use the "Hoffmann Machining Calculator" [30] to calculate the most relevant cutting and performance data for turning, drilling, and milling methods. It can calculate the speed and feed rate for the production methods of turning, drilling, and milling by entering the corresponding parameters and subsequently readjusting them to the machine parameters. FSWizard Machinist Calculator, developed by Zero_Divide [31], automatically uses the recommended cutting speed and chip loading for machining with various indexed and solid face mills, drills, cranes, etc. "Machining Calculator" [32], developed by Soft Engineer Developer, is a valuable solution for process engineers, CNC programmers, operators, and machine tool adjusters. The calculator allows for calculating the technological parameters of machining (milling, turning, drilling, and other technological parameters by selecting the material to be processed and entering various machining parameters). The advantage of these technological solutions for calculating cutting modes is the speed of obtaining results. However, the disadvantage is the impossibility of transferring the received data for further calculations of fixture layouts.

3 Research Methodology

The method of analysis, process-oriented approach, object modeling, and programming will be used to solve the task. The system was modeled using system analysis, namely IDEF0 and the UML language. According to the IDEF0 standard, a process-oriented model has been developed that fully reflects the system‘s structure, the processes being executed, and the information flows transmitted by these functions. Diagrams of use cases and sequences were constructed using the UML language. The obtained diagrams made it possible to present the system so that it could be easily converted into a program code, which significantly facilitates further program implementation. The program code was developed using the Visual Basic language in the Microsoft Visual Studio environment. The conceptual model of the database is created using ER modeling.

The final solution is a subsystem analyzing the selected fixture layout. The main stages of the project include the creation of the technical task; subsystem modeling (creating an ER model of the database, developing IDEF0 and UML diagrams, and modeling the user interface); subsystem development; testing and debugging; development of documentation. Thus, a specialized software module was developed and integrated into the CAFD system.

In the general case, the calculation scheme for hole making (drilling, reboring, and countersinking) is presented in Fig. 1a. When performing calculations for the boring process, and it is also used the scheme mentioned above, excluding the "Axial force calculation" stage. When calculating the milling process, use the algorithm presented in Fig. 1b.

Therefore, for the reengineering of the CAFD system, the following software modules must be developed: a module for working with database tables necessary for the organization of design and a module for determining the force characteristics of milling and drilling operations, in which calculations are performed for three basic manufacturing steps, namely drilling, milling, and boring.

Since the subsystem of engineering analysis is a component of the CAFD, a common database will be used. Tables are added to the existing structure, which is designed to store coefficients for calculating cutting modes, data on tools and materials, and save an archive of performed calculations. Figure 2 presents a conceptual model of the added part of the database. The table of materials already exists, so it is not considered when upgrading the database (the corresponding entity in Fig. 2 is not filled).

Model elements describe the following entities:

- coefficients necessary for calculations (table Koef) – coefficient name, value, type of technological machining, formula;
- calculation results for drilling operations (table Sverdl_rez) – (diameter, a diameter of the existing hole, depth, type of machining, cutting depth, feed, cutting speed, spindle rotation frequency, torque, axial force, power, material, tool);

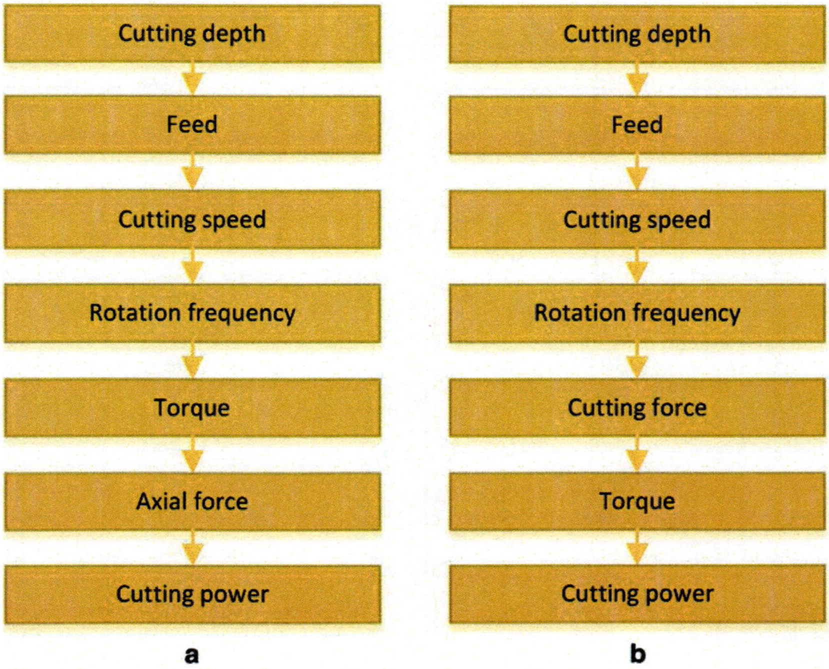

Fig. 1 Calculation scheme for drilling (**a**) and milling (**b**) operations

- calculation results for milling operations (table Frez_rez) – milling width, milling depth, cutter width, cutter depth, cutter length, cutter diameter, cutter type, cutting speed, spindle speed, cutting force, torque, power, machined material, tool;
- tools for performing operations (table Instrument) – name, the tool material, type of machining.

Relationships between entities are established based on the following rules:

- the material determines the coefficients. The coefficient is determined by the material;
- the tool determines the coefficients. The coefficient is determined by the tool;
- the result of calculations for drilling operations depends on the workpiece material. The workpiece material affects the result of calculations for drilling operations;
- the result of calculations for drilling operations depends on the tool. The tool affects the result of calculations for drilling operations;
- the result of calculations for milling operations depends on the workpiece material. The workpiece material affects the result of calculations for milling operations;
- the result of calculations for milling operations depends on the tool. The tool affects the result of calculations for milling.

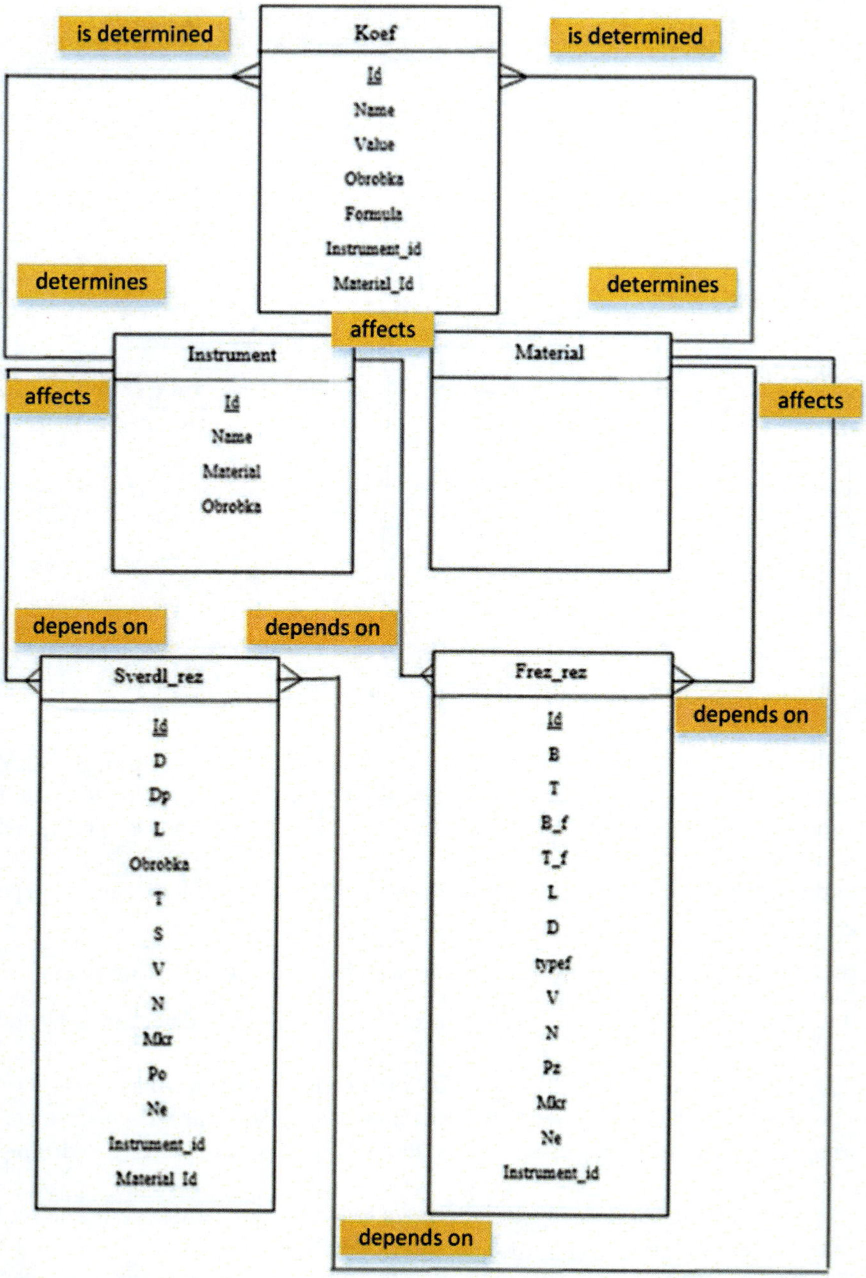

Fig. 2 ER model of the subject area

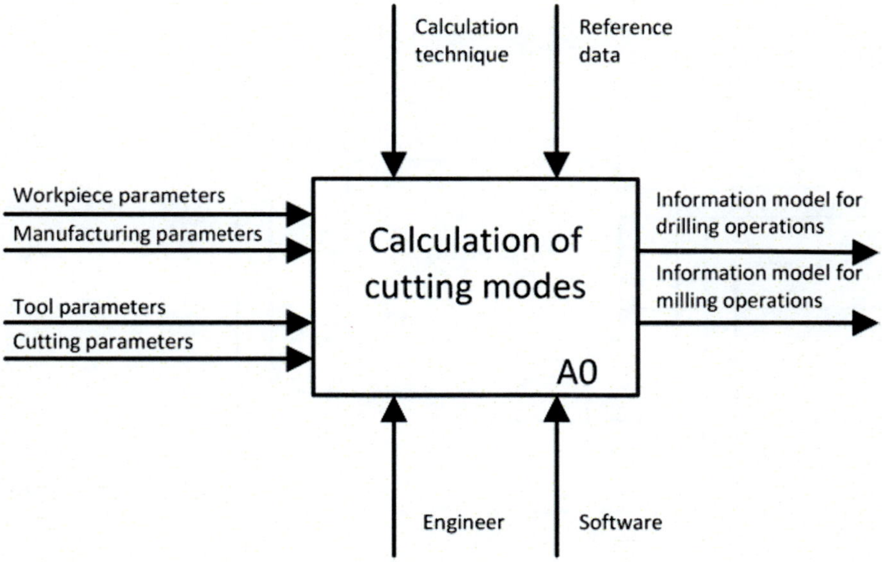

Fig. 3 IDEF0 diagram of calculation of cutting modes

4 Results

At the beginning of the creation of subsystems, it is essential to understand all the processes that should take place in the engineering analysis module and that need to be automated. Therefore, a process-oriented model has been developed that visually describes the execution of cutting mode calculations.

The calculation of cutting modes includes the design of drilling operations and the design of milling operations. Figure 3 presents an IDEF0 diagram of the calculation of cutting modes. The decomposition of the calculation of cutting modes for drilling and milling operations is illustrated in Fig. 4.

Cutting modes calculation is presented in detail in the diagrams of lower-level decompositions. The design of drilling operations includes determining the coefficients, cutting depth, cutting speed, torque, spindle rotation frequency, torque, axial force, cutting power, and creating an information model of drilling operations (Fig. 5). The milling operation is decomposed into the determination of milling coefficients, calculation of cutting speed, spindle rotation frequency, cutting force, torque, power, and creation of an information model of milling operation (Fig. 6).

The general view of user interaction with the subsystem is described in the form of a UML model, which represents the possible options for using the software modules of the subsystem and sequence diagrams depicting the flow of events within the framework of a specific use case.

The following options for using software modules are provided.

Option "View/Change Coefficients" allows the user to view, add, change, and delete the coefficients used in calculating cutting modes.

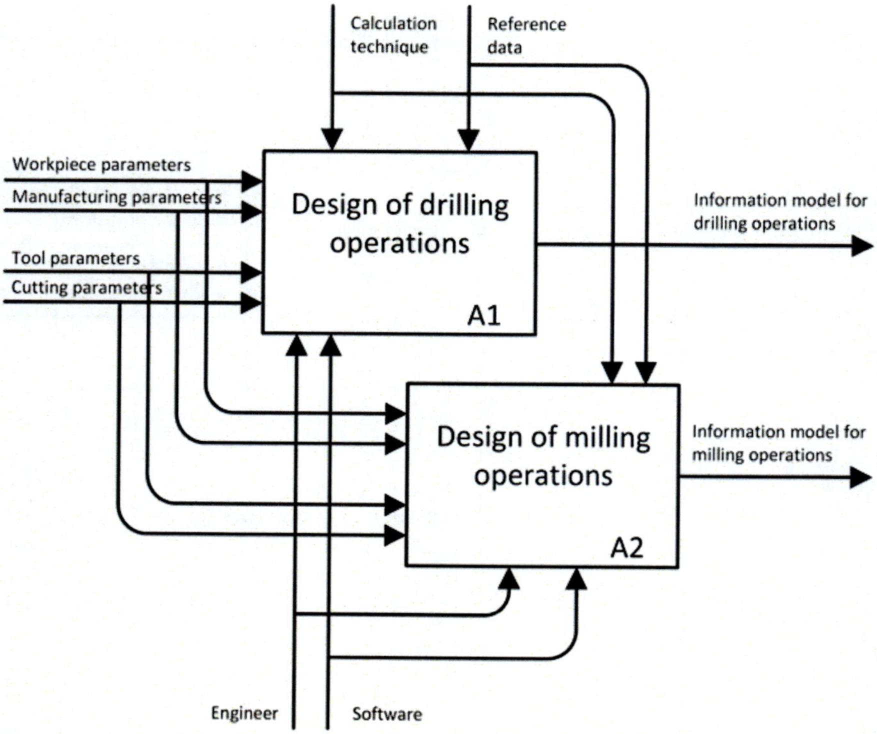

Fig. 4 Decomposition of the calculation of cutting modes

Process: the user selects the menu item "Library of cutting mode coefficients," and the form for viewing/editing cutting mode coefficients appears. The user fills in the required fields in the form. Depending on the entered data and the user's choice, the coefficients used in the calculation of cutting modes are reviewed, added, changed, or deleted. If a coefficient is changed, added, or deleted, the table of coefficients in the database is updated.

Option "Calculation of cutting modes" allows the user to design drilling and milling operations.

Process: the user selects the menu item "Cutting mode calculation," and the cutting mode calculation form appears. The user fills in the required fields and clicks the "Calculate" button. Calculation data is recorded in the database. At the user's request, an MS Excel report is generated with the results of the calculations.

According to the built models, appropriate software modules were developed and integrated into the general CAFD. The windows' design style and the developed software's architecture model coincide with those adopted in the basic software.

The module for working with the database (Fig. 7) is intended for viewing, editing, adding, and deleting coefficients for calculating cutting modes.

The user enters the parameters necessary for performing calculations when filling in the input data (Fig. 8).

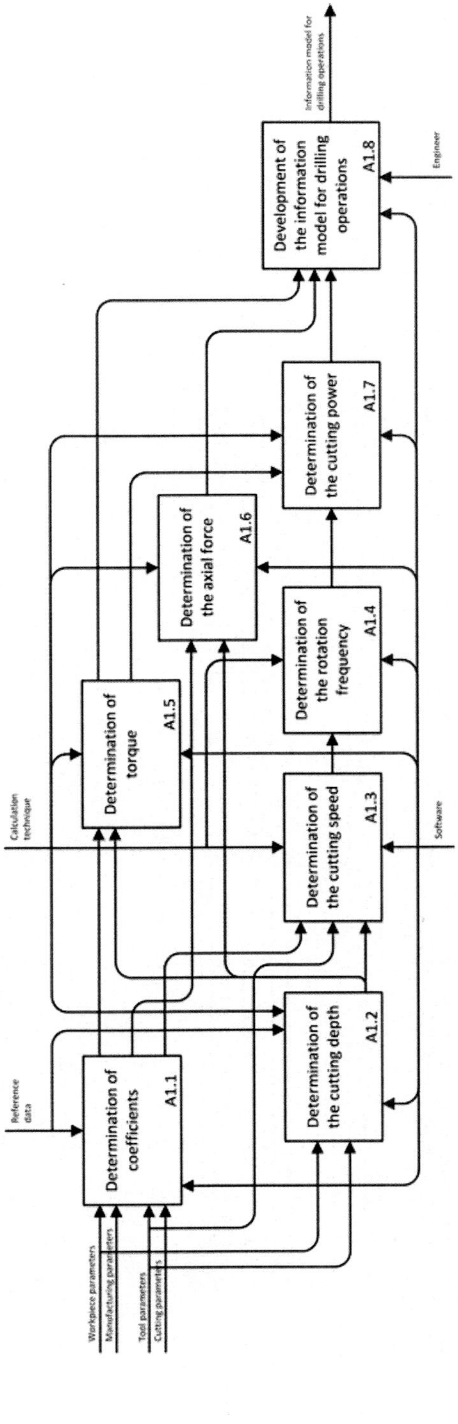

Fig. 5 Decomposition of the design of drilling operations

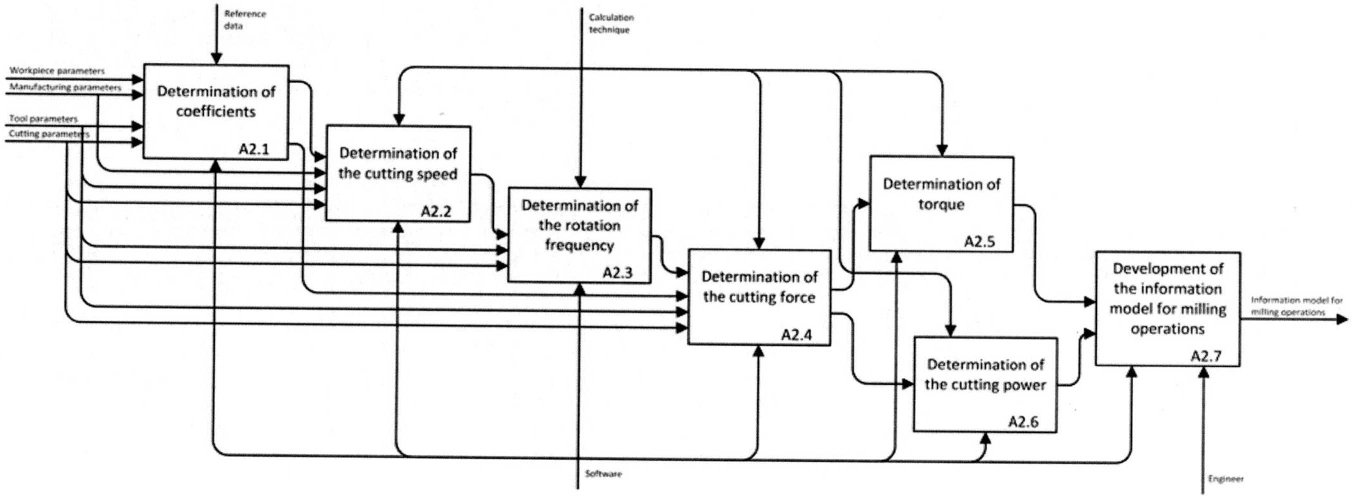

Fig. 6 Decomposition of designing a milling operation

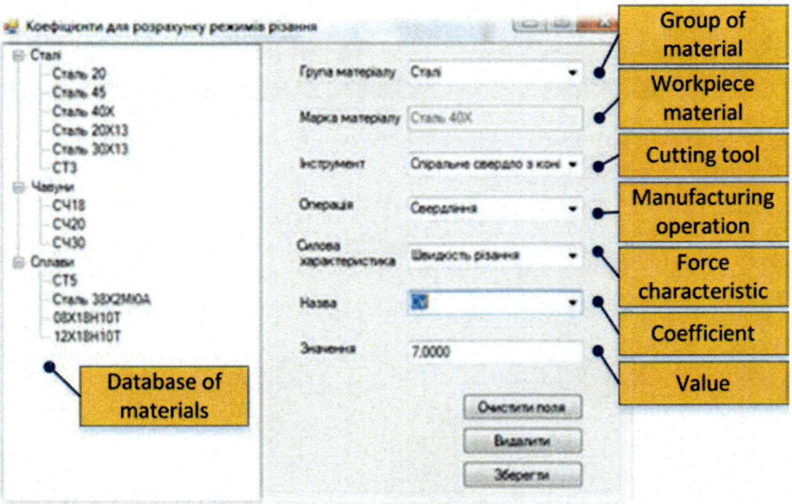

Fig. 7 Database module

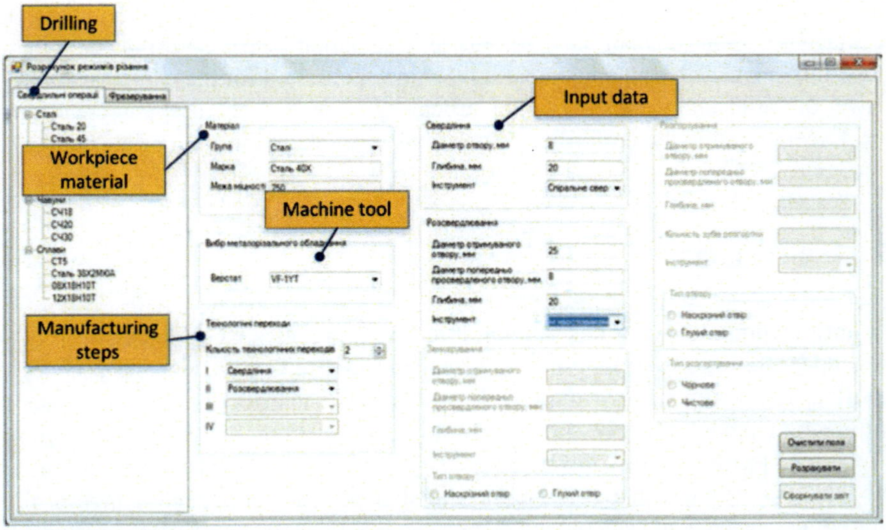

Fig. 8 Filling in input data

The module for calculating cutting modes (Fig. 9) is intended for designing drilling and milling operations. The form has two tabs that allow you to perform the corresponding functions.

All modules are checked for completeness of fields, correct input, or compatibility of individual parameters. In case of an error, a corresponding message is displayed. If all fields are filled in correctly, the cutting modes are calculated, a record is automatically created in the database containing the calculation results, and a

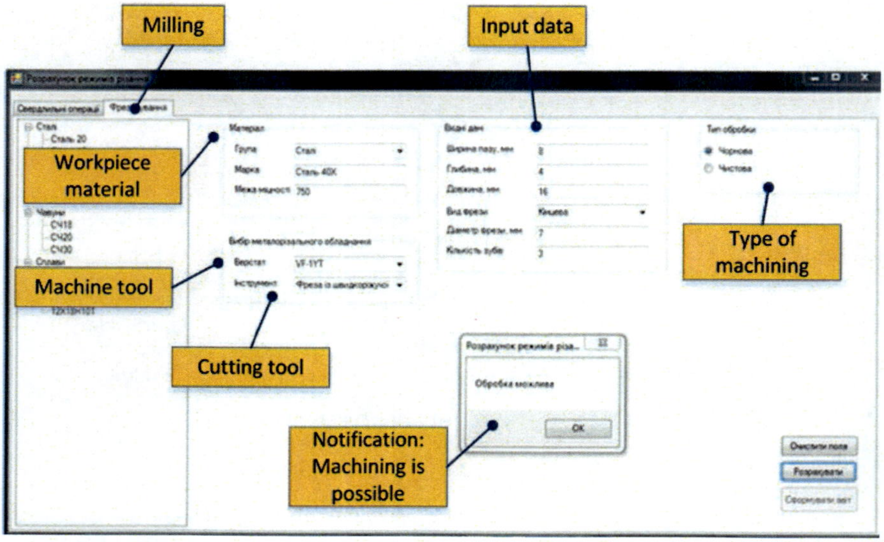

Fig. 9 Module for calculation of cutting modes

message is displayed about the possibility of calculation. The conclusion about this possibility is made based on comparing the power required for machining with the power of the machine considering the efficiency. If desired, the user can create an MS Excel report on the performed calculations by clicking the appropriate button.

5 Conclusions

An information model for determining force characteristics when calculating cutting modes for drilling-milling-boring operations has been developed, and a subsystem that is a component of the CAFD system has been developed. The created information models clearly show performing calculations and the subsystem's structure, functions, and software implementation. The subsystem consists of a module for working with a database and a module for calculating cutting modes. To store reference data and calculation results, appropriate tables have been created in a database shared with the CAFD system.

The implemented subsystem of the fixture‘s layout analysis allows the calculation of the cutting modes in an automated manner, saving the obtained results in the electronic archive and generating a report. The application of the developed subsystem speeds up the implementation of project procedures. The software implementation of the subsystem expanded the functionality of the CAFD system.

The practical value of the work is in increasing the efficiency of the process of designing manufacturing processes using automated calculations of force characteristics during drilling, milling, and boring operations.

Acknowledgments This research was funded by the research project "Improvement of the Production Planning by Implementation of the Computer-Aided Fixture Design System" within the Joint Ukrainian–Slovak R & D Projects for the period of 2022–2023 granted by the Slovak Research and Development Agency (SK-UA-21-0060) and the Ministry of Education and Science of Ukraine (State reg. no. 0122 U002657). This publication is the result of the Project implementation "Development of excellent research capacities in the field of additive technologies for the Industry of the 21st century" (ITMS: 313011BWN5), supported by the Operational Program Integrated Infrastructure funded by the ERDF. The research was partially supported by the Research and Educational Center for Industrial Engineering (Sumy State University) and the International Association for Technological Development and Innovations.

References

1. Trojanowska, J., Pająk, E.: Using the theory of constraints to production processes improvement. In: Proceedings of the 7th International DAAAM Baltic Conference: Industrial Engineering, 2010, p. 109816. Tallinn (2010)
2. Dobrotvorskiy, S., Basova, Y., Dobrovolska, L., Sokol, Y., Kazantsev, N.: Big challenges of small manufacturing enterprises in industry 4.0. In: Ivanov, V., Trojanowska, J., Pavlenko, I., Zajac, J., Peraković, D. (eds.) Advances in Design, Simulation and Manufacturing III. DSMIE 2020 Lecture Notes in Mechanical Engineering, pp. 118–127. Springer, Cham (2020). https://doi.org/10.1007/978-3-030-50794-7_12
3. Reis, L., Varela, M.L.R., Machado, J.M., Trojanowska, J.: Application of lean approaches and techniques in an automotive company. Roman. Rev. Precis. Mechan. Optics Mechatron. **2016**(50), 112–118 (2016)
4. Varela, M.L.R., Madureira, A.M., Dantas, J.D., Santos, A.S., Putnik, G.D., Trojanowska, J., Machado, J.: Collaborative paradigm for single-machine scheduling under just-in-time principles: Total holding-tardiness cost problem. Manage. Prod. Eng. Rev. **9**(1), 90–103 (2018). https://doi.org/10.24425/119404
5. Bakhovskyy, P., Yevsiuk, M., Zabolotnyi, O., Cagáňová, D., Tkachuk, A.: Stages of the virtual technical functions concept networks development. In: Cagáňová, D., Horňáková, N., Pusca, A., Cunha, P.F. (eds.) Advances in Industrial Internet of Things, Engineering and Management. EAI/Springer Innovations in Communication and Computing, pp. 119–135. Springer, Cham (2021). https://doi.org/10.1007/978-3-030-69705-1_7
6. Chetverzhuk, T., Zabolotnyi, O., Sychuk, V., Polinkevych, R., Tkachuk, A.: A method of body parts force displacements calculation of metal-cutting machine tools using CAD and CAE technologies. Annal. Emerg. Technol. Comput. **3**(4), 37–47 (2019). https://doi.org/10.33166/AETiC.2019.04.004
7. Hashemi, H., Shaharoun, A.M., Izman, S., Ganji, B., Namazian, Z., Shojaei, S.: Fixture design automation and optimization techniques: review and future trends. Int. J. Eng. Transact. Appl. **27**(11), 1787–1794 (2014). https://doi.org/10.5829/idosi.ije.2014.27.11a.17
8. Ivanov, V., Vashchenko, S., Rong, Y.: Information support of the computer-aided fixture design system. In: 12th international conference on ICT in education, research and industrial applications, ICTERI 2016, vol. 1614, pp. 73–86. CEUR Workshop Proceedings (2016)
9. Ivanov, V., Pavlenko, I., Vashchenko, S., Zajac, J.: Information system for computer-aided fixture design. In: Knapčíková, L., Balog, M. (eds.) Industry 4.0: Trends in Management of Intelligent Manufacturing Systems. EAI/Springer Innovations in Communication and Computing, pp. 121–132. Springer, Cham (2019). https://doi.org/10.1007/978-3-030-14011-3_11
10. Grabis, J., Haidabrus, B., Protsenko, S., Protsenko, I., Rovna, A.: Data science approach for IT project management. Vide. Tehnologija. Resursi Environ, Technol. Res. **2**, 51–55 (2019). https://doi.org/10.17770/etr2019vol2.4163

11. Yakovenko, I., Permyakov, A., Prihodko, O., Basova, Y., Ivanova, M.: Structural optimization of technological layout of modular machine tools. In: Tonkonogyi, V., et al. (eds.) Advanced Manufacturing Processes. InterPartner 2019 Lecture Notes in Mechanical Engineering, pp. 352–363. Springer, Cham (2020). https://doi.org/10.1007/978-3-030-40724-7_36
12. Sokolov, V., Porkuian, O., Krol, O., Stepanova, O.: Design calculation of automatic rotary motion electrohydraulic drive for technological equipment. In: Ivanov, V., Trojanowska, J., Pavlenko, I., Zajac, J., Peraković, D. (eds.) Advances in Design, Simulation and Manufacturing IV. DSMIE 2021 Lecture Notes in Mechanical Engineering, pp. 133–142. Springer, Cham (2021). https://doi.org/10.1007/978-3-030-77719-7_14
13. Yakovenko, I., Permyakov, A., Ivanova, M., Basova, Y., Shepeliev, D.: Lifecycle management of modular machine tools. In: Tonkonogyi, V., Ivanov, V., Trojanowska, J., Oborskyi, G., Pavlenko, I. (eds.) Advanced Manufacturing Processes III. InterPartner 2021 Lecture Notes in Mechanical Engineering, pp. 127–137. Springer, Cham (2022). https://doi.org/10.1007/978-3-030-91327-4_13
14. Hovorun, T.P., Pylypenko, O.V., Berladir, K.V., Dyadyura, K.O., Dunaeva, M.N., Vorobiov, S.I., Panda, A.: Physical-mechanical properties and structural-phase state of nanostructured wear-resistant coatings based on nitrides of refractory metals Ti and Zr. Funct. Mater. **26**(3), 548–555 (2019). https://doi.org/10.15407/fm26.03.548
15. Hovorun, T.P., Berladir, K.V., Bilous, O.A., Liubych, O.Y., Vorobiov, S.I.: Development of alloy resistant in conditions of abrasive wear. Funct. Mat. **28**(1), 170–177 (2021). https://doi.org/10.15407/fm28.01.170
16. Wu, H., Zheng, H., Li, X., Wang, W., Xiang, X., Meng, X.: A geometric accuracy analysis and tolerance robust design approach for a vertical machining center based on the reliability theory. Measure. J. Int. Measure. Confeder. **161**, 107809 (2020). https://doi.org/10.1016/j.measurement.2020.107809
17. Denysenko, Y., Kysylevska, A., Panchenko, O., Zaloga, V., Dynnyk, O.: Decision-making based on prediction of oil quality indicators in the enterprise's information system. In: Advanced Manufacturing Processes. InterPartner 2019. Lecture Notes in Mechanical Engineering, pp. 25–31. Springer, Cham (2020). https://doi.org/10.1007/978-3-030-40724-7_3
18. Krol, O., Sokolov, V.: Modelling of spindle nodes for machining centers. J. Phys. Conf. Ser. **1084**(1), 012007 (2018). https://doi.org/10.1088/1742-6596/1084/1/012007
19. Kushnirov, P., Denysenko, Y., Ostapenko, B., Zhyhylii, D., Stupin, B.: Improvement of the milling effectiveness by application of composite milling heads. In: Ivanov, V., Trojanowska, J., Pavlenko, I., Rauch, E., Peraković, D. (eds.) Advances in Design, Simulation and Manufacturing V. DSMIE 2022 Lecture Notes in Mechanical Engineering, pp. 293–301. Springer, Cham (2022). https://doi.org/10.1007/978-3-031-06025-0_29
20. Dobrotvorskiy, S., Kononenko, S., Basova, Y., Dobrovolska, L., Edl, M.: Development of optimum thin-walled parts milling parameters calculation technique. In: Ivanov, V., Trojanowska, J., Pavlenko, I., Zajac, J., Peraković, D. (eds.) Advances in Design, Simulation and Manufacturing IV. DSMIE 2021 Lecture Notes in Mechanical Engineering, pp. 343–352. Springer, Cham (2021). https://doi.org/10.1007/978-3-030-77719-7_34
21. Lv, L., Deng, Z., Meng, H., Liu, T., Wan, L.: A multi-objective decision-making method for machining process plan and an application. J. Clean. Prod. **260**(1), 121072 (2020). https://doi.org/10.1016/j.jclepro.2020.121072
22. Fayzimatov, S., Xusanov, Y., Valixonov, D.: Optimization conditions of drilling polymeric composite materials. Am. J. Eng. Technol. **3**(02), 22–30 (2021). https://doi.org/10.37547/tajet/Volume03Issue02-04
23. Varghese, V., Ramesh, M.R., Chakradhar, D.: Experimental investigation and optimization of machining parameters for sustainable machining. Mater. Manuf. Process. **33**(16), 1782–1792 (2018). https://doi.org/10.1080/10426914.2018.1476760
24. Seco Assistant. https://play.google.com/store/apps/details?id=com.secotools.assistant. Last accessed 2022/11/26

25. Machining Calculator. https://play.google.com/store/apps/details?id=com.sandvik.material-calculator.activities. Last accessed 2022/11/26
26. Walter Machining Calculator. https://play.google.com/store/apps/details?id=com.walter_tools.wmc. Last accessed 2022/11/26
27. TaeguTec Tool CutZZ. https://play.google.com/store/apps/details?id=com.taegutec.catalog. Last accessed 2022/11/26
28. ISCAR Cutting Info. https://play.google.com/store/apps/details?id=com.iscar.cuttinginfo. Last accessed 2022/11/26
29. Speed & Feed. https://play.google.com/store/apps/details?id=com.kennametal.Pilot. Last accessed 2022/11/26
30. Hoffmann Machining Calculator. https://play.google.com/store/apps/details?id=de.hoffmann.zr&hl=en. Last accessed 2022/11/26
31. FSWizard Machinist Calculator. https://play.google.com/store/apps/details?id=com.beta.fswizard_lite&hl=en. Last accessed 2022/11/26
32. Machining Calculator. https://play.google.com/store/apps/details?id=com.millingcalculator.millingcalculator&hl=en. Last accessed 2022/11/26

Improving the Performance of Grey-Box Model of 3-DOF Compliant Robotic Arm with Fluidic Muscles

Tomáš Čakurda, Monika Trojanová, Alexander Hošovský, and Oleksandr Sokolov

1 Introduction

Terms such as robot and manipulator are already self-evident and an integral part of today's not only industrial practice. These robotic wholes can help people with difficult tasks/actions, or replace a person where the working conditions are not suitable for his health. However, simultaneously, with the development, research, and application of soft actuators comes another dimension of robotics, which is characterized by the phrase safe interaction. When using rigid actuators, a person's movement in the robot's space is limited to prevent accidents and damage to health. Using soft actuators makes human interaction with the machine (robot/manipulator) safer. Such soft actuators also include pneumatic artificial muscles, which have different properties depending on the design of the muscles, the materials used, or the operating parameters. A more detailed overview of the different types of pneumatic artificial muscles, including the originator, time classification, and brief description, is given in [1]. The basic characteristic feature of pneumatic artificial muscles is their property – contraction (contraction/expansion/rotation) as a result of the action of a certain stimulus (change in pressure, tension, current, etc.). The resulting desired movement of the manipulator, driven by this type of muscles, is, therefore (among other things) dependent on a detailed examination of the individual properties of the actuator; kinematics, and dynamics of the given drive, but also of the entire device.

T. Čakurda (✉) · M. Trojanová · A. Hošovský · O. Sokolov
Faculty of Manufacturing Technologies with a seat in Prešov, Technical University of Košice, Prešov, Slovakia
e-mail: tomas.cakurda@tuke.sk

M. Balog et al. (eds.), *EAI International Conference on Automation and Control in Theory and Practice*, EAI/Springer Innovations in Communication and Computing, https://doi.org/10.1007/978-3-031-31967-9_24

The field of kinematics and dynamics modeling is described in many research papers (e.g. [2] or [3]), as it represents an important part of the subsequent control of systems. Well-known books that describe dynamic and kinematic modeling of systems are works by [4, 5]. If it is necessary to describe the relationship between two parameters of the system, it is advisable to use dynamic modeling and description using a dynamic model. The latter can be derived using two basic principles [6]. The first is the Newton-Euler formulation, separately described parts. The more complex the system, the more demanding the process of dynamic modeling using this method becomes. The second principle is the Euler-Lagrange formulation. Compared to the first method, this method describes the system as a whole using the system's total kinetic and potential energy. This method is very often used in the field of mechanics to derive dynamic models of systems. For example, in one of the latest works [7] the Euler-Lagrange method was used to obtain a decoupled position and orientation fractional-order dynamic model of a two-links robotic manipulator. In another work [8], the authors model the dynamics of a robotic manipulator, which has a horizontal position of the base upwards and has four degrees of freedom. Both principles of dynamic modeling (Euler-Lagrange and Newton-Euler) were used and described in [9] for the derivation of motion equations during different walking stages of a biped robot. Similarly, to derive the equations of motion of the four-legged robot prototype (Mini-Bot), the authors in [10] used the Euler-Lagrange method, tracking the position and angular velocity for each leg of the robot.

In this article, the authors describe the dynamic model of an experimental manipulator and then the process of validation of the created model. Since the system is powered by pneumatic artificial muscles that have specific properties, the overall modeling process was all the more challenging. The system model consists of a dynamic model of the active members and a dynamic model of the movement links. Newton's second law was used to describe the dynamics of the drive, and Lagrangian formalism was used to describe the dynamics of the arm. The Matlab Simulink software was used to create the simulation model, while the model contained the parameters of the system, which were obtained from the 3D CAD model of the given manipulator. As part of the model validation process, the conformity of the simulated outputs (joint rotation angle) with the real ones from the system was monitored using the NRMSE and MAE statistical criteria. To ensure the highest degree of agreement between the simulated and the real output, the parameters of the model were optimized using the Trust Region Algorithm for the nonlinear optimization method of least squares. The article is formally divided into five chapters, where the introduction is followed by an explanation of the theoretical starting points oriented toward the description of the dynamic model of the investigated system. The third part of the article describes the investigated experimental system itself. The fourth chapter deals with model validation, and the fifth is devoted to parameter estimation.

2 Design of the Dynamic Model

The experimental system represents a robotic arm formed by three links with three rotating joints. The scheme of basic structural components, their location, and their designation are shown in Fig. 1.

2.1 Dynamic Model of Manipulator Arm

The Lagrange equations were used to derive the dynamic model of arm movement. The general formulation of the Lagrange equations for experimental manipulator is given in Eq. (1) [12]:

$$\frac{d}{dt}\left[\frac{\partial L}{\partial \dot{\varphi}_i}\right]-\frac{\partial L}{\partial \varphi_i}=\tau_i \qquad i=1,2,3 \tag{1}$$

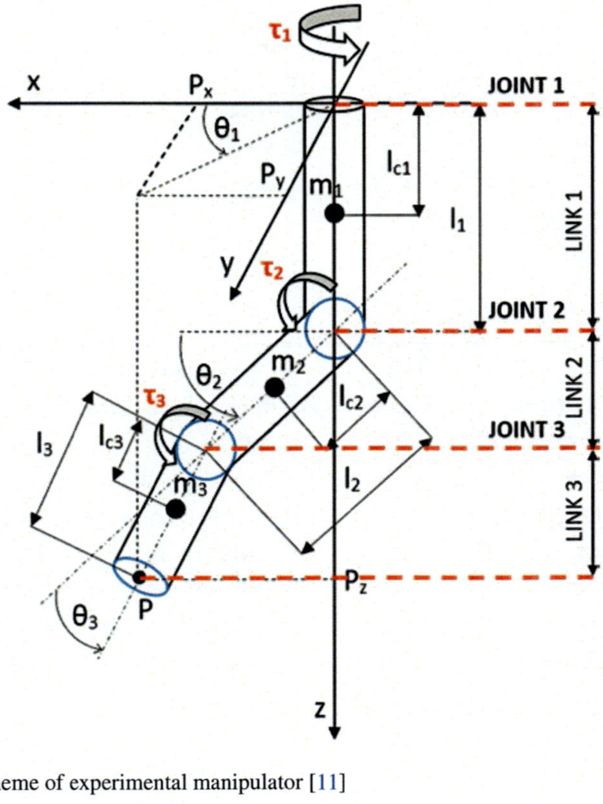

Fig. 1 The scheme of experimental manipulator [11]

where L is the Lagrangian, φ_i is the joint position, derivation of the joint position is joint velocities, and τ_i is the force term.

The dynamics of the arm movement are described by the equations that were derived based on the Lagrangian calculation and the simultaneous application of Eq. (1). A summary overview of variables, their designations, and units is presented in Fig. 2. The final forms for the force components τ_1, τ_2, and τ_3 of Link 1, 2, and 3 are determined by Eqs. (2, 3, and 4) [13]

$$\begin{aligned}\tau_1 = &\left[\begin{array}{l}\left(I_{y2}-I_{z2}-m_2 l_{c2}^2\right)c_2 s_2+\left(I_{y3}-I_{z3}\right)c_{23}s_{23}-m_3\left(l_2 c_2+l_{c3}c_{23}\right)\\ \left(l_2 s_2+l_{c3}s_{23}\right)\end{array}\right]\dot{\theta}_1\dot{\theta}_2+\\ &\left[\begin{array}{l}\left(I_{y2}-I_{z2}-m_2 l_{c2}^2\right)c_2 s_2+\left(I_{y3}-I_{z3}\right)c_{23}s_{23}-m_3\left(l_2 \mathrm{c}_2+l_{c3}c_{23}\right)\\ \left(l_2 s_2+l_{c3}s_{23}\right)\end{array}\right]\dot{\theta}_1\dot{\theta}_2+\\ &\left[\left(I_{y3}-I_{z3}\right)c_{23}s_{23}-m_3 l_{c3}s_{23}\left(l_2 c_2+l_{c3}c_{23}\right)\right]\dot{\theta}_1\dot{\theta}_3+\\ &\left[\left(I_{y3}-I_{z3}\right)c_{23}s_{23}-m_3 l_{c3}s_{23}\left(l_2 c_2+l_{c3}c_{23}\right)\right]\dot{\theta}_1\theta_3+\\ &\left[I_{y2}s_2^2+I_{y3}s_{23}^2+I_{z1}+I_{z2}c_2^2+I_{z3}c_{23}^2+m_2 l_{c2}^2 c_2^2+m_3\left(l_2 c_2+l_{c3}c_{23}\right)^2\right]\end{aligned} \quad (2)$$

$$\begin{aligned}\tau_2 = &\left[I_{x3}+m_3 l_{c3}^2+m_3 1_2 l_{c3}c_3\right]\ddot{\theta}_3+\\ &\left[\begin{array}{l}\left(I_{z2}-I_{y2}+m_2 l_{c2}^2\right)c_2 s_2+\left(I_{z3}-I_{y3}\right)\\ c_{23}s_{23}+m_3\left(l_2 c_2+l_{c3}c_{23}\right)\left(l_2 s_2+l_{c3}s_{23}\right)\end{array}\right]\dot{\theta}_1^2\\ &-\left[l_2 m_3 l_{c3}s_3\right]\dot{\theta}_2\dot{\theta}_3-\left[l_2 m_3 l_{c3}s_3\right]\dot{\theta}_2\dot{\theta}_3-\left[l_2 m_3 l_{c3}s_3\right]\dot{\theta}_3^2\\ &+\left[I_{x2}+I_{x3}+m_3 l_2^2+m_2 l_{c2}^2+m_3 l_{c3}^2+2m_3 l_2 l_{c3}c_3\right]\\ &-\left(m_2 g l_{c2}+m_3 g l_2\right)c_2-m_3 l_{c3}c_{23}\end{aligned} \quad (3)$$

Variable	Symbol	Unit	Variable	Symbol	Unit
Link moment of inertia	I	kg.m^2	Absolute muscle pressure	P_m	Pa
Atmospheric pressure	P_a	Pa	Muscle volume	V_m	m^3
Mass to move	m	kg	Muscle contraction	ζ	%
Muscle pair torque	τ	N.m	Volume of compressed air	V_a	m^3
Link length	l	m	Muscle force	F	N
Link center of mass	l_c	m	External muscle force	F_e	N
Joint angle	θ	rad	Muscle damping coefficient	μ	m^2·s
Sonic conductance	C	m^3·s^{-1}·Pa^{-1}	Sprocket diameter	d	m
Gravity force	F_g	N	Gravity constant	g	m.s^{-2}

Fig. 2 Variables used in dynamic model [12]

$$\tau_3 = \left[I_{x3} + m_3 l_{c3}^2 + m_3 l_2 l_{c3} c_3\right]\ddot{\theta}_2 + \left[\left(I_{z3} - I_{y3}\right)c_{23}s_{23} + m_3 l_{c3} s_{23}\left(l_2 c_2 + l_{c3} c_{23}\right)\right]$$
$$\dot{\theta}_1^2 + l_2 m_3 l_{c3} s_3 \dot{\theta}_2 + \left(I_{x3} + m_3 l_{c3}^2\right) - m_3 g l_{c3} c_{23} \quad (4)$$

In a compact form, the dynamic model expressed by three nonlinear differential equations can be expressed as Eq. (5) [14]. When the manipulator moves, friction occurs in the joints, which must be added to Eq. (5). Equation 6, supplemented by the friction component, represents the complete dynamic model base on Lagrangian formalism [14]. Equation 7 represents the friction force, where the friction model used is based on the augmented Coulomb friction model [15]. The used friction model is a combination of the Coulomb model, and the viscous friction model and also includes a term describing the Stribeck effect. Constants f_{ci}, f_{di}, f_{ei}, f_{gi} where i = 1, 2, 3 are parameters obtained base on gravity tests and $\dot{\theta}$ is the angular velocity [16]

$$M(\theta)\ddot{\theta} + C\left(\theta,\dot{\theta}\right)\dot{\theta} + G(\theta) = \tau \quad (5)$$

$$M(\theta)\ddot{\theta} + C\left(\theta,\dot{\theta}\right)\dot{\theta} + G(\theta) + F\left(\dot{\theta}\right) = \tau \quad (6)$$

$$F\left(\dot{\theta}\right) = f_{ci}\,\mathrm{sgn}\left(\dot{\theta}\right) + f_{di}\dot{\theta} + f_{ei}\dot{\theta}^2\,\mathrm{sgn}\left(\dot{\theta}\right) + f_{gi}\dot{\theta}^3 \quad (7)$$

2.2 Dynamic Model of Actuators

To supplement the dynamic model of the system, it is necessary to derive a dynamic model of pneumatic artificial muscles representing the used type of actuator of links. Newton's second law was applied to derive the dynamic model of contraction representing by Eq. (8), where m – moved mass, y – muscle displacement, $F_D\left(\dot{\zeta},P_m\right)$ – nonlinear damper force, $F_S(\zeta, P_m)$ – nonlinear spring force, P_m – muscle pressure, ζ – muscle contraction, F_e – external (opposing) muscle force and F_g is gravity force [17]

$$m\ddot{y} + F_D\left(\dot{\zeta},P_m\right) + F_S\left(\zeta,P_m\right) = F_e - F_g \quad (8)$$

The rotary joint, which is driven by two pneumatic artificial muscles in an antagonistic connection under the action of gravity, represents the connection of two members next to each other, $F_D\left(\dot{\zeta},P_m\right)$ and $F_S(\zeta, P_m)$, where one acts against the gravitational force and the other acts in its direction. Based on this statement, it is possible to determine two equations for muscle movement (Eqs. 9 and 10). The effect of gravity does not affect joint 1; therefore, the relevant equations are applied only during the movement of Joint 2 and Joint 3. [18]

$$m\ddot{y}_1 + F_{D1}\left(\dot{\zeta}, P_m\right) + F_{S1}\left(\zeta, P_m\right) = F_{D2}\left(\dot{\zeta}, P_m\right) + F_{S2}\left(\zeta, P_m\right) + F_g \quad (9)$$

$$m\ddot{y}_2 + F_{D2}\left(\dot{\zeta}, P_m\right) + F_{S2}\left(\zeta, P_m\right) = F_{D1}\left(\dot{\zeta}, P_m\right) + F_{S1}\left(\zeta, P_m\right) - F_g \quad (10)$$

The nonlinear spring force in the given equations corresponds to the function for the PAM static force. This force depends on the pressure in the muscle and its contraction. This function was approximated using MLP (A multilayer perceptron - a feedforward network of simple neurons that transforms specified inputs into neural network outputs). The output is calculated by the perceptron, using the nonlinear activation function given by Eq. (11), where w is weights vector, x is the input vector and b is the bias. The basis for this calculation is a linear combination of real inputs. φ in Eq. (11) represents the activation function. The hyperbolic tangent represented the type of used activation function. The approximated function is then represented by Eq. (12). [19]

$$y = \varphi(z) = \varphi\left(\sum_{i=1}^{n} w_i x_i + b\right) \quad (11)$$

$$y = \varphi(z) = \tan h(z) = \frac{2}{\left(1 + e^{-2z}\right)^{-1}} \quad (12)$$

Equation 13 represents the calculation method of the nonlinear damper force, which expresses that the force is the product of pressure and the speed with which the muscle contracts. μ in Eq. (13) represent a damping coefficient [$m^2 \cdot s$].

$$F_D\left(\dot{\zeta}, P_m\right) = \mu P_m \dot{\zeta} \quad (13)$$

The nonlinear terms in Eq. (8) is associated with pressure, which implies the need to describe the dynamics of the pneumatic and its behavior over time. Assuming a gas in a closed space, the Boyle-Marriott law can be used to derive the dynamics. This terms is given in Eq. (14), where P_a – atmospheric pressure, V_a – volume of compressed air within muscle, P_m – absolute muscle pressure and V_m – muscle volume. Solving for P_m and taking the time derivative gives Eq. (15). [20]

$$P_a V_a = P_m V_m \quad (14)$$

$$\dot{P}_m = P_a \frac{\dot{V}_a}{V_m} - P_m \frac{\dot{V}_m}{V_m} \quad (15)$$

For the possibility of determining Eq. (15), it is necessary to define the volume of the muscles $V_m(\zeta)$ given by Eq. (16). Rate of change of volume $\dot{V}_m\left(\zeta, \dot{\zeta}\right)$ is given by Eq. (17). Equation 17 represents the time derivative of Eq. (16) [21]. Figure 3

	20-400N	40-400N	20-350N
a_1	$-1.778*10^{-3}$	$-1.535*10^{-2}$	$-1.543*10^{-3}$
a_2	$1.583*10^{-5}$	$1.581*10^{-3}$	$-5.312*10^{-7}$
a_3	$7.815*10^{-4}$	$2.994*10^{-3}$	$6.832*10^{-4}$
a_4	$1.256*10^{-4}$	$5.036*10^{-4}$	$1.099*10^{-4}$

Fig. 3 Coefficients from the curve fitting toolbox [11]

lists the values of the polynomial coefficients, which we determined using the Curve Fitting Toolbox in Matlab.

$$V_m(\zeta) = a_1\zeta^3 + a_2\zeta^2 + a_3\zeta + a_4 \tag{16}$$

$$\dot{V}_m(\zeta,\dot{\zeta}) = 3a_1\zeta^2\dot{\zeta} + 2a_2\zeta\dot{\zeta} + a_3\dot{\zeta} \tag{17}$$

The last necessary parameter for the derivation of a complete dynamic model of pressure in the muscle is the volume flow of air through the valve $\dot{V}_a$ is given by Eqs. (18 and 19), where P_1 is upstream pressure, P_2 is downstream pressure, T_0 – temperature of air at reference conditions, T_1 – upstream temperature of air, b – critical ratio, C – sonic conductance. [22]

$$\dot{V}_a = P_1 C\sqrt{\frac{T_0}{T_1}}\sqrt{1-\left(\frac{\frac{P_2}{P_1}-b}{1-b}\right)^2}, if\ \frac{P_2}{P_1} > b \tag{18}$$

$$\dot{V}_a = P_1 C\sqrt{\frac{T_0}{T_1}}, if\ \frac{P_2}{P_1} \le b \tag{19}$$

Based on the given model, we express the mass flow using two parameters of the valve, namely sonic conductance and critical ratio.

A pair of muscles in an antagonistic connection develops a torque during their movement, and the dynamics of muscle contraction depend on two nonlinear terms (Eq. 8). The torque is then given by Eq. (20), where F_1, F_2 are forces in muscles and d is sprocket diameter.

$$\tau = (F_1 - F_2) * d/2 \tag{20}$$

After the muscles are deflected, then stabilized in a certain position and taking into account Eq. (20), the muscles generate the difference between the forces to compensate for the torque under the influence of the gravitational force. The relevant forms are given in Eqs. (21 and 22), where $k = g/r$ and g – gravity constant and r – sprocket diameter.

$$\Delta F_1 = \left[\left(m_3 * l_2 + m_2 * l_{c2} \right) \cos\theta_2 + m_3 l_{c3} \cos\left(\theta_2 + \theta_3\right) \right] k \tag{21}$$

$$\Delta F_2 = m_3 l_{c3} \cos\left(\theta_2 + \theta_3\right) k \tag{22}$$

3 Description of the Experimental Manipulator

The investigated experimental system is a robotic manipulator arm that has three degrees of freedom and is powered by an unconventional type of soft actuators – fluid muscles. The movement of the arm is ensured by three pairs of muscles, which are produced by FESTO. Muscles of the DMSP type (with pressed connection) are used in three sizes (DMSP 20-400N-RM-CM, DMSP 40-400N-RM-CM, FESTO DMSP 20-350N-RM-CM), on the basis of which the muscles have different properties and work scopes. If the fluid muscle is used alone, a tensile force is developed during its operation, which is dependent on the pressure in the muscle and on the contraction of the muscle. Muscle contraction occurs when a medium is supplied to the muscle-compressed air. However, if the muscles are engaged antagonistically and the arm includes joints and the pulley-toothed belt subsystem, then the resulting movement of such a system is in the form of turning the arm, as the pulling force of the muscles is manifested as a torque due to the involved components. The experimental system is shown in Fig. 4.

The AIRSTAR AC 401/50 compressor is used for infusing the muscles of the mentioned system, and the pressure in the muscles is regulated by MATRIX ERP50-type pressure regulators. The advantage of these pressure regulators is that they also have built-in pressure sensors. In the experimental system, it is possible to monitor the connection of three joints and three links, while a load is placed at the end of the third link. The entire arm of the manipulator is attached to the upper platform of the structure. which is assembled from aluminum profiles. The arm is controlled using the Matlab Simulink software environment, while the power supply is provided by the KORAD KD3005D source and the control signals are processed by the I/O PCI card Humusoft MF624.

4 Validation of Dynamic Model

In the second chapter, a complete dynamic model of the experimental manipulator with three degrees of freedom of movement was derived. For the needs of validation, this model was exported to the Matlab Simulink program environment, where a simulation scheme was created, representing the real behavior of the relevant

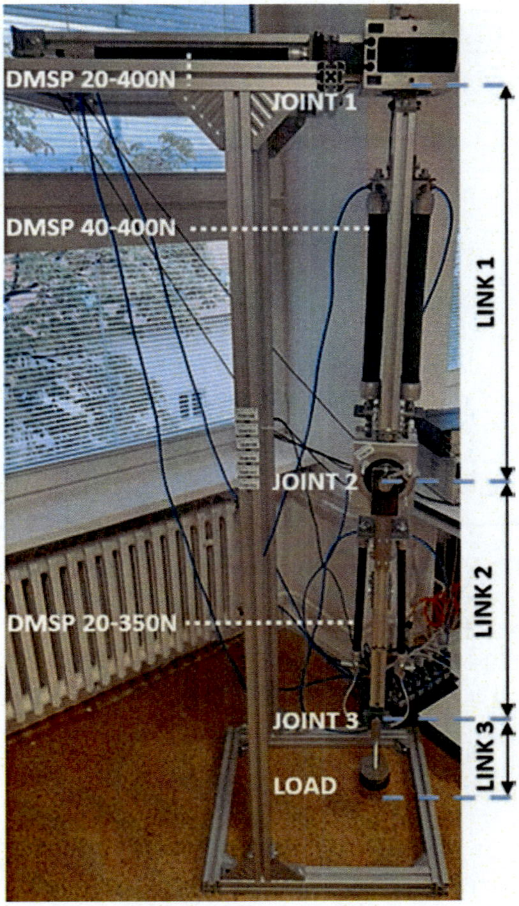

Fig. 4 Experimental robotic arm [11]

investigated system. To supplement the model, the relevant physical parameters were taken from the CAD software used to model the individual components of the manipulator and their connection into a complete 3D model of the system. The individual coefficients of the friction model used were obtained experimentally on the basis of gravity tests. The critical ratio and sonic conductance of the pressure regulators in the muscles correspond to the technical parameters specified by the manufacturer. The relevant parameter values are listed in Fig. 5, and their designations are explained in Fig. 2.

In the validation process, the simulation outputs, which represented the matrices from the Simulink environment, and the data processed on the basis of the measurements were compared. The comparison of simulation outputs (model) and system outputs (measurements) was made on the basis of two statistical indicators, namely NRMSE (Normal Root Mean Square Error) and MAE (Mean Absolute Error). The difference between them lies in the fact that the first mentioned indicator expresses the extent to which the compared outputs match. The mathematical formulation of

Designation	Value	Unit	Designation	Value	Unit
I_{x2}	0.0558	$kg.m^2$	f_{d1}	0.2248	N.m
I_{x3}	0.0234	$kg.m^2$	f_{e1}	0.0117	$N.m.s.rad^{-1}$
I_{y2}	0.0035	$kg.m^2$	f_{g1}	0.0812	N.m
I_{y3}	0.0003	$kg.m^2$	f_{c2}	0.0247	$N.m.s.rad^{-1}$
I_{z1}	0.0030	$kg.m^2$	f_{d2}	0.2248	N.m
I_{z2}	0.0035	$kg.m^2$	f_{e2}	0.0117	$N.m.s.rad^{-1}$
I_{z3}	0.0003	$kg.m^2$	f_{g2}	0.0812	N.m
l_2	0.8300	m	f_{c3}	0.0149	$N.m.s.rad^{-1}$
m_2	11.750	kg	f_{d3}	0.0523	N.m
m_3	4.2000	kg	f_{e3}	0.0100	$N.m.s.rad^{-1}$
l_{c2}	0.4150	m	f_{g3}	0.0100	N.m
l_{c3}	0.2300	m	b	0.4330	-
f_{c1}	0.0247	$N.m.s.rad^{-1}$	C	12.300	$m^3 \cdot s^{-1} \cdot Pa^{-1}$

Fig. 5 List of parameter values

NRMSE is given in Eq. (23). The most optimal value of the relevant indicator is 100%. On the other hand, the MAE indicator represents the created error in numerical expression, which arose during the simulation in comparison with the measured values (Eq. 24). [23]

$$NRMSE = \left(1 - \frac{\sqrt{\sum_{j=1}^{n} \left[y_j - \hat{y}_j \right]^2}}{\sqrt{\sum_{j=1}^{n} \left[y_j - \frac{1}{n} \sum_{j=1}^{n} y_j \right]^2}} \right) \times 100\% \tag{23}$$

$$MAE = \frac{1}{n} \sum_{j=1}^{n} \left| y_j - \hat{y}_j \right| \tag{24}$$

4.1 Muscle Dynamics Model Validation

To compare the created model to simulate the movement of a real system, it was necessary to obtain a representative sample of measurements from the experimental manipulator. The movement of the arms is ensured by three pairs of pneumatic artificial muscles based on the control signal from the Matlab program, respectively, its Signal Builder tool. During the measurements, the initial value of the pressure in the muscles was chosen to be 550 kPa, while subsequently, for the given time intervals, one of the pair of muscles was deflated by the preset pressure value, and after the expiration of the set time interval, it was pressurized again to the initial value. Under the specified conditions, ten measurements were made at a time interval of one measurement of 120 seconds. Based on the measurements, we obtained the observed parameter, which represented the course of the deviation of the angle of joints. The arithmetic mean was determined from the measurements taken and then compared with the output of the simulation based on the selected statistical indicators.

The control signal for the excitation of the upper pair of muscles driving Link 1 consisted in deflating the right muscle by 400 kPa in the twentieth second of the measurement and then pressurizing it in the fiftieth second to the original pressure value. The same procedure was repeated for deflating and inflating the left muscle in the seventieth and hundredth seconds of the measurement. For the movement of Link 3, we used the same excitation method as for Link 1. The average value of pressure in the muscles obtained by measurement and the simulated course of pressure using the model is shown for Link 1 in Fig. 6 and for Link 3 in Fig. 7.

Unlike the upper and lower pairs of muscles, the control signal for the excitation of the pair of muscles driving Link 2 was to release the left muscle by 300 kPa at the twentieth and return to the original pressure at the fiftieth second. Inflation and deflation of the right muscle by 300 kPa followed at the seventieth and hundredth seconds of the measurement. The average value of the pressure in the muscles obtained by measurement and the simulated course of the pressure using the model is shown for Link 2 in Fig. 8.

The results of the validation of the pneumatic part of the dynamic model of the experimental manipulator are presented in Fig. 9. The value of the statistical indicator NRMSE was around 90%. The values of the indicator ranged within the interval, with the lowest value of 87.83% for the right muscle of Link 2 and the highest value of 91.27% for the left muscle of Link 3. The comparison of the measured pressure

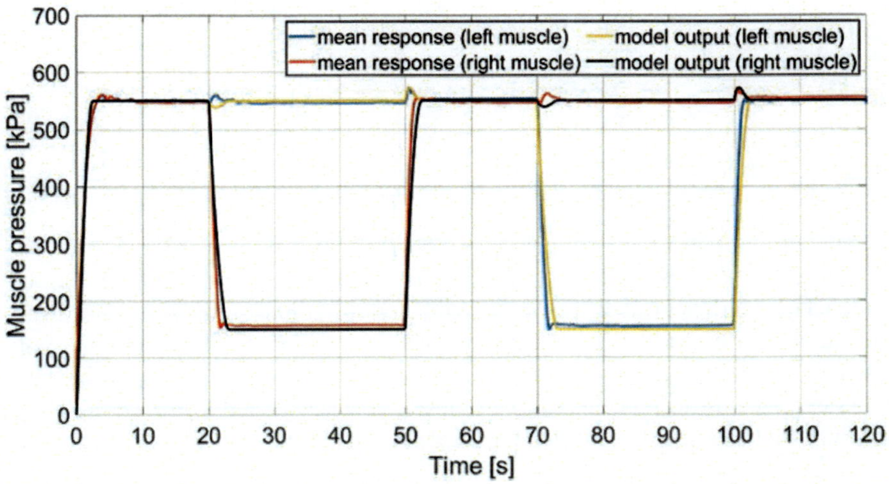

Fig. 6 Measured and simulated value of muscle pressure for Link 1

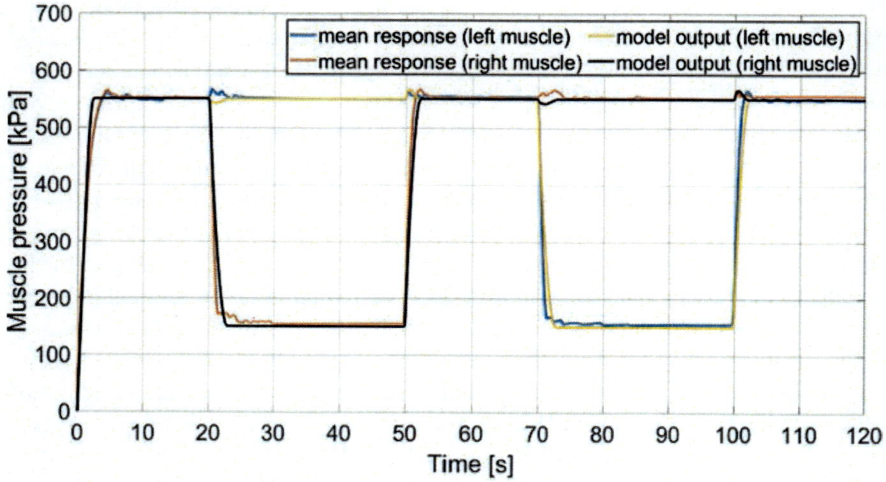

Fig. 7 Measured and simulated value of muscle pressure for Link 3

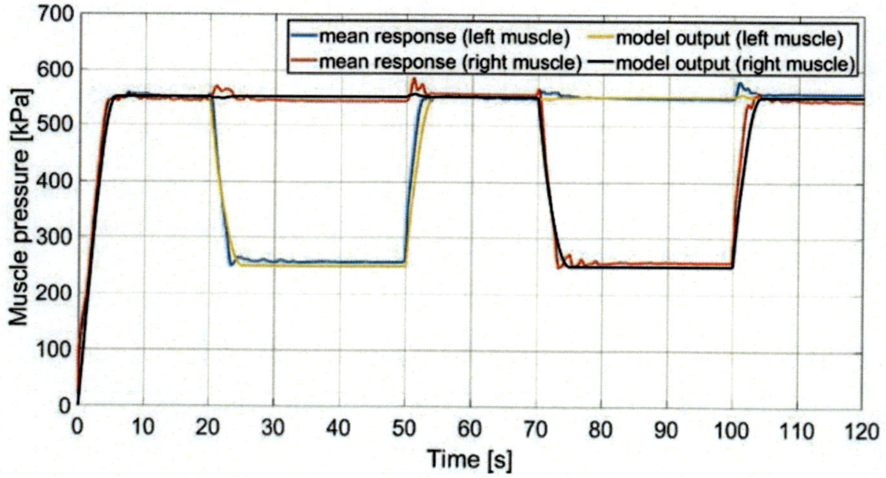

Fig. 8 Measured and simulated value of muscle pressure for Link 2

courses in the muscles and the simulation output showed that the pneumatic part is modeled on a sufficient level. The model satisfactorily simulates the real pressurization and depressurization of pneumatic muscles with minor deviations from the real system. Possible sources of discrepancy between the real system and the model are errors in modeling airflow dynamics, or muscle contraction and its time derivative, on which the volume of the artificial muscle depends.

Link 1			
Right muscle		Left muscle	
NRMSE	90.64 %	NRMSE	90.89 %
MAE	7.4142	MAE	6.4681
Link 2			
Right muscle		Left muscle	
NRMSE	87.83 %	NRMSE	89.56 %
MAE	9.9946	MAE	7.9934
Link 3			
Right muscle		Left muscle	
NRMSE	91.16 %	NRMSE	91.27 %
MAE	7.4877	MAE	6.2126

Fig. 9 Comparison of simulated and measured courses based on selected indicators

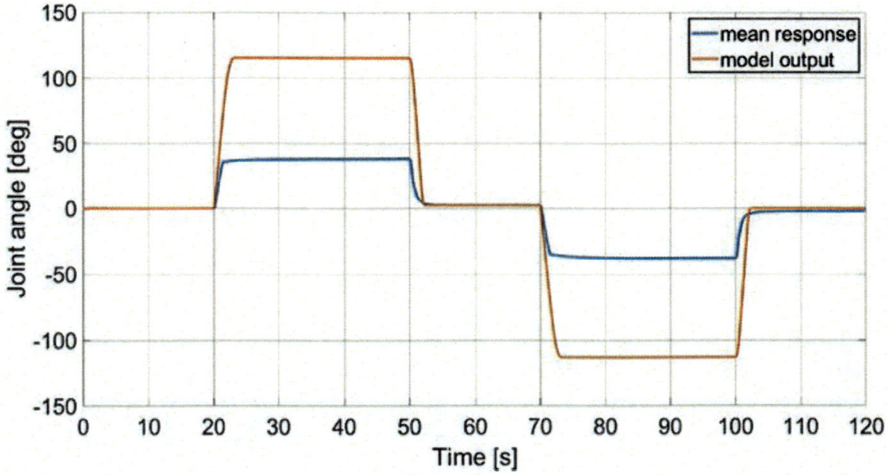

Fig. 10 Measured and simulated angle value for Link 1

4.2 *Complete Model Validation*

In the process of confirm of the model of the system, work was done with simultaneous propulsion by all three pairs of pneumatic artificial muscles. The control signals were identical to the previous chapter, and the investigated output parameter was monitored.

System and model outputs are illustrated in Figs. 10, 11, and 12.

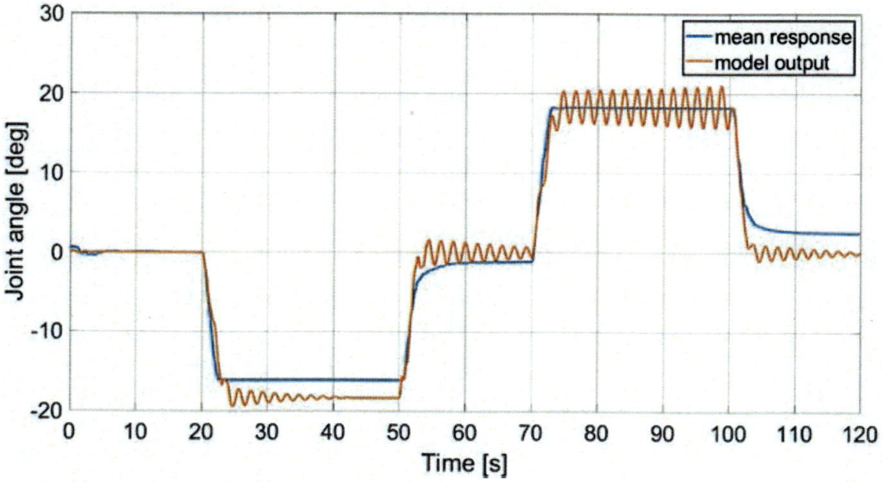

Fig. 11 Measured and simulated angle value for Link 2

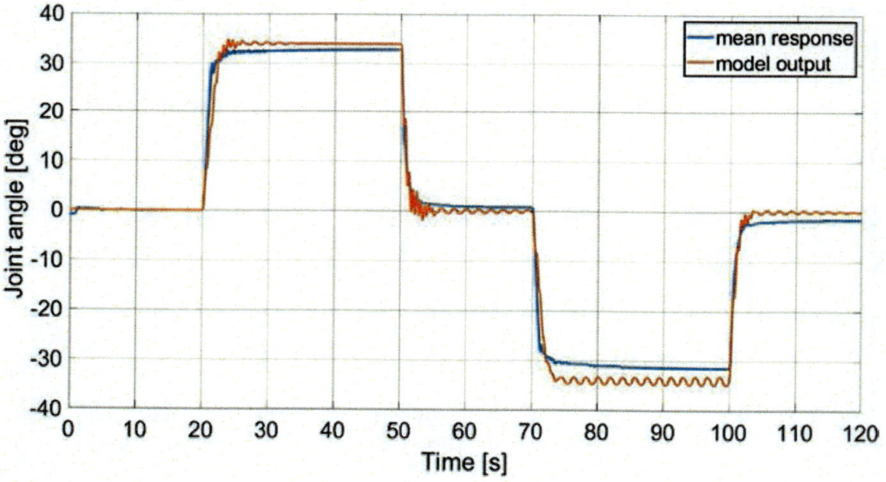

Fig. 12 Measured and simulated angle value for Link 3

The size of joint 1, which is simulated by the created scheme, has a higher value than the measured data. This indicates that the coefficients of friction are low to compensate for the torque exerted by the upper pair of muscles without the simultaneous action of gravity, as in Joint 2 and Joint 3, on the links of which gravity acts during movement.

The size of joint 2 and joint 3, which are the output of the simulation, correspond sufficiently with the measured sizes. Oscillations in their courses indicate the need to adjust friction coefficients compared to gravity tests.

Link 1		Link 2		Link 3	
NRMSE	32.96 %	NRMSE	84.81 %	NRMSE	90.8 %
MAE	38.5972	MAE	1.5986	MAE	1.6535

Fig. 13 Comparison of simulated and measured courses of angles based on selected indicators

The results of the validation based on the selected statistical indicators are presented in Fig. 13. The statistical indicators have the worst values for Joint 1 of Link 1, respectively NRMSE is equal to 32.93%, and MAE has a value of 38.5972. On the other hand, we achieved the best values for Joint 3 of Link 3. The NRMSE is 90.8%, and the MAE is 1.6535.

Based on the insufficient results of the confirm of model, taking into account the selected statistical indicators, in comparison with the measured values, parameter estimation in the Matlab Simulink environment was used to achieve better results. The estimator serves to describe the behavior of the model.

5 Parameter Estimation for Designed Model

To estimate the parameters of the dynamic model, we have to formulate it as an optimization problem. The optimization takes place according to the following equation [21]:

$$\begin{gathered}\min f(x)\\ x \in R^n\end{gathered} \tag{25}$$

It is necessary to determine the optimization algorithm and correctly define the objective function. A sum-squared type of error function is defined as [21]:

$$(\lambda) = \sum_{k=1}^{K}\left[\bar{\theta}(k) - \theta(k,\lambda)\right]^T \left[\bar{\theta}(k) - \theta(k,\lambda)\right] \tag{26}$$

where λ is a vector of measured values of joint angle, K is the number of samples and $\bar{\theta}$ is a vector of model parameters.

Numerical iterative methods are used to solve optimization tasks, that is, the approximate solution of the function at the point x_k is in the k-th iteration. The new point (approach to the solution) x_{k+1} is determined computationally based on an algorithm. This procedure is applied until we get an acceptable solution. We used the Trust Region Algorithm for Nonlinear Least Squares as an algorithm in the Matlab Simulink environment. The given algorithm considers the model to be reliable if the region is close to the solution. The given region is referred to as the credible region and is modified at each subsequent iteration, i.e., if in a certain iteration the approximate model is not good enough (it contains a bad point), the credible region will not increase, but on the contrary, will decrease.

When estimating the parameters, we focused on the parameters listed in the simulation scheme of the real system. To create the experiment, it was necessary to compile the matrices of the desired outputs, which represented the measured values of the angle of rotation of individual joints, depending on time. An important preparation step was also the selection of parameters that should be estimated.

Physical parameters exported from the 3D CAD model and individual coefficients of the friction model were selected for estimation. Individual parameters and bounds values are listed in Fig. 14. The basis for parameter estimation is the optimization method and algorithm, which determines the sequence of steps in which the estimation takes place and the criteria for parameter estimation itself. For estimation, we used the Parameter Estimator tool in Matlab Simulink. Based on parameter estimation in the Matlab Simulink program tool (Parameter Estimator), we received new, estimated parameter values. The simulation scheme of the experimental manipulator was modified by replacing it with estimated parameters that were reused in the validation of the complete system model. The new values are shown in Fig. 15.

Based on the validation of the complete dynamic model of the investigated system, we obtained different values of the selected statistical indicators. The new values are shown in Fig. 16. In contrast to the initial validation, we achieved the desired improvement of indicators, especially for Link 1. The value of the NRMSE indicator for Link 1 and Link 2 reached an almost identical number. Within the MAE statistical indicator, Link 3 achieved the best value of 1.0969.

The improvement of the model outputs compared to the measured data is mainly shown by the graphical curves after the parameter estimation process. A comparison

Designation	Parameter Estimation Bounds	Designation	Parameter Estimation Bounds
I_{x2}	0.001 – 0.1 kg.m^2	f_{d1}	0.01 – 5 N.m
I_{x3}	0.001 – 0.1 kg.m^2	f_{e1}	0.01 – 5 N.m.s.rad^{-1}
I_{y2}	0.0001 – 0.01 kg.m^2	f_{g1}	0.01 – 5 N.m
I_{y3}	0.0001 – 0.01 kg.m^2	f_{c2}	0.01 – 5 N.m.s.rad^{-1}
I_{z1}	0.0001 – 0.01 kg.m^2	f_{d2}	0.01 – 5 N.m
I_{z2}	0.0001 – 0.01 kg.m^2	f_{e2}	0.01 – 5 N.m.s.rad^{-1}
I_{z3}	0.0001 – 0.01 kg.m^2	f_{g2}	0.01 – 5 N.m
m_2	11 – 13 kg	f_{c3}	0.01 – 5 N.m.s.rad^{-1}
m_3	3.5 – 5 kg	f_{d3}	0.01 – 5 N.m
l_{c2}	0.35 – 0.5 m	f_{e3}	0.01 – 5 N.m.s.rad^{-1}
l_{c3}	0.19 – 0.25 m	f_{g3}	0.01 – 5 N.m
f_{c1}	0.01 – 5 N.m.s.rad^{-1}		

Fig. 14 The list of the bounds of the parameters

Designation	Value	Unit	Designation	Value	Unit
I_{x2}	0.0546	$kg.m^2$	f_{d1}	1.0640	N.m
I_{x3}	0.0246	$kg.m^2$	f_{e1}	0.8156	$N.m.s.rad^{-1}$
I_{y2}	0.0010	$kg.m^2$	f_{g1}	2.7832	N.m
I_{y3}	0.0010	$kg.m^2$	f_{c2}	0.3254	$N.m.s.rad^{-1}$
I_{z1}	0.0019	$kg.m^2$	f_{d2}	0.0699	N.m
I_{z2}	0.0056	$kg.m^2$	f_{e2}	0.0364	$N.m.s.rad^{-1}$
I_{z3}	0.0003	$kg.m^2$	f_{g2}	2.6858	N.m
l_2	0.8300	m	f_{c3}	0.0257	$N.m.s.rad^{-1}$
m_2	12.522	kg	f_{d3}	0.0192	N.m
m_3	4.3007	kg	f_{e3}	0.2558	$N.m.s.rad^{-1}$
l_{c2}	0.4676	m	f_{g3}	0.1249	N.m
l_{c3}	0.2220	m	b	0.4330	-
f_{c1}	0.0631	$N.m.s.rad^{-1}$	C	12.300	$m^3 \cdot s^{-1} \cdot Pa^{-1}$

Fig. 15 Parameter values after parameter estimation

Link 1		Link 2		Link 3	
NRMSE	92.61 %	NRMSE	83.13 %	NRMSE	92.5 %
MAE	1.1763	MAE	1.3919	MAE	1.0969

Fig. 16 Comparison of courses of angles based on selected indicators after estimation

of system outputs (measured waveforms) and simulation outputs (model) after parameter estimation is shown in Figs. 17, 18, and 19. The estimated model is able to satisfactorily simulate the real movement of the experimental manipulator with smaller deviations from the measured data. The created model describes the behavior of the real system.

6 Conclusion

The goal of the relevant article was the validation of a dynamic model of an experimental manipulator. The goal was to create a dynamic model that would satisfactorily simulate the real movement of the manipulator. The simulation scheme was inserted in the Matlab Simulink with the addition of parameter values from the 3D CAD model and friction coefficients obtained experimentally. The dynamic model consisted of the dynamics of the movement of individual links obtained on the basis of the Lagrangian formalism and the dynamic model of the pneumatic part base on Newton's second law and Boyle–Mariott law. Based on the performance of a series of measurements, we obtained representative courses of rotation angles of

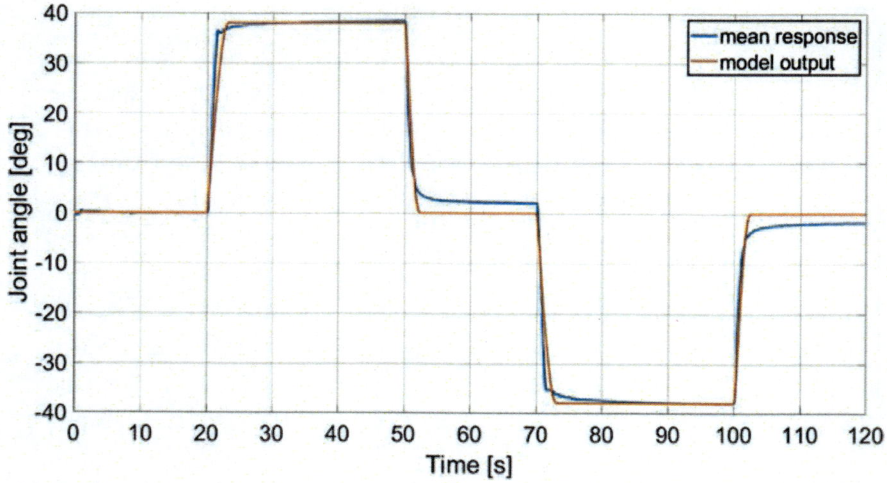

Fig. 17 Measured and simulated angle value after estimation for Link 1

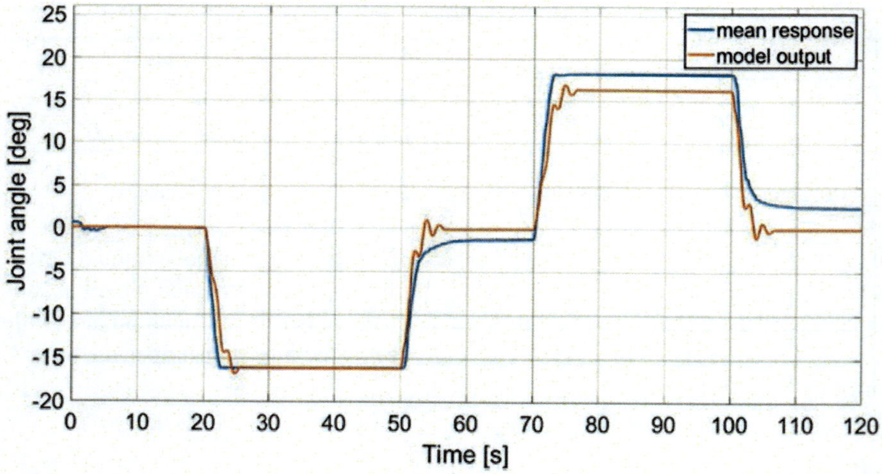

Fig. 18 Measured and simulated angle value after estimation for Link 2

individual links, which we compared with the simulation outputs. The comparison was made based on two statistical indicators NRMSE (Normal Root Mean Square Error) and MAE (Mean Absolute Error). The comparison showed that the pneumatic part is modeled at a sufficient level, but we improved the simulation of the movement of the arm based on the estimation of the selected parameters. A complete dynamic model with substituted parameter values simulates the movement of the experimental manipulator at a usable level with smaller deviations from the measured data. The obtained and represented results show that the created and subsequently estimated dynamic model describes the behavior of the real manipulator

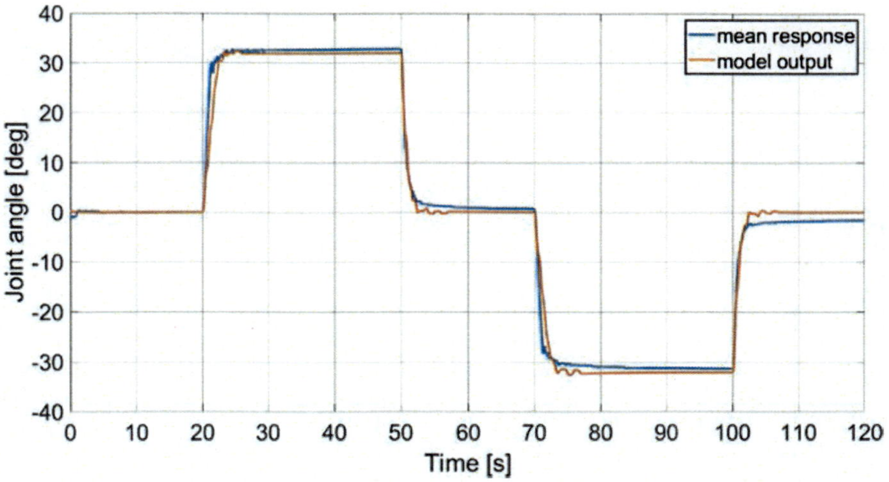

Fig. 19 Measured and simulated angle value after estimation for Link 3

at a sufficient level with the possibility of further applications of the simulation scheme.

Acknowledgments This work was supported by the Slovak Research and Development Agency under contract No. APVV-19-0590, and also by the projects VEGA 1/0061/23, KEGA 022TUKE-4/2023 granted by the Ministry of Education, Science, Research, and Sport of the Slovak Republic.

References

1. Takosoglu, J.E., Laski, P.A., Blasiak, S., Bracha, G., Pietrala, D.: Determining the static characteristics of pneumatic muscles. Meas. Control. **49**, 62–71 (2016)
2. Nezhad, M.N., Korayem, A.H.: Dynamic modeling of industrial manipulator Hyundai HS165 in order to determine the dynamic load-carrying capacity for a specified trajectory. In: 2019 7th International Conference on Robotics and Mechatronics (ICRoM). pp. 538–543 (2019).
3. Varela-Aldás, J., Andaluz, V.H., Chicaiza, F.A.: Modelling and control of a mobile manipulator for trajectory tracking. In: 2018 International Conference on Information Systems and Computer Science (INCISCOS). pp. 69–74 (2018).
4. Siciliano, B., Khatib, O.: Springer Handbook of Robotics. Springer (2016)
5. Kurfess, T.R.: Robotics and Automation Handbook. CRC Press (2018)
6. Spong, M.W., Hutchinson, S., Vidyasagar, M.: Robot Modeling and Control. Wiley (2020)
7. Azar, A.T., Serrano, F.E., Kamal, N.A., Kumar, S., Ibraheem, I.K., Humaidi, A.J., Gorripotu, T.S., Pilla, R.: Fractional-order euler–lagrange dynamic formulation and control of asynchronous switched robotic systems. In: Poonia, R.C., Singh, V., Singh Jat, D., Diván, M.J., Khan, M.S. (eds.) Proceedings of Third International Conference on Sustainable Computing, pp. 479–490. Springer Nature, Singapore (2022)
8. Al-Qahtani, H.M., Mohammed, A.A., Sunar, M.: Dynamics and control of a robotic arm having four links. Arab. J. Sci. Eng. **42**, 1841–1852 (2017)

9. Al-Shuka, H.F.N., Corves, B., Zhu, W.-H.: Dynamics of Biped Robots During a Complete Gait Cycle: Euler-Lagrange vs. Newton-Euler Formulations. School of Control Science and Engineering, Shandong University (2019)
10. Naing, S.Y., Rain, T.: Analysis of position and angular velocity of four-legged robot (mini-bot) from dynamic model using Euler-Lagrange method. In: International Conference on Industrial Engineering, Applications and Manufacturing (ICIEAM). pp. 1–4 (2019).
11. Cakurda, T., Trojanova, M., Hosovsky, A.: Complete dynamic model validation of 3- Dof compliant manipulator with fluidic muscles. MM Sci. J., 6139–6146 (2022)
12. Kelly, R., Davila, V.S., Perez, J.A.L.: Control of Robot Manipulators in Joint Space. Springer, London (2005)
13. Murray, R.M., Li, Z., Sastry, S.S.: A Mathematical Introduction to Robotic Manipulation. CRC Press, Boca Raton (1994)
14. Ashagrie, A., Salau, A.O., Weldcherkos, T.: Modeling and control of a 3-DOF articulated robotic manipulator using self-tuning fuzzy sliding mode controller. Cogent Eng. **8**, 1950105 (2021)
15. Liu, Y., Li, J., Zhang, Z., Hu, X., Zhang, W.: Experimental comparison of five friction models on the same test-bed of the micro stick-slip motion system. Mech. Sci. **6**, 15–28 (2015)
16. Hao, L., Pagani, R., Beschi, M., Legnani, G.: Dynamic and friction parameters of an industrial robot: identification, comparison and repetitiveness analysis. Robotics. **10**, 49 (2021)
17. Serres, J.L.: Dynamic Characterization of a Pneumatic Muscle Actuator and Its Application to a Resistive Training Device (2008).
18. Pitel, J., Tóthová, M.: Dynamic modeling of PAM based actuator using modified Hill's muscle model. In: Proceedings of the 14th International Carpathian Control Conference (ICCC). pp. 307–310 (2013).
19. Awad, M., Khanna, R.: Efficient Learning Machines: Theories, Concepts, and Applications for Engineers and System Designers. Springer (2015)
20. Tipler, P.A., Mosca, G.: Physics for Scientists and Engineers: Extended Version. W. H. Freeman, New York (2003)
21. Kecskeméthy, A., Liem, K., Singh, M.D.: A Fluidic Muscle-Driven Force-Controlled Parallel Platform for Physical Simulation of Virtual Spatial Force-Displacement Laws. Riga Technical University (2008)
22. Beater, P.: Pneumatic Drives: System Design, Modelling and Control. Springer, London (2007)
23. Trojanová, M., Čakurda, T., Hošovský, A., Krenický, T.: Estimation of grey-box dynamic model of 2-DOF pneumatic actuator robotic arm using gravity tests. Appl. Sci. **11**, 4490 (2021)

Development of a Virtual Bench for Simulation and Research of a Mobile Robot Motor

Božek Pavol and Peterka Jozef

1 Introduction

The actual problem in technical universities is the development of virtual benches, which is especially important in connection with the development of distance learning, increasing the cost of laboratory equipment. Visual simulation systems are used as a tool environment for creating such simulators, the most popular of which are SimInTech, and MatLab/Simulink. The possibility of further modification of virtual benches and extension of their libraries (virtual laboratories, tutorials, and self-testing) is also taken into account. Visualization requirements are particularly important in the application to the educational process.

The research aims to develop a virtual simulation bench to simulate a mobile robot in SimInTech software. This paper uses simulation modeling approaches and methods of model-oriented design of robotic systems.

2 Literature Review

The paper [1] shows how much everyday life can be impacted by the value of virtual stands, and information systems. The application of information technology in real contemporary situations must be the focal point, therefore, examples of governance in the digital world are given. Based on student preference, there are some intriguing examples presented that are equally relevant to students' interests and concerns,

B. Pavol · P. Jozef (✉)
Slovak University of Technology in Bratislava, Bratislava, Slovakia
e-mail: pavol.bozek@stuba.sk; jozef.peterka@stuba.sk

M. Balog et al. (eds.), *EAI International Conference on Automation and Control in Theory and Practice*, EAI/Springer Innovations in Communication and Computing, https://doi.org/10.1007/978-3-031-31967-9_25

highlighting the reasons why virtual and information systems have become central to governance in the digital world, such as reconnaissance via drones.

In [2], a virtual system for monitoring the technical condition of electromechanical equipment with a neural network-based decision-making unit was described. An asynchronous gearing drive was operated as a control object. The decisions of the control unit were executed based on a combined analysis of vibration data (from the gearbox) and induction motor current consumption. The diagnostic attributes for this purpose were selected as follows: velocity, vibration acceleration, and current in the stator winding phases of the drive motor. In [3] the development of architecture and software decision-making units is based on neural networks, as well as is its training and testing. To this date, one of the main drawbacks of education in higher education institutions is the decreasing level of interactivity between students and the subjects of their study. To increase this level of interest and potentially solve the arising issues, is the modeling and development of special software, such as use of virtual benches, capable of performing a simulation of various technical systems, in particular, asynchronous motors. Simultaneously, improving the control quality of induction motors due to their presence in conditions of information uncertainty is a relatively demanding task, which is considered in the paper. We examine the static lateral stability of machine-tractor units by using the methods of multibody dynamics (MBD) [4]. The simulation model was created using SolidWorks Motion, CAD (Computer Aided Design), and CAE (Computer Aided Engineering). Based on a virtual stand approach, a method has been developed for analyzing the costs of computing resources for access control system implementation. In order to provide a better simulation environment, virtual stands use computing complex at every level of the access control process [5]. The paper [6] debates a range of obstacles connected with designing and developing the virtual test stand. A very important task is to develop better control algorithms based on discrete models, given that almost all of the tasks of controlling CNC (Computer Numeric Control) machines and robot drives cannot be accurately described by linear models, and a general solution does not exist for nonlinear models. A discrete model of a nonlinear system is one in which the state, control, and measurement matrices are variable, resulting in an infinite number of variants. For this reason, fitting tools are crucial to determining whether mathematical models and real objects are suitable. A review paper [7] focuses on theoretical issues related to studying the dynamics of CNC machines and robots and modeling their drives. The paper [8] presents virtual testing unit characteristics for investigating unsteady flow phenomena in a centrifugal compressor using an ANSYS 12.0 system.

The paper [9, 10] describes how to identify a DC motor by specifying motor parameters such as armature winding resistance, inductance, stator magnetic flux, and viscous friction coefficient in drive supports. When using a quasi-optimal nonlinear control algorithm, these parameters are components of the state matrix and determine the magnitude of the voltage. The change of these parameters due to degradation or under certain conditions of drive operation causes a discrepancy between the model state and the true one, resulting in an increase in power consumption and transient time. A methodology for calculating the identification criterion for a nonlinear control system in discrete form is proposed. The determinant of the measurement matrix is calculated at each step of discrete time. Their analysis shows that motor identification is possible in transient modes. When the armature winding resistance of the motor deviates from the nominal, the transient time and the amount of overshoot increase significantly. When the armature resistance is decreased by 25% below the nominal value, the determinant of the motor measurement matrix reaches the threshold value of the identifiability criterion. Thus, the loss of identifiability indicates the presence of a defect. The obtained results of the study can be used as a fault detection in such drives.

In the paper [11] the mathematical model construction consists of the motion of the output link of an industrial robot with six degrees of freedom as is considered in the study of the accuracy of its positioning. The mathematical robotic model solves the inverse kinematics problem and gives the solution to the direct dynamics problem, and as well the positioning problem of the robot working point. Significant influence on the control systems of complex mechatronic object characteristics, such as manipulation robots, is provided by their kinematic and dynamic parameters. A detailed mathematical description of these parameters is crucial for the use in the implementation.

Students studying simulation and model-oriented design of mechatronic systems and control devices will benefit from the virtual bench. There is a review of existing virtual stands. The MR drive model in state space is constructed [12].

An optimally tuned PID (proportional-integral-derivative) control for an induction motor-based transport robot actuator was created. The actuator's mathematical models were obtained in both discrete and continuous forms. The PID controller's parametric synthesis was accomplished using continuous and discrete actuator models. SimInTech was used to run numerical simulations for the adaptive regulator while considering cargo weight (from empty to fully loaded). The scheme of automatic adjustment of actuator PID coefficients taking cargo weight into account was proposed. The framework of automatic classification of PID regulator coefficients for an actuator based on cargo weight was proposed. The optimal values of the amplification coefficients of a discrete PID control law were determined as a result of its parametric synthesis. There was no over-control, and the transient time met the initial criteria for optimizing the mathematical algorithm by angular velocity [13].

An occurrence of an extensive variety of virtual benches is described in [14–20]. The volume of the article does not allow us to review and analyze them all.

3 Research Methodology

The research methodology is based on the representation of a mathematical model of an electric motor in the state space and simulation modeling. The model of the MR motor consists of a model of control object dynamics, a model of control system (regulator), a signal base, and a control panel model, containing control buttons, indication elements, and technical animation for reflection of current parameters of MR.

The MR motor is represented by differential Eqs. (1 and 2) of control and observation in the state space:

$$\frac{dx(t)}{dt} = A(t)x(t) + B(t)u(t) \tag{1}$$

$$y(t) = C(t)x(t) \tag{2}$$

where **x**(t) is a state vector of the MR drive; **u**(t) is a control vector; **y**(t) is an output vector; **A**(t) is a state matrix; **B**(t) is the control matrix; **C**(t) is the measurement matrix.

The following first-order differential equations are known for a DC motor (3)÷(7):

$$u = R\left(\frac{L}{R}\frac{di}{dt} + i\right) + e \tag{3}$$

$$J\frac{d\omega}{dt} = T_E - k_f\omega - T_L \tag{4}$$

$$\omega = \frac{d\theta}{dt} \tag{5}$$

$$e = k_E\omega \tag{6}$$

$$T_E = k_T i \tag{7}$$

where u, i, e are voltage, electric current, and EMF of the motor armature; L, R, are inductance, the resistance of the motor armature; ω, T_E, T_L, and θ are angular speed, electromagnetic torque of the motor, load torque and shaft rotation angle of the motor; J is a moment of inertia of the motor rotor and load; k_E, k_T are coefficients which are design constants of the motor; k_f is coefficient of viscous friction of the motor.

In formula (4) the load torque is constant, $T_L = 0.45$ N•m for this example. If necessary, it can be specified as a random variable generator.

Equation (3) is written in Cauchy form (8):

$$L\frac{di}{dt} = -Ri - k_f\omega + u \tag{8}$$

The pair of Eqs. (4) and (8), on the basis of which the vector-matrix model of a DC motor is built, is used for the state space model of a DC motor. The control vector is a scalar quantity – the supply voltage u, proportional to the given speed.

Since the drive controller is to provide torque and speed control, the armature current I and the armature speed ω are generalized coordinates. The control is the armature voltage u and the disturbance is the load torque T_L. Parameters of the model are active resistance and inductance of the circuit and armature, denoted R and L respectively, a reduced moment of inertia J, design constants k_E and k_T, and viscous friction coefficient k_f. Solving the initial system with respect to the first derivatives, the equation of the DC motor in the state space (1) in vector–matrix form is obtained (9), (10):

$$\dot{x} = \left[\dot{I}\,\dot{\omega}\right] = \left[-\frac{R}{L} - \frac{k_E}{L}\frac{k_T}{J} - \frac{k_f\omega + T_L}{J\omega}\right]\left[I\,\omega\right] + \left[\frac{1}{L}0\right]u \tag{9}$$

$$y = \left[I\,\omega\right] = \left[1001\right]x \tag{10}$$

The virtual simulator consists of two main subsystems (MR model, control unit) and one auxiliary subsystem for plotting.

The parameters of the DC motor are shown in Table 1.

The gearbox parameters are shown in Table 2.

Figure 1 shows 4 models of the DC motors with reducing armature winding and digital twin of DC motor. The model is implemented in the SimInTech software. Figure 2 shows a diagram of the electric current residual of four models of the DC motors with reducing armature winding and digital twin of DC motor with a linear change in the input voltage.

Figure 3 shows a diagram of the electric current residual of four models of the DC motors with reducing armature winding and digital twin of DC motor at stepped input voltage.

The electric current residual increases as the resistance of the DC motor armature winding decreases. With a stepped voltage input, the electric current residual is larger.

Table 1 Parameters of the DC motor

Parameter	Value
Moment of inertia applied to the rotor shaft, kg•m^2	1.000
Rotor circuit inductance, H	0.001
Rotor circuit resistance, ohm	1.900
Coefficient of anti-EMF	0.450
Torque constant	0.360

Table 2 Gearbox parameters

Parameter	Value
Friction coefficient, Nms	1.000
Load inertia moment, kgm^2	200
Elastic stiffness factor, nm	900,000
Gearing ratio	140
Half of backlash, rad	0.002

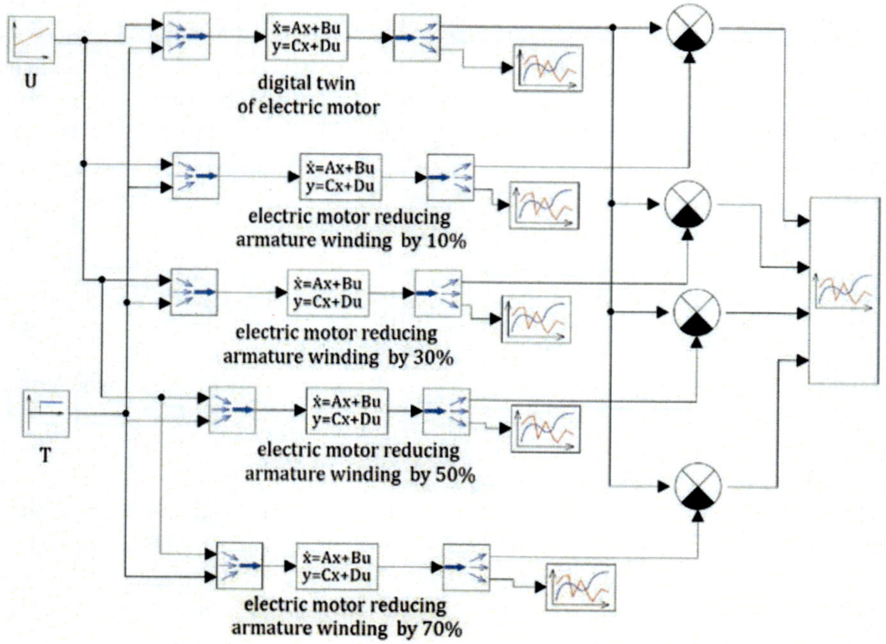

Fig. 1 Models of the DC motor with reducing armature winding and digital twin of DC motor

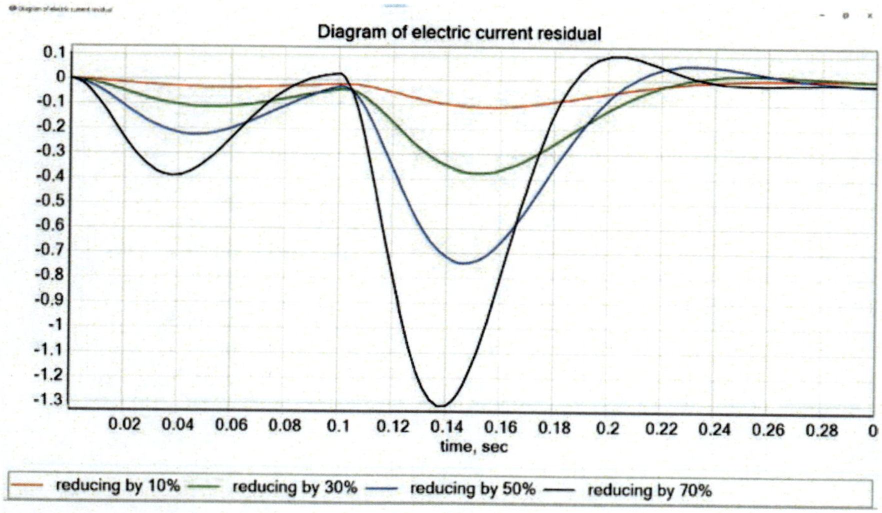

Fig. 2 Diagram of electric current residual of four models of the DC motors with reducing armature winding and digital twin of DC motor with a linear change in the input voltage

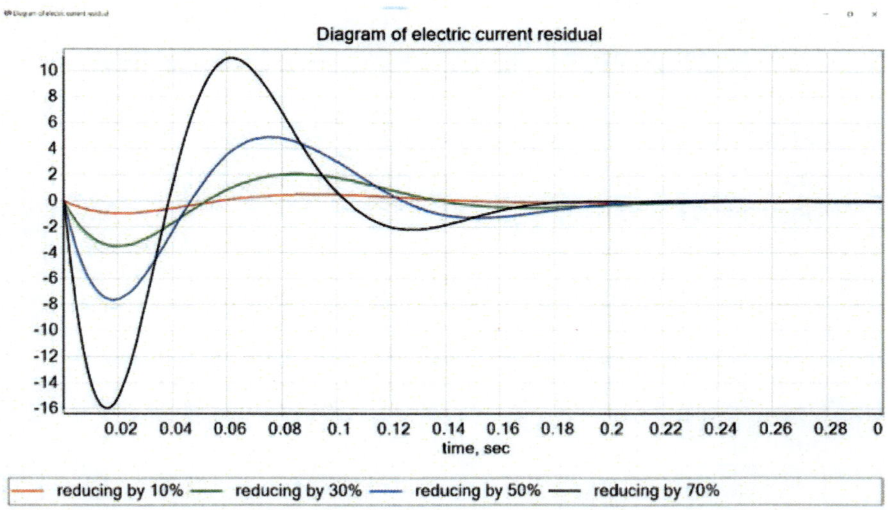

Fig. 3 Diagram of electric current residual of four models of the DC motors with reducing armature winding and digital twin of DC motor at stepped input voltage

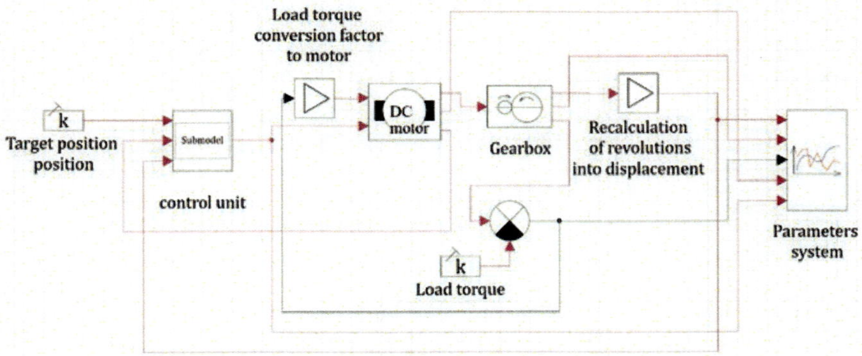

Fig. 4 Model of a mobile robot motor, X-coordinate movement

Figure 4 shows the model of the MR driver, which consists of an MR control unit, DC motor, gearbox, and unit for conversion of gearbox shaft revolution into linear motion of MR.

Figure 5 shows the submodel of the control unit to generate a PWM (Pulse Width Modulation) signal for the DC motor. Figure 6 shows PWM signal for the DC motor.

Figure 7 Shows the MR Motor model Parameters: MR Motor Displacement, Gearbox Angular Velocity, Torque, MR Motor Speed, and control Signal

The simulation results show the following MR model parameters: MR displacement 2.5 m, maximum gearbox angular velocity 0.2 rads^{-1}, maximum torque on the gearbox output shaft 15 Nm, maximum DC motor angular velocity 30 rads^{-1}.

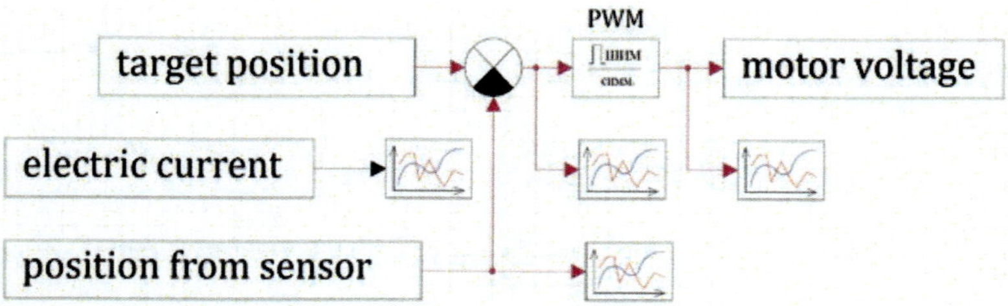

Fig. 5 Submodel of control unit to generate a PWM signal for the DC motor

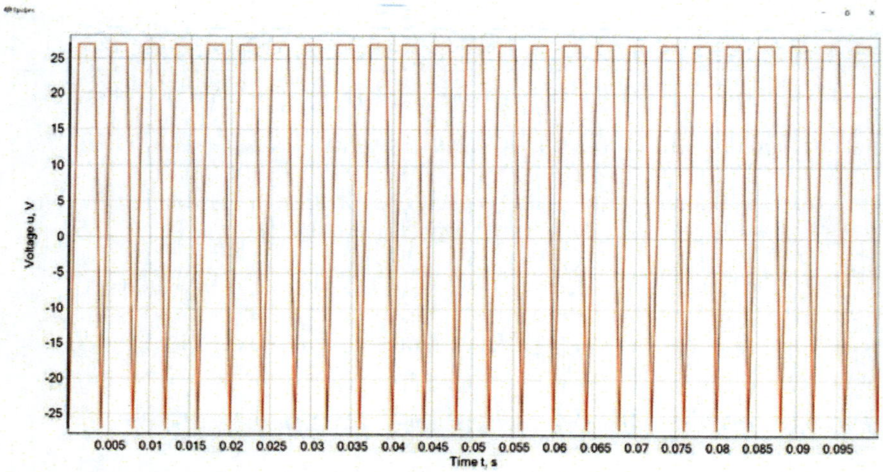

Fig. 6 PWM signal for the DC motor

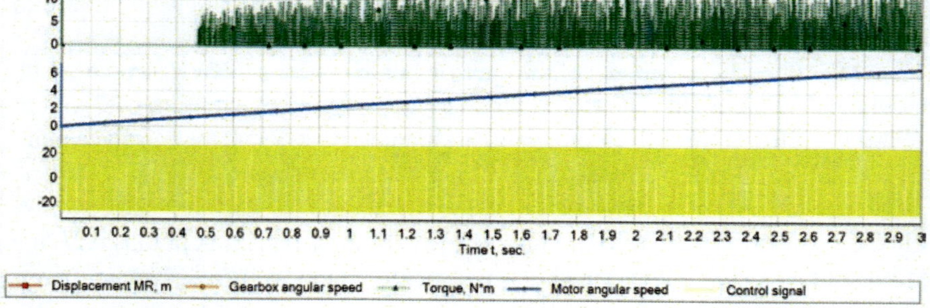

Fig. 7 Simulation Results of the Mobile Robot: Mobile robot Movement, Gearbox Angular Velocity, Torque, Mobile robot Speed, and control Signal

4 Results and Discussion

There are quite a lot of computer-controlled laboratory stands on the market, which have a high cost. The analysis of available laboratory equipment and software allows us to draw the following conclusions.

Negative tendencies in the innovation of laboratory stands have been recognized: laboratory equipment obsoleteness; significant cost; and a limited range of equipment. Positive trends in their development have been identified, including the advancement of software capabilities (technically advanced languages), enabling quick and easy creation of any applications; the presence and enhancement of specialized software for creating multimedia application features, along with customized software designed for engineering automation (AutoCAD, KOMPAS, Altium Designer, and so on) and mathematical operations (Matlab, MathCAD, etc.). Software tools, including SimInTech, MatLab/Simulink, followed by procedure-oriented programming languages such as VB, Delphi, Java, etc., as well as problem-oriented programming languages LISP, C++, C#, DBMS languages can be engaged in the process. In this regard, it is preferable to surrender traditional laboratory benches in favor of broader use of simulation modeling and computational analysis methods.

The basis for such a transition should be virtual laboratory benches.

5 Conclusion

A review of the literature showed the high relevance of developing virtual benches. As a result of this research, a virtual bench for mobile robot motors has been developed. A state space DC motor model has been developed. Experiments with 4 DC motor models with the reduced armature winding have been performed. Experimental results are compared with the results of a digital twin motor. At a decrease of armature winding resistance from 10% to 70%, the electric current residual has increased from 0.1 A to 1.3 A, i.e., has become 13 times bigger at the linear change of voltage signal. With a 10% to 70% reduction in armature resistance, the electric current residual has increased from 0.9 A to 16 A, 18 times the incremental voltage signal.

Modeling of MR has been performed in SimInTech software. As a result of this simulation, the following parameters were obtained: MR displacement on 2.5 m, maximum gearbox angular speed of 0.2 rads-1, maximum torque on the gearbox output shaft of 15 Nm, the maximum angular speed of the DC motor 30 rads-1. The advantage of the proposed solution is that the motor can be quickly investigated and identified. For example, it is possible to determine how much the armature resistance of a real motor differs from the nameplate values.

Acknowledgments The article was written with the support of projects KEGA 006STU-4/2021, and KEGA 001STU-4/2022.

References

1. Valacich, J., Schneider, C.: Information systems today: Managing the digital world. Pearson. 560 (2017)
2. Peterka, J., Bozek, P., Nikitin, Y.: Diagnostics of automated technological devices. MM Sci. J., 4027–4034 (2020)
3. Stepanov, P., Bozek, P., Lagutkin, S., Nikitin, Y.: Development of a decision box of diagnostic system for electric drives. MM Sci. J. 4954–4961 (2021)
4. Lysych, M.N.: A study of the static lateral stability of a tillage machine-tractor unit on a virtual stand. J. Phys. Conf. Ser. **1515**, 042033 (2020)
5. Magomedov, S., Ilin, D., Nikulchev, E.: Resource analysis of the log files storage based on simulation models in a virtual environment. Appl. Sci. 11(11) (2021)
6. Averbukh, V.L., Averbukh, N.V., Vasev, P., Gajniyarov, I.: The Tasks of Designing and Developing Virtual Test Stands. Proceedings. Global Smart Industry Conference, GloSIC 2020, Chelyabinsk, September, pp. 49–54 (2020)
7. Trefilov, S., Bozek, P., Nikitin, Y., Martinovic, J., Peterka, J.: Diagnostics of actuators of machine tools drives according to the identifiability criterion by the state space. MM Sci. J. 5291–5296 (2021)
8. Lebedev, A., Kozhukhov, Y., Ivanov, V., Kartashov, S., Nguyen, M.H., Lebedeva, S., Lopulalan, Kh., Aksenov, A.: Unsteady processes in a centrifugal compressor: From a physical experiment to a virtual stand. IOP Conference Series: Materials Science and Engineering, Rostov-on-Don. P. 012005 (2020). https://doi.org/10.1088/1757-899X/1001/1/012005
9. Karandeeva, I: Application of virtual stands in educational process in order to improve the quality and effectiveness of education. Modern Sci. Res. Innovat. In Russ. 1(13) (2020)
10. Lekomtsev, P.V., Nikitin, U.R., Trefilov, S.A.: Identification of DC motor on the basis of quasi-optimal nonlinear control algorithm. Vestnik IzhGTU imeni M. T. Kalashnikova. **24**(2), 68–76 (2021). https://doi.org/10.22213/2413-1172-2021-2-68-76
11. Turygin, Y.V., Zubkova, Y.V., Speranskikh, T.N.: Development of mathematical model of robot output link motion based on positioning error analysis. Vestnik IzhGTU imeni M. T. Kalashnikova. 21(1) 19–22 (2018). https://doi.org/10.22213/2413-1172-2018-1-19-22. (in Russ.)
12. Nikitin, Yu.R., Zubkova, Yu.V., Sosnovich, E.V., Masanov P.A.: Development of a virtual stand for modeling and research of a mobile robot in the SimInTech program. Intelligent systems in production. 20(2) 78–85 (2020). (in Russ.). https://doi.org/10.22213/2410-9304-2022-2-78-85
13. Božek, P., Nikitin, Y.: The development of an optimally-tuned PID control for the actuator of a transport robot. Actuators. **10**(8), 1951–1913 (2021)
14. Nikitin, Y., Bozek, P., Peterka, J.: Logical-linguistic model of diagnostics of electric drivers with sensors support. Sensors. **20**(16), 1–19 (2020)
15. Božek, P., Nikitin, Y., Krenicky, T.: Diagnostics of Mechatronic Systems. Series: Studies in Systems. Decision and Control 345. Springer Nature, Switzerland AG (2021)
16. Božek, P., Nikitin, Y., Krenicky, T.: Methods, models, algorithms for diagnostics of mechatronic systems. In: diagnostics of mechatronic systems. Studies in systems. Decis. Control. **345**, 17–26 (2021)
17. Božek, P., Nikitin, Y., Krenicky, T.: Model Systems for Diagnosticing of mechatronic objects. In diagnostics of mechatronic systems. Studies in systems. Decis. Control. **345**, 27–61 (2021)
18. Kuric, I., Klačková, I., Nikitin, Y.R., Zajačko, I., Císar, M., Tucki, K.: Analysis of diagnostic methods and energy of production systems drives. Processes 9(843) (2021)
19. Božek, P., Krenicky, T., Stepanov, P., Nikitin, Y., Gupalov, B., Lagutkin, S.: Development of a system for predictive diagnostics of traction asynchronous electric drive for trains. MM Sci. J., 6123–6126 (2022)
20. Krenicky, T., Nikitin, Y., Božek, P.: Model-based design of induction motor control system in MATLAB. Appl. Sci. 12(11957) (2022)

Monte Carlo Simulation in a Digital Calibration Certificate

George Sammarah, Martin Halaj, Lukáš Bartalský, and Jakub Palenčár

1 Introduction

Any measuring system necessarily shows slight changes throughout time. These variances are mostly brought on by how mechanical or electronic components are affected by the environment and how they are used.

To reduce or eliminate such adverse effects, each meter should undergo regular calibration. The calibration must be performed repeatedly at suitable intervals; the length of these intervals depends on a variety of factors (required uncertainty, frequency of use, type of use, stability of the equipment, etc.). When a measuring device needs to be calibrated is not a set rule. Instruments should generally be calibrated:

- prior to the first use of new, uncalibrated instruments,
- as often as the manufacturer or internal company guidelines specify,
- if there is cause to suspect that the instrument does not produce reliable findings, such as if it is broken or does not match other equipment,
- if the tool has been returned from repair and the calibration is no longer accurate [1].

The output at the end of a calibration process is a calibration certificate that provides the results of the calibration.

The measurement disparities between the instrument being tested (the measurand) and the standard values are documented in a calibration certificate. The document must specify the measurement uncertainties for each calibrated point. The

G. Sammarah (✉) · M. Halaj · L. Bartalský · J. Palenčár
Slovak University of Technology in Bratislava, Faculty of Mechanical Engineering, Institute of Automation, Measurement and Applied Informatics, Bratislava, Slovakia
e-mail: george.sammarah@stuba.sk

M. Balog et al. (eds.), *EAI International Conference on Automation and Control in Theory and Practice*, EAI/Springer Innovations in Communication and Computing, https://doi.org/10.1007/978-3-031-31967-9_26

digital calibration certificate brings the possibility of providing a lot of additional data, which can be advantageously used in measurement.

Definition. Definition of calibration is provided in internationally recognized vocabulary of metrology – VIM [2]. VIM in clause 2.39 defines calibration as follows: Calibration – "operation that, under specified conditions, in a first step, establishes a relation between the quantity values with measurement uncertainties provided by measurement standards and corresponding indications with associated measurement uncertainties and, in a second step, uses this information to establish a relation for obtaining a measurement result from an indication" [2].

NOTE 1 A calibration may be expressed by a statement, calibration function, calibration diagram, calibration curve, or calibration table. In some cases, it may consist of an additive or multiplicative correction of the indication with associated measurement uncertainty.

NOTE 2 Calibration should not be confused with adjustment of a measuring system, often mistakenly called "self-calibration," nor with verification of calibration.

NOTE 3 Often, the first step alone in the above definition is perceived as being calibration. [2].

This definition describes calibration as a tool that realizes the correct dissemination of the units vertically along the traceability chain within the calibration hierarchy from national standards to working measuring instruments [2]. The reader shall note that the definition comprises two stages, carried out one after the other:

- determination of the calibrated instrument performance, done in the laboratory,
- obtaining the measurement results by a calibrated instrument, which is carried out by the user after the instrument is calibrated.

The submitted paper deals with this relationship among the calibration work and the practical use of the calibrated instrument, seen from the Digital Calibration Certificate (DCC) point of view.

Reporting Calibration Results A calibration result may be expressed as a statement, calibration function, calibration diagram, calibration curve, or calibration table, as stated in the specification of the calibration procedure [3]. In most cases, the calibration is carried out in several discrete points (values of the quantity generated by the measurement standard). As a result, a calibration table with calibration results in those particular points is stated in the calibration certificate (see Table 1).

Besides the tabled results, an important statement is written in the calibration certificate, below the table with results:

The reported expanded uncertainty of measurement is in accordance with EA 4/02 stated as the coverage factor $k = 2$ times the typical measurement uncertainty. That translates to a coverage probability of about 95% for a normal distribution.

The above-described case deals with calibration in discrete points. But most measuring instruments belong to the indicative measuring instrument, which provides a continuous indication of the measured quantity value. In that case, the

Table 1 Calibration result in discrete points

Point	Temperature – standard / °C	Temperature – calibrated device / °C	Expanded uncertainty U / mK
1	10,00	10,02	48
2	20,00	19,98	48
3	30,00	30,01	48
4	40,00	40,02	48
5	50,00	49,98	48

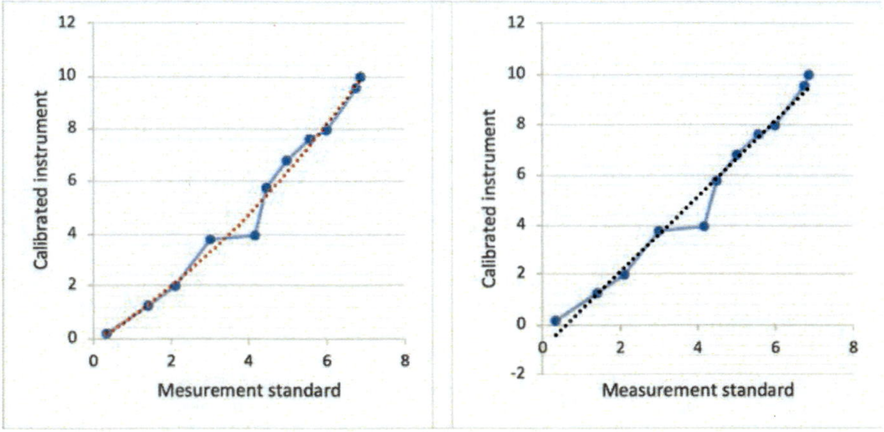

Fig. 1 Example of a calibration curve (**a**) polynomial, (**b**) a straight line

calibration curve is a more appropriate expression of a calibration result (see Fig. 1) using different measuring points [4].

Often only the calibration line is used, so the linear dependence between the indication of the meter and the value of the measured quantity (see Fig. 1).

As written earlier, the calibration consists of two steps, having two different purposes:

- determination of behavior of a measuring instrument, which is conducted in a laboratory and which may result in drafting a calibration curve,
- use of information obtained from the first stage to get the measured quantity value. In fact, the indication of a measuring instrument shall be transformed into a quantity value, using the so-called transformation curve (see Fig. 2). The transformation curve is a mirror reflection of a calibration curve, and its parameters can be determined from the parameters of a calibration curve. This is an especially easy task for a straight line calibration curve. Essentially, this is a reverse task compared to the first stage of a calibration process.

The provided examples on Figs. 1 and 2 do not contain and uncertainty range, which may encompass the estimate of a calibration (transformation) curve.

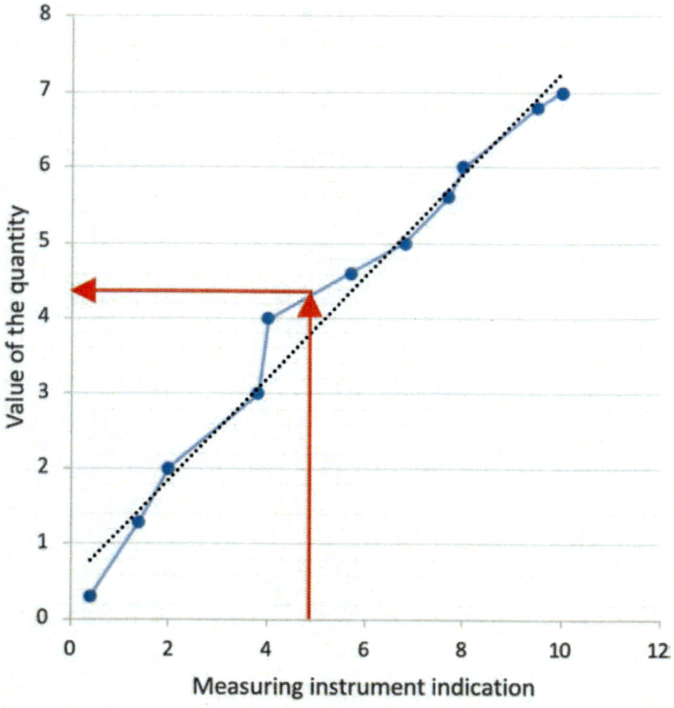

Fig. 2 Example of a transformation curve

2 Digital Calibration Certificate

The Digital Calibration Certificate (DCC) is used to transmit calibration findings uniformly and for electronic storage, authentication, encryption, and signing of data. Future processing of this data will be made possible by a notion that is currently being developed [5]. The target group is all devices around the world that need proof of the metrological continuity and the outcomes of their measurements. There are various industrial facilities that demand evidence of their quality management systems, such as metrology institutions, designated institutes, national calibration offices, calibration laboratories, and many others. Despite the fact that the data gathered during the calibration is labor-intensive and prone to errors, the analog calibration certificate has so far produced excessive value for the business. This inherent shortcoming of its analog cousin is made up for by DCC. Processes for digitally supporting manufacturing and quality monitoring are greatly aided by the machine's readability. This generates a lot of additional value for the DCC company [6].

A much more advanced way to create digital certificates is to use a format based on XML. The certificate is created in XML format rather than just as a document. In this instance, each specific certificate detail is encoded as a meaningful XML field or tag. Both humans and machines can read XML files. To ensure its legitimacy and integrity, it might be digitally signed and encrypted [1]. The benefit of an XML format is the ability to analyze and process a much larger quantity of information than we can in nondigital formats.

General Features The design of the DCC is crucial. General information, such as a calibration gauge, conforms to a prescribed form. This establishes a common foundation that national calibration offices may control. The measurement results, however, only apply to particular expert groups. Companies frequently have several data exchange formats that are well-established in the industry and must to be preserved [5]. The DCC's design is essentially a technical issue for which metrology institutes are responsible for the first time.

Leaving the paper calibration certificate in its original digital document format, like pdf, or image format, like jpg, is the most natural way to digitize it. Because certificates are frequently prepared digitally before printing, this technique is straightforward. It can be widely utilized in a variety of settings for electronic certificate exchange, storage, and archiving. However, like with paper certificates, such a system is vulnerable to fabrication and necessitates additional manual processes to validate the certificate's authenticity and integrity. Furthermore, it is impossible to process and read the contents of such a certificate mechanically.

Standard Requirements ISO / IEC 17025: 2017 is the basic standard for the implementation and assessment of a quality management system in calibration and testing laboratories. Therefore, it is necessary to examine what requirements the standard states for the issued certificates and how they could be met when issuing digital certificates. Some requirements of this standard are defined by [7]:

- *Confidentiality.* Rules for ensuring the confidentiality of laboratory operations.
- *Authorization of personnel.* Requirements for defining the authorization of personnel to perform metrological activities and to approve issued documents.
- *Records of equipment used for calibration.* Records of laboratory equipment maintenance.
- *Metrological traceability.* Demonstration of the metrological traceability of the equipment used for calibration.
- *Technical records.* Calibration certificates are considered technical records, therefore, the general requirements for records apply also to them.
- *Reporting of results.* Details of the structure and content of reports, including calibration certificates.
- *Laboratory information management system.* These are the requirements that are placed in the laboratory to work with information.

The structure of the digital calibration certificate must contain necessary information; however, the format may differ based on the organization and measuring region. In general, depending on the viewpoint, a calibration certificate's content can be categorized in a variety of ways. For the purposes of this paper, we divided a digital calibration certificate's contents into four groups:

- *Administrative information.* Information of primary interest is found in an extremely controlled area. Data fields must be filled out and have a specific structure.

Calibration Results (Regulated Area) One of the hardest problems to overcome using DCC is the representation of measurement results.

Comments (Unregulated Area) Individual details regarding the measuring process are included in the comments section, which adds details about the measurement results.

Additional Information (Unregulated Area) An electronic copy of the calibration certificate can be stored here in PDF format, which can then be translated to XML. This makes it possible for DCC users to switch to the digital world while still viewing a picture of their typical calibration certificate [8]. With DCC, this area will be highly required after and used to control customer measurements.

DCC creation Putting digital certificates into practice is a complex process. DCCs should contain data in accordance with relevant data sources and terminology for metrology. Such resources in the international framework are:

- Guidance on the expression of GUM uncertainties [9],
- Technical standard ISO / IEC 17025 [7],
- Constant values according to CODATA tables,
- International Metrology Dictionary – VIM [2],
- International System of Units SI.

At the same time, as digital certificate data is consistent with these sources, the information in DCC is divided into four sections – (1) administrative data, (2) measured results, (3) comments, (4) documents. And this structure must be followed when creating digital certificates (see Fig. 3). The figure describes the different procedures for creating DCC on both sides, staff and machines, and outlines what they should develop during each step of the process.

According to Fig. 3 the procedure of creating a DCC is consisted of the following steps:

- The administrative staff member accepts the calibration request in paper form, processes the calibration identification data and uploads it to the computer. This data will be stored in local databases for further processing, and if the administrative staff entered any additional data, the data in that particular customer's file would be updated.

During calibration, the technician can manually enter some parameters of the measurement process, which he must enter as input data. This data is automatically stored and processed. After the measurement, the measured results are obtained and stored in a database, where the technician can select them for further processing.

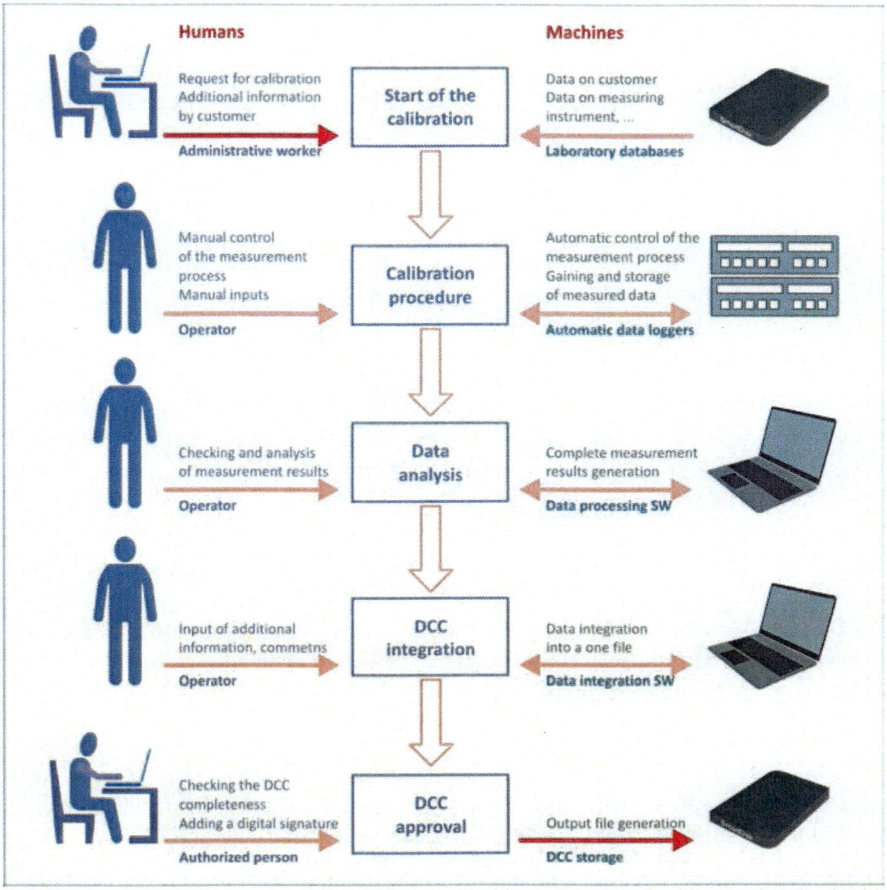

Fig. 3 Schematic procedure for creating DCC

Data analysis is in progress, the software generates the values measured on the standard and on the calibrated scale, and the technician checks whether the measured results meet the requirements, in that case, he would move them further to the DCC integration.

Additional files (videos, photos, records, ...), comments, and opinions of the technician on the scale would be entered here; if necessary, the software would integrate all the necessary mentioned data.

The last step in making a DCC is to check the whole correctness of the digital certificate. After the inspection, the authorized person must approve and sign the digital certificate with a digital signature. The system would create an output file containing DCC, simulation of measured results, videos, photographs, records, and other necessary information about the calibration procedure. In the end, this file is sent to the applicant of this metrology service.

3 Methods for Evaluation of Measurement Result

As written in part 1 of this paper (and as stated in the definition of calibration), the calibration result must always contain an expressed uncertainty of measurement. Despite the form (calibration table in discrete points, a continuous calibration curve), a measurement uncertainty represents an inevitable parameter that always must be stated on a calibration certificate.

Calculation of measurement uncertainty is a widely discussed research topics with many theoretical works and practical guidelines available. Guide for Expression of Uncertainty in Measurement (GUM) is the fundamental one [9].

The Bayesian probabilistic approach of GUM introduces two principles:

- The shape of a measurement result can be represented in terms of a probability density function (PDF), which specifies the value to be reported (often the distribution's mean) and a coverage interval where the value of the measurand is believed to lie with a specified probability [10]. It means we get an estimate of the measurement result and corresponding uncertainty. And an evaluation of a coverage interval is sometimes causing problems when evaluating calibration results with a requested probability,
- PDF encodes and transmits belief in the range of values that can be assigned to a measured quantity [10]. We can estimate the behavior of future observations of the same measurand (future measurements) (Fig. 4).

This means – we have some information on measurand (PDF0) prior to the measurement, measurement results refine that information (PDFm). For example, we assume Gaussian probability distribution in its general form. After measurement, we obtain the particular Gaussian PDF with calculated mean and dispersion.

GUM uncertainty Framework Just to remember the classical analytical approach – the propagation and summarizing stages of the GUM uncertainty framework constitute the following computational steps:

Fig. 4 Normal (Gaussian) probability distribution function

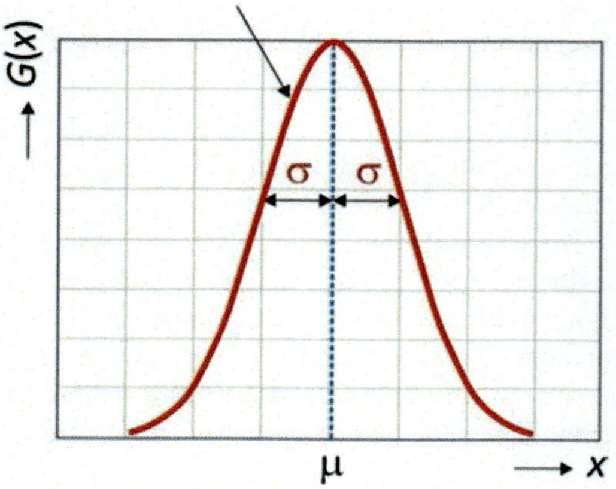

- For the input quantities $X = (X_1, ..., X_N)^T$ obtain the expectations $x = (x_1, ..., x_N)^T$ and the standard deviations (standard uncertainties) $u(x) = [u(x_1), ..., u(x_N)]^T$.

Set the degrees of freedom (infinite or finite) associated with each $u(x_i)$.

For each pair i, j for which X_i and X_j are not independent, obtain the covariance (mutual uncertainty) $u(x_i, x_j)$ [11].

Form the partial derivatives of the first order of $f(X)$ with respect to X.

Calculate y, which is the model evaluated at X equal to x.

Calculate the model sensitivity coefficients, which are partial derivatives according to point 4., evaluated at x [12].

Calculate the standard uncertainty $u(y)$ by combining $u(x)$, the $u(x_i, x_j)$, and the model sensitivity coefficients.

Calculate ν_{eff}, the effective degrees of freedom associated with $u(y)$.

Calculate the expanded uncertainty U_p, and hence a coverage interval (for a stipulated coverage probability p) for Y, regarded as a random variable, by forming the appropriate multiple of $u(y)$ [11]. This is the approach widely used for evaluation of measurement results, obviously also for evaluation of calibration result.

Monte Carlo method (MCM) Clause 5.4.5 of Supplement 1 to GUM [9] states: „In reality, the propagation of distributions may be done without approximations only for simple instances. One approximation method is used by the GUM uncertainty framework, and another by the Monte Carlo Method (MCM). The GUM uncertainty framework is accurate for a small but significant fraction of issues. Although MCM is never completely accurate, it is more reliable than the GUM uncertainty framework for a wide range of issues [11].

MCM as an implementation of the propagation of distributions can be stated as a step-by-step procedure:

- Select the number M of Monte Carlo trials to be made:

(a) Generate M vectors, by sampling from the assigned PDFs, as realizations of the set of N input quantities X_i, $i = 1, ..., N$,

(b) For each such vector, form the corresponding model value of $Y = f(M)$, i.e. calculate M model values y_r, $r = 1, \ldots M$, [13].

(c) Sort these M model values into strictly increasing order, using the sorted model values to provide G.

(d) Use G to form an estimate y of Y and the standard uncertainty $u(y)$ associated with y.

(e) Use G to form appropriate coverage interval for Y, for a required coverage probability p [14].

- Figure 5 presents a simple comparison of principles of the classical GUM approach (left) and the Monte Carlo method (right).

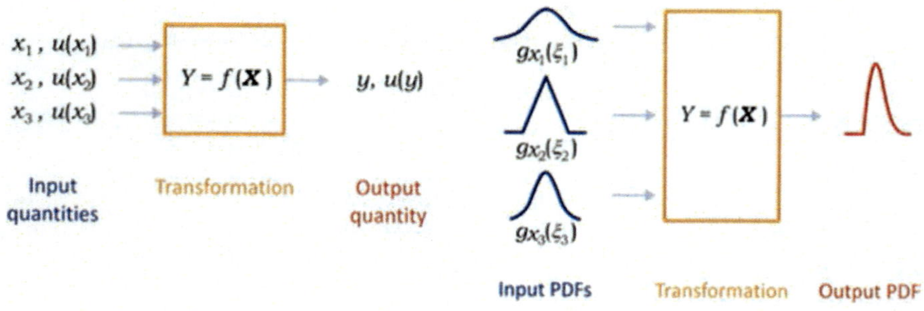

Fig. 5 Schematic representation of a law of distributions propagation

4 Utilization of MCM for Calibration

Estimates of the calibration curve parameters, their standard and expanded uncertainties (coverage intervals), and related probability distributions are obtained when introducing a digital calibration certificate using the original nonlinear measurement model and the Monte Carlo method (MCM) based on the law of propagation distributions (LPD), as described in the GUM Appendix (9). In this way, we can also obtain probability distributions and coverage intervals for quantities measured using a calibrated instrument [15].

Process Description We will use the MCM simulation in two phases:

- in the laboratory when determining the dependence of the instrument indication and the value of the measured quantity,
- when measuring is done by a calibrated instrument.

The data obtained by the simulation in the first phase are stored in DCC as additional data. Let's describe the whole process in more details:

1. In the first phase, the measurement model corresponding to a specific calibration must be determined. Also the input quantities must be listed, like the number of measurements and measured values is determined and all influencing quantities for our specific measurement must be determined or in terms of mean, range, or distribution of individual quantities. The measured data from the standard as well as the data from the calibrated instrument are generated. Coefficients *a, b* are calculated and stored in a file from which these values would be used further. This calculation is repeated *M* times, which we store to certain file.
2. In the second step, the customer would receive a DCC containing *M* pairs *a, b*. In his simulation, the customer should enter his input values as measured values and influencing quantities and set up the measurement model so that x_0 can be calculated. They generate random quantities $x_{0\,mer1}$ of the calibrated instrument

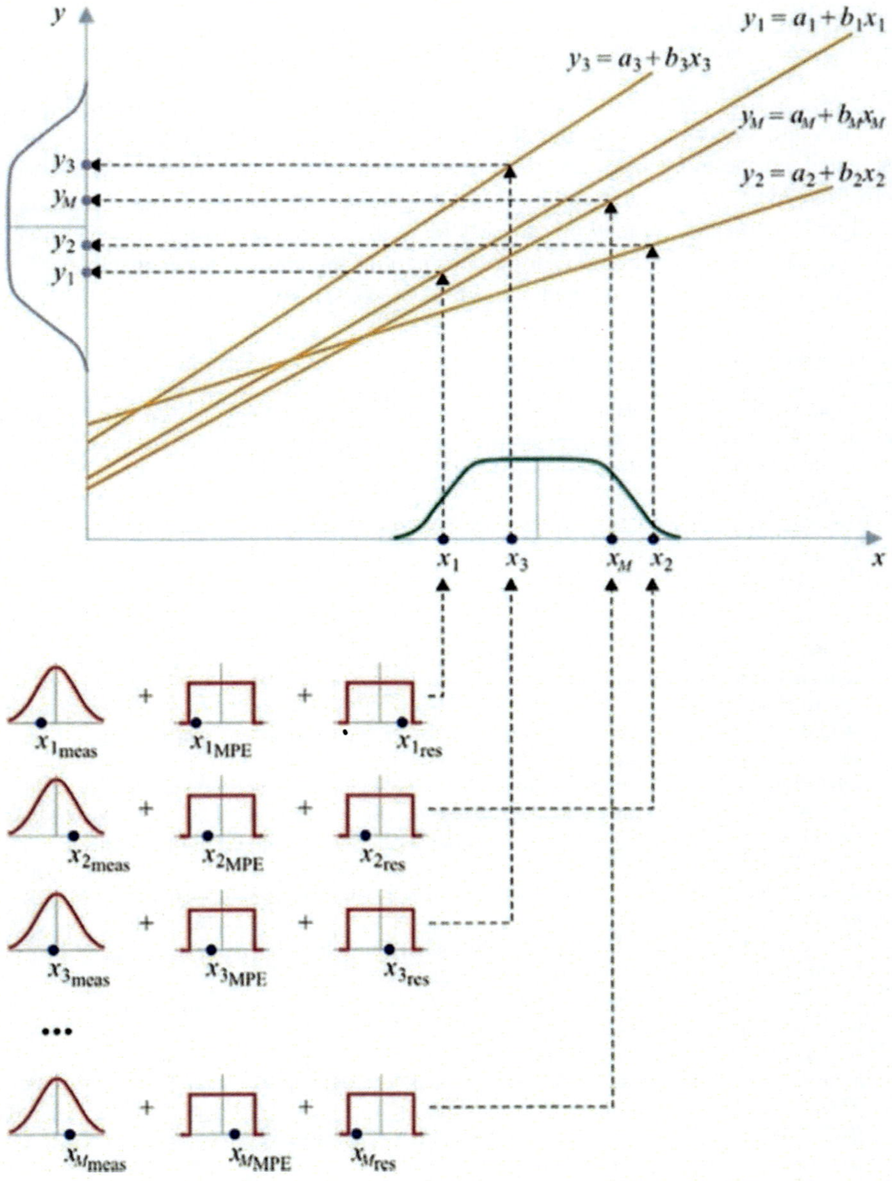

Fig. 6 Schematic representation of multiple calibration lines

as well as influencing quantities. x_{01} is calculated and using the pairs a_1, b_1 in the certificate, y_{01} could be calculated, and thus we would have the first line y_1. And this is repeated M times, and we get y_1 .... to y_M (see Fig. 6). These results are arranged from the smallest to the largest y, so we get a histogram of the values of the measured quantity.

5 Conclusion

The Monte Carlo approach has lately gained popularity, particularly for nonlinear models and situations where the assumption of a normal distribution may not be satisfied. Furthermore, the uncertainty propagation process only requires symmetric coverage intervals, which in some cases it may not be satisfactory. By incorporating a digital certificate, we may employ MCM in the calibration of meters where we need to identify the calibration (or transformation) function. This certificate can be linked to both the calibration and measurement assessment programs.

Acknowledgments This paper was supported by the KEGA grant No. 013STU-4/2021 Strengthening of activating teaching methods in the field of technical measurement, KEGA grant No. 014STU-4/2020 Introduction of progressive methods for increasing the level of the educational process of subject design and evaluation of measurements, Slovak Research and Development Agency, project APVV-21-0216.

References

1. Mohammed, S., Gadelrab, R.A.: Abouhogail: Towards a new generation of digital calibration certificate: analysis and survey. In: Measurement, vol. 181, pp. 1–10 (2021). https://doi.org/10.1016/j.measurement.2021.109611
2. JCGM 200: 2012. International vocabulary of metrology – basic and general concepts and associated terms (VIM). 3rd edn. (2008)
3. SvantekAcademy, https://svantek.com/academy/calibration/. Last accessed 2023/01/29
4. Grosinger, P., Rybář, J., Dunaj, Š., Ďuriš, S., Hučko, B.: A New Payload Swing Angle Sensing Device and Its Accuracy. Sensors. **21**(6612) (2021). https://doi.org/10.3390/s21196612
5. Physikalisch-technische bundesanstalt. https://www.ptb.de/cms/fileadmin/internet/publikationen/ptb_mitteilun- gen/mitt2017/Heft4/PTB-Mitteilungen_2017_Heft_4.pdf/. Last accessed 2023/02/01
6. Harmes-Liedtke, U., Jose Oteiza, J.: Global quality infrastructure index. Report 2020. GOII (2020). https://doi.org/10.13140/RG.2.2.27782.50249
7. ISO/IEC 17025:2017. General requirements for the competence of testing and calibration laboratories, /www.iso.org/. Last accessed 2023/02/04
8. Sammarah, G., Halaj, M.: Enhanced capabilities of a digital calibration certificate. In: Measurement Management and Metrological Order in the Organization. 177 (2022). ISBN 978-1-7399378-1-2
9. JCGM 100: 2008. Evaluation of measurement data — guide to the expression of uncertainty in measurement (GUM). 1st edn. (2008)
10. Tyler Estler, W., Brynn Hibbert, D.: A new guidance document on measurement uncertainty and conformity assessment. OLIML Bull. **2**(Vol. LIV), 14–16 (2013)
11. JCGM 101: 2008. Evaluation of measurement data —guide to the expression of uncertainty in measurement. GUM 1995. (2008)
12. Warwick. http://wrap.warwick.ac.uk/2775/. Last accessed 2023/02/02

13. Špetlík, M., Březina, J.: Groundwater contaminant transport solved by Monte Carlo methods accelerated by deep learning meta-model. Appl. Sci. **12**(15), 7382 (2022). https://doi.org/10.3390/app12157382
14. Wen, X., et al.: Uncertainty estimation of robot geometric parameters and end-effecter position based on new generation GPS. Mathematical Probl. Eng. **2019**(11), 1 (2019). https://doi.org/10.1155/2019/7830489
15. Palenčár, J., et al.: ISO linear calibration and measurement uncertainty of the result obtained with the calibrated instrument. Measure. Sci. Rev. **22**(6), 293–307 (2022)

Control of Temperature Field in the Secondary Cooling Zone of Continuous Casting Process Based on Robust Internal Model Control (IMC) Approach

Sohaibullah Zarghoon, Lukáš Bartalský, and Cyril Belavý

1 Introduction

The continuous casting process consists of parts described as follows: ladle (1) is used to move and store molten metal into a tundish during the operation. When the ladle is being switched, the tundish (2) enables a reservoir of metal to supply the water-cooled mold (4) through the submerged entry nozzle (3), and the first solidification of metal takes place While the casting core is still liquid (5), the solidified shell of the casting strand is developing at the mold walls. As the steel strand passes through the secondary cooling zone (6), which is divided into a number of subzones (6.1,6.2, 6. n), the liquid steel gradually solidifies, continuing a process that began in the water-cooled mold. The metallurgical length in the continuous casting machine is the distance from the meniscus level to the point of complete solidification (8), [1–8] (Fig. 1).

1.1 Modeling and Identification of Secondary Cooling Zone as Distributed Parameter System (DPS)

In the mathematical theory of control of distributed parameters systems (DPS), they are called a system represented by a partial differential equation, in which the input and output relation describe distributed input and distributed output system which shows in Fig. 2.

S. Zarghoon (✉) · L. Bartalský · C. Belavý
Institute of Automation, Measurement and Applied Informatics, Faculty of Mechanical Engineering, Slovak University of Technology in Bratislava, Bratislava, Slovakia
e-mail: sohaibullah.zarghoon@stuba.sk

M. Balog et al. (eds.), *EAI International Conference on Automation and Control in Theory and Practice*, EAI/Springer Innovations in Communication and Computing, https://doi.org/10.1007/978-3-031-31967-9_27

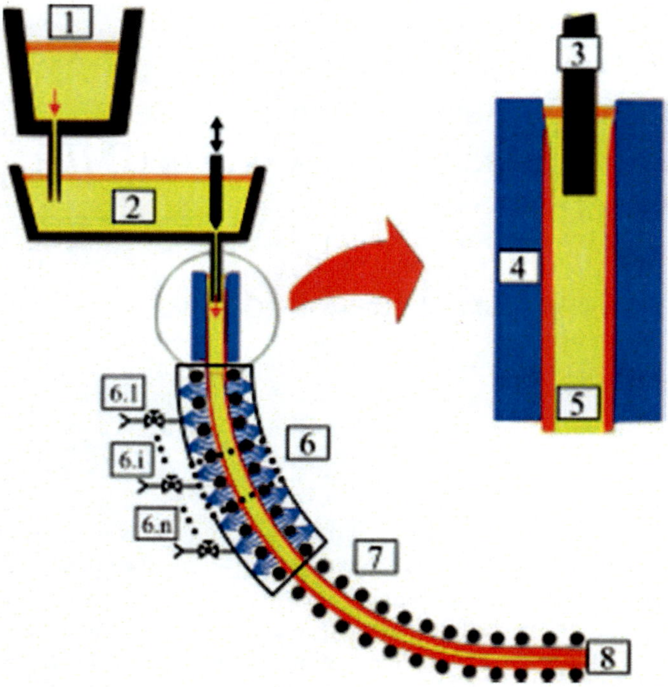

Fig. 1 Steel casting process

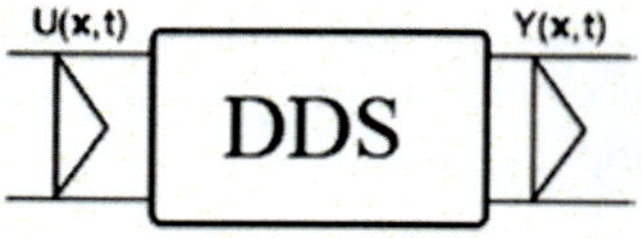

Fig. 2 Distributed input, output system [2]

Where $U(x, t)$ is distributed input $Y(x, t)$ is distributed output, and DDS is Distributed input and distributed output system. In engineering practice after connecting actuators and generators of distributed input quantities to DDS, systems with distributed parameters are interpreted as systems with lumped input and distributed output systems (LDS) Fig. 3, an example of a such system we can say is a secondary cooling system of continuous casting technology which lumped inputs are water cooling nozzle and distributed output is temperature filed [2–7, 9].

HLDS – shows the system with lumped input and distributed output, $U_n(k)$ – lumped input vector components in continuous form, and $Y(x, t)$ is distributed output (Fig. 4).

Where LDS – lumped input and distributed output system, $\{SA_i\}_{i=1,n}$ is actuating components of a lumped input variable, $\{GU_i\}_{i=1,n}$ are generators of the distributed input variable, $U(t) = \{U_i(t)\}_i$ is LDS's lumped input variable vector, $\{U_iA_i(t)\}_i$ is lumped input variables for $\{GU_i(t)\}_i$, $\{U_i(\xi, t)\}_i$ is DDS distributed input variables and $\{U_i(\xi, t)\}_i$ is DDS distributed input variable as a result [6]. At zero initial and

Fig. 3 LDS with a zero-order hold unit [3–6]

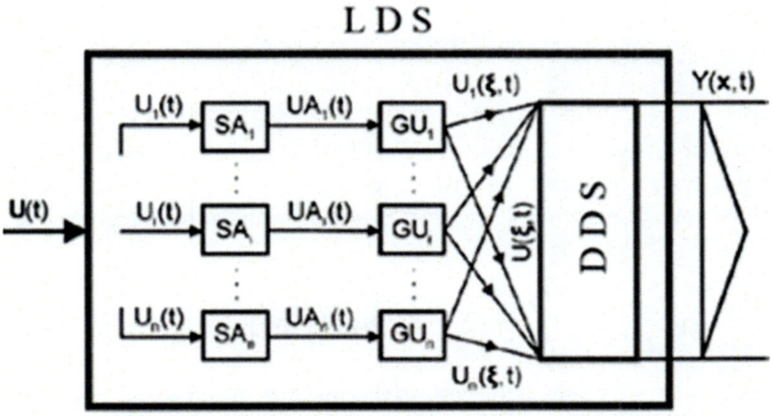

Fig. 4 Structure of LDS [4–6]

boundary conditions, the result distributed output quantity in the input-output relationship is determined in the continuous form below:

$$Y(x,t) = \sum_{n}^{i=1} Y_i(x,t) = \sum_{n}^{i=1} \mathcal{G}_i(x,t) \otimes U_i(t) = \sum_{n}^{i=1} \int_{t}^{0} \mathcal{G}_i(x,t-\tau) U_i(\tau) d\tau \tag{1}$$

Where its discrete form is:

$$Y(x,\mathrm{k}) = \sum_{n}^{i=1} Y_i(x,\mathrm{k}) = \sum_{n}^{i=1} \mathcal{G}H_i(x,\mathrm{k}) \oplus U_i(\mathrm{k}) = \sum_{n}^{i=1} \sum_{k}^{q=0} \mathcal{G}H_i(x,k-q) U_i(q) \tag{2}$$

In the above equation $\otimes$ is the convolution product and $\oplus$ is the convolution sum, $\{U_i(t)\}_i$ and $U(t) = \{U_i(k)\}_i$ are continuous and discrete lumped input quantities, $\{G_i(x, t)\}_i$ are distributed impulse characteristics between the inputs and outputs of LDS, and $\{G_i(x, k)\}_i$ are their discrete form with zero-order hold unit HLDS Fig. 3.

Where:

$$\{Y_i(x,k)\}_i = \{\mathcal{G}H_i(x,k) \oplus U_i(k)\}_i = \left\{ \sum_{k}^{q=0} \mathcal{G}H_i(x,k-q) U_i(q) \right\}_i \}_i \tag{3}$$

$$\{Y_i(x,t)\}_i = \{\mathcal{G}_i(x,t) \otimes U_i(t)\}_i = \left\{ \int_{t}^{0} \mathcal{G}_i(x,t-\tau) U_i(\tau) d\tau \right\}_i \tag{4}$$

are distributed continuous and discrete partial output variables belonging to individual lumped input variables.

We get the result by using the Laplace transformation:

$$Y(x,s) = \sum_{n}^{i=1} Y_i(x,s) = \sum_{n}^{i=1} S_i(x,s)U_i(s) \tag{5}$$

$$\{Y_i(x,s) = S_i(x,s)U_i(s)\}_i \tag{6}$$

Next, consider the partial courses of discrete partial transient characteristics also in the space dependence steady-state, which means when time $k \to \alpha$. Following that, the reduced courses in the set $\{\mathcal{H}H_i(x,\alpha)\}_i$ are determined for the given relations:

$$\left\{\mathcal{H}HR_i(x,\alpha) = \frac{\mathcal{H}H_i(x,\alpha)}{\mathcal{H}H_i(x_i,\alpha)}\right\}_i \tag{7}$$

The dynamic characteristics at the basis of discrete distributed partial transient characteristics are divided into time and space components based on the above information time component are transfer function $\{SH_i(x_i, z)\}_i$, and the space components are $\{\mathcal{H}H_i(x,\alpha)\}_i$ [5–7, 10]. The secondary cooling zone model simulation has been done in Procast software.

The system is classified by the DPS theory as an LDS (lumped input and distributed output system), with its inputs being a configuration of water flow rates from nozzles dispersed over the surface of the strand and divided into three zones, and its output is a temperature field which is shown in Fig. 5.

Identification purpose needed to obtain the space components of dynamics in the steady-state $\{\mathcal{H}H_i(x,\alpha)\}_i$ from the data which is measured in 300 points. Time components of the dynamics are obtained by identifying the responses $\{Y_i(x, k)\}_i$, where the point is chosen according to the highest gain. The obtained transient characteristic in the time and space domain for zone #1 of SCZ are shown in Fig. 6. The location of the measured points with the fastest dynamics in the individual zone of the secondary cooling zone is shown in Table 1:

At these points, three transfer functions were obtained from the simulated responses using the System Identification Toolbox. The models were searched for in the form of a process model with its parameter nominal values:

$$G_s(s) = K_P \times \frac{1}{1 + T_P s} \tag{8}$$

Three transfer functions in the form of (8) with their nominal parameters were obtained and next, they were transformed into discrete-time transfer functions with sample time 5 s and it is included in Table 2.

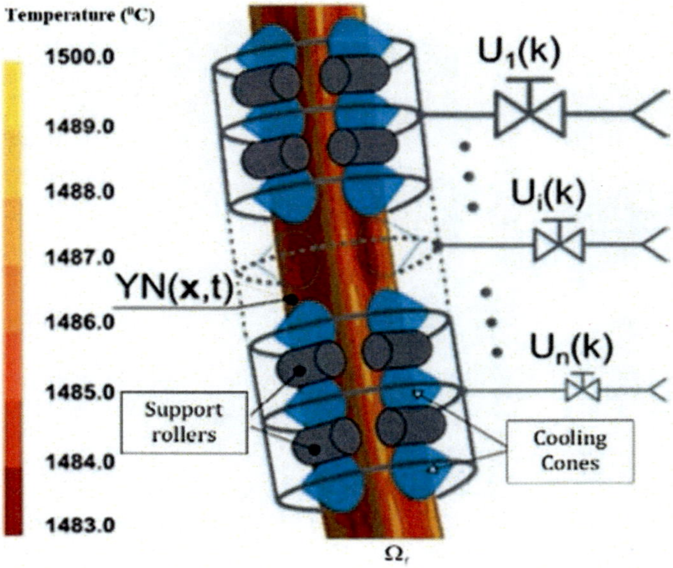

Fig. 5 FEM model of the secondary cooling zone [6]

2 Robust IMC-Based Control Circuit of DPS

It is possible to construct and solve control synthesis for various control circuit structures utilizing HLDS models to explain the dynamic of DPS. The decomposition of the dynamic of the DPS control synthesis in this case is decomposed to control synthesis in space and time domains. The IMC-based control circuit for the lumped input and distribution output system is shown in Fig. 7.

In the CS part of the control synthesis in the SS block, based on the reduced steady-state distributed transient characteristics $\left\{\mathcal{H}HR_i(x,\alpha)\right\}_i$ solve the approximation task of the distributed control variable and distributed control variable $W(\overline{x},k)$.

$$\min_{\breve{Y}_i} \; Y(x,k)-\sum_{n}^{i=1} Y_i(x_i,k)\mathcal{H}HR_i(x,\infty) \;=\; Y(x,k)-\sum_{n}^{i=1}\breve{Y}_i(k)\mathcal{H}HR_i(x,\infty) \tag{9}$$

$$\min_{\breve{W}_i} \; W(x,k)-\sum_{n}^{i=1} W_i(x_i,k)\mathcal{H}HR_i(x,\infty) \;=\; W(x,k)-\sum_{n}^{i=1}\breve{W}_i(k)\mathcal{H}HR_i(x,\infty) \tag{10}$$

The solution of the above approximation task is a vector of approximation coefficients with components $\left\{\breve{Y}_i(k)\right\}_i$ and $\left\{\overline{W}_i(k)\right\}_i$, the difference between these components is control error $\overline{\overline{E}}(k)$.

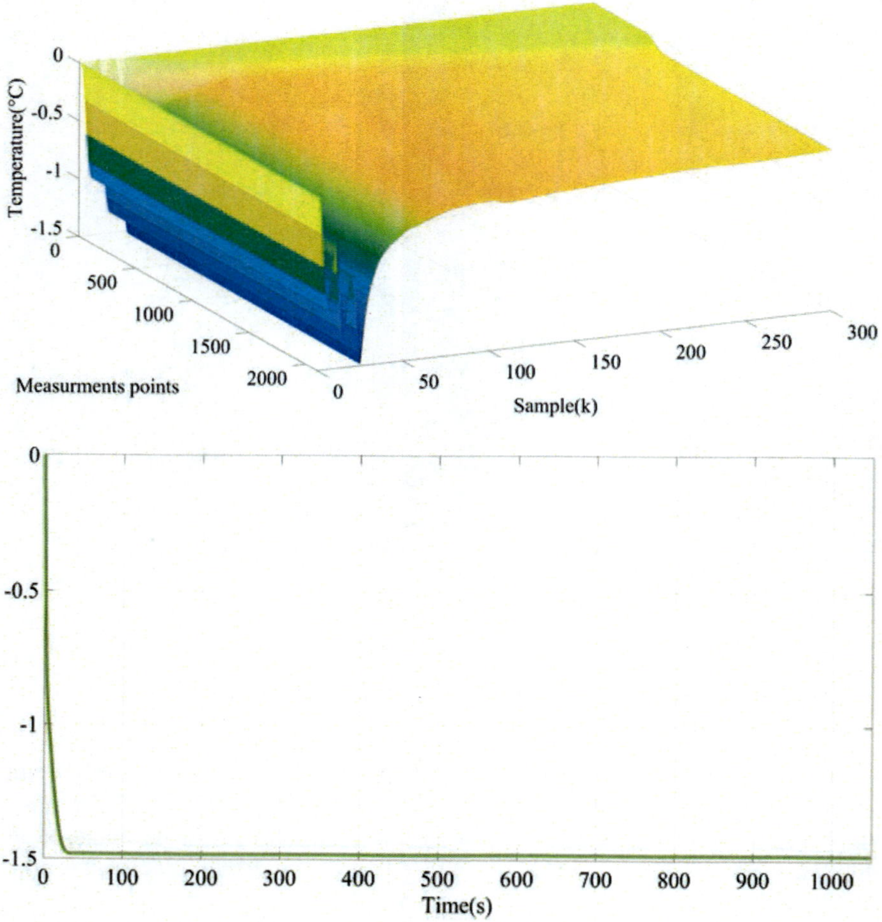

Fig. 6 Transient characteristics in zone #1, up in the space domain, down in the time domain

Table 1 Location of fast dynamics in every single zone of SCZ

Zone #1	Zone #2	Zone #3
X1 = 25	X1 = 166	X1 = 298

Table 2 Transfer function parameters values

Zone #1		Zone #2		Zone #3	
K_{p1}	T_{p1}	K_{p2}	T_{p2}	K_{p3}	T_{p3}
−0.1023	0.9309	−0.09454	0.9888	−0.0123	0.983

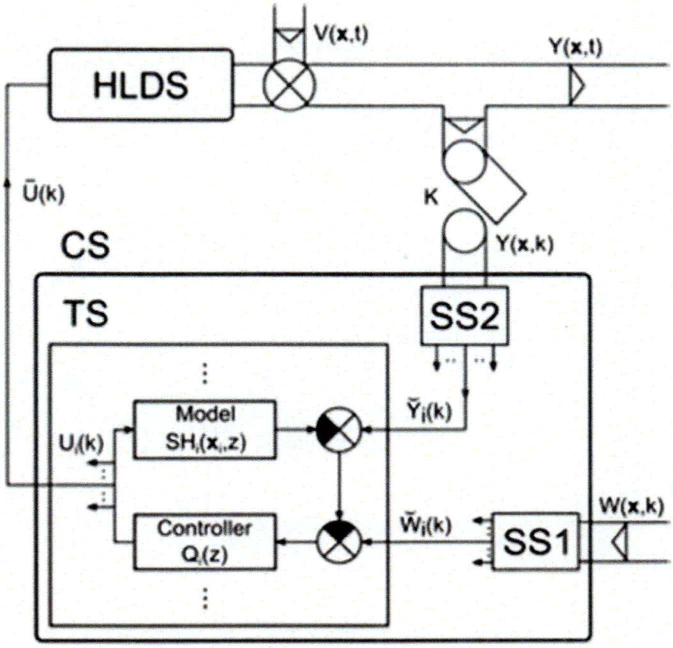

Fig. 7 Time-space control loop [7, 9–13]

$$\breve{E}(k) = \{\breve{E}_i(k)\}_i = \{\breve{W}_i(k) - \breve{Y}_i(k)\}_i \tag{11}$$

In the time synthesis, the control error $\breve{E}(k)$ will be treated and these controllers generate the components $\{U_i(k)\}_i$ of the control variable vector $U(k)$ [6, 7, 9].

3 Analysis of Uncertainties of Identified Transfer Function

The transfer functions $\{SH_i(\overline{x}_i, z)\}_{i=1,3}$ are lumped as models of the controlled system in the control synthesis's time component. The process of modeling and identification, in this case, provides the key sources of uncertainty. The obtained transfer function for the zone was analyzed according to their parameters. The function ureal is the simplest function to define an uncertain real parameter in Robust Control Toolbox®. The following lines can be used to set the uncertain parameters of zone #1 of the secondary cooling system using the function ureal [10]:

S=K_{p1}*(1)/(T_{p11}*s + 1).
$K_{p1} = -1.4808$.
$T_{p11} = 6.985$.
K_{p1} = ureal ('K_{p1},' -1.4808,' Mode', 'Range', 'Percentage,' [−20 20]).
T_{p11} = ureal ('T_{p11}', 6.985,' Mode', 'Range', 'Percentage,' [−20 20]).

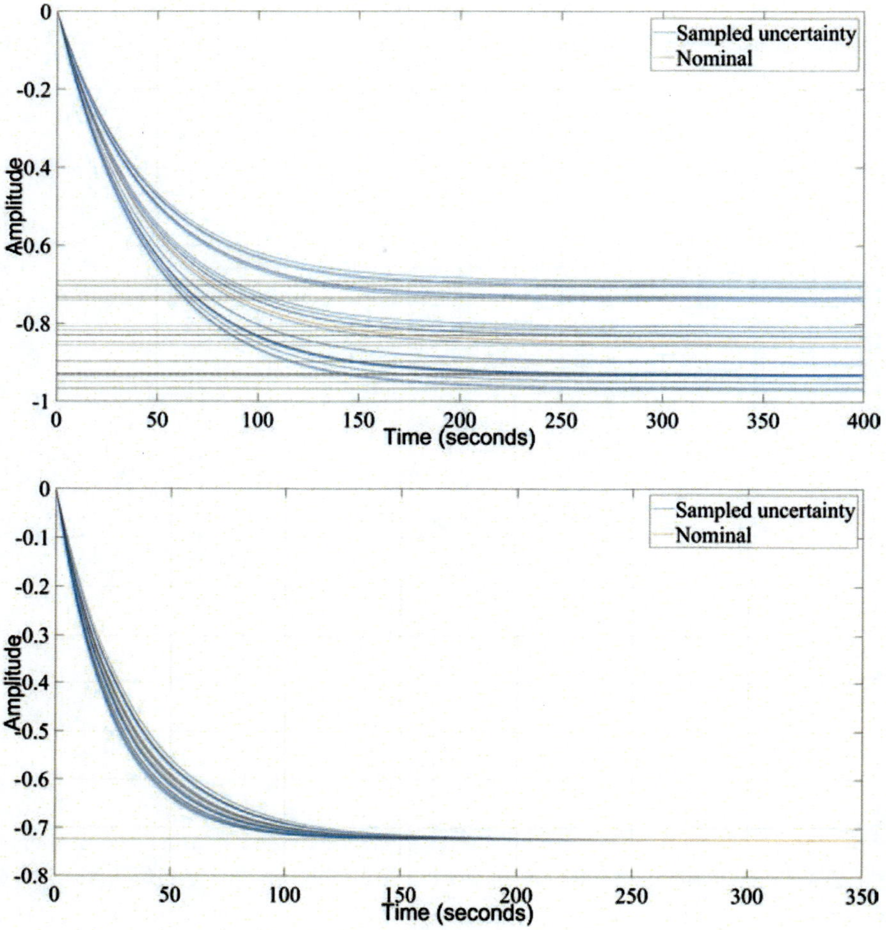

Fig. 8 Uncertainty of transfer function parameters of Zone #1

The following data is displayed as a result:

The uncertain real parameter "K_{p1}" with nominal value −1.48 and a range [−1.78, −1.18].

The uncertain real parameter "T_{p11}" with a nominal value of 6.99 and a range [5.59, 8.38].

The resulting set of transient properties is depicted in Fig. 8, where the gain k variability affects the dynamics of the transient property in the steady state, and the time constant Tp1 variability in the denominator of the transfer function's dynamics affects the transient property's value [10].

4 Design of Robust IMC Control

Uncertainties in the examined system must be handled when controlling a system with lumped input and distributed output. Additionally, there are more uncertainties when it comes to space domain control synthesis. Let's investigate these uncertainties'sources in the feedback-distributed parameter control system. Operations in the distributed parameter control system, beginning with the feedback-distributed parameter control loop [9], are shown in Fig. 7. The control time synthesis block Fig. 7 describes the design technique for IMC $Q_i(z)$ controllers in control synthesis with IMC structure. Individual control circuits use the identified transfer functions $\{SH_i(x_i, z)\}_i$ as nominal transfer functions describing the process model. H2 optimal controllers $Q_i(z)$ for input $\breve{W}(k)$ in the form of a unit step function $\gamma(z) = \frac{z}{z-1}$ are determined by minimizing the criterion:

$$\min_{Q_i(z)} E_i(x_i, z)_2 = \min_{Q_i(z)} (1 - SH_i(x_i, z) Q_i(z)) \gamma(z)_2 \tag{12}$$

The stable nominal transfer functions of the system $\{SH_i(x_i, z)\}_i$ are factorized in the first stage.

$$\mathrm{SH}_i(x_i, z) = \mathrm{SH}_{iN}(x_i, z)\ \mathrm{SH}_{iM}(x_i, z) \tag{13}$$

Where contains unstable zeros of the transfer function $\mathrm{SH}_i(x_i, z)$. Then the discrete IMC controller $Q_i(z)$ is in the following form:

$$Q_i(z) = \mathrm{SH}_{iM}(\mathrm{x}_i, z)^{-1} \tag{14}$$

In the second stage add a filter $F_i(z)$ with an adjustable parameter α to the controller $Q_i(z)$.

$$F_i(z) = \frac{(1-\alpha_i)}{z-\alpha_i} \tag{15}$$

The final formulation of the IMC controller $Q_{iF}(z)$ will be in the form:

$$Q_{iF}(z) = Q_i(z) F_i(z) = Q_i(z) \frac{(1-\alpha_i)}{z-\alpha_i} \tag{16}$$

The discrete IMC controller $Q_{iF}(z)$ has only one adjustable parameter α. We identify it by solving an optimization problem, ensuring that the requirements for robust stability and performance are met [12].

$$\left|F_i\left(e^{j\omega T}\right)\right| < \left[|S_i\left(x_i, j\omega\right) Q_i\left(e^{j\omega T}\right)| \bar{L}_{mi}(\omega)\right]^{-1}, \ 0 \le \omega \le \frac{\pi}{T} \tag{17}$$

$$\left|Q_i\left(j\omega\right)\right| \bar{L}_{ai}\left(\omega\right) + \left|1 - S_i\left(x_i, j\omega\right)\right| G_{wi}\left(\omega\right) < 1, \ 0 \le \omega \le \frac{\pi}{T} \tag{18}$$

4.1 Optimization of Controller Parameter α

The robust controller's alpha parameter is the only tuning parameter that can affect the quality. The quality of the control system is determined by the measure of control error in space and time.

$$E\left(x, k\right) = W\left(x, k\right) - Y\left(x, k\right) \tag{19}$$

It is appropriate to express this in quadratic norm:

$$E\left(x, k\right) \ = \ W\left(x, k\right) - Y\left(x, k\right) \tag{20}$$

The course of the quadratic norm of the control error $E\left(x, k\right)$ is therefore showing the quality of the control, which is not only dependent on space x but also on time dependence k. This can be used to define the criterion function for the optimization of the controller parameters, in this case, the tuning parameter $\{\alpha_i\}_{i=1,3}$ of the robust controller is in the following shape:

$$J = \min_{\alpha_i} \sum_{N}^{k=0} W\left(x, k\right) - Y\left(x, k\right) \tag{21}$$

For this purpose, in the MATLAB & Simulink library from the Simulink design optimization toolbox in the model verification section, a check custom bounds block has been chosen, and it is connected to the mean control loop Fig. 11. Simulink Design Optimization operates based on boundary definition; these boundaries serve as barriers between which a controlled signal must be maintained. Simulink Design Optimization allows entering the upper and lower limit of the boundaries based on the output signal. The process of the boundary's selection is clearly shown in Fig. 9 [13-6].

Simulink Design Optimization automatically selects the maximum and minimum (α, $-\alpha$) if the user does not use the option to limit the values when selecting parameters. This prolongs the calculation time. The optimization process will start to see Fig. 10, then Simulink Design Optimization automatically writes the values to the variables in the workspace when the optimization operation is completed, making them immediately available for further use in MATLAB and Simulink software.

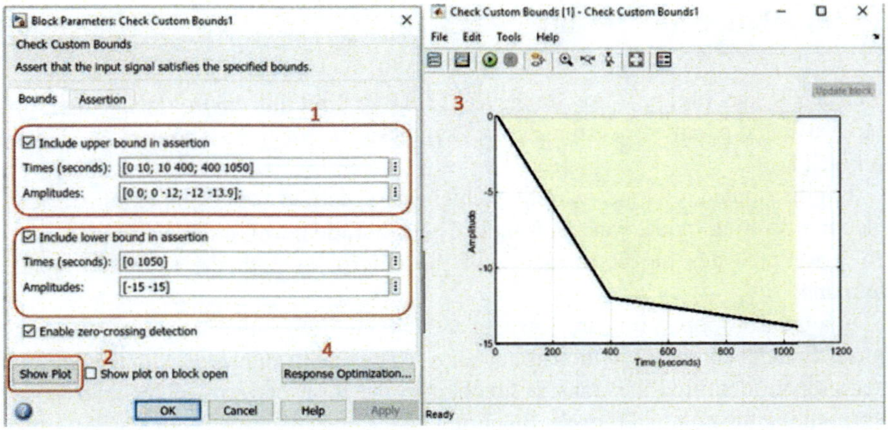

Fig. 9 Boundaries selection and its output

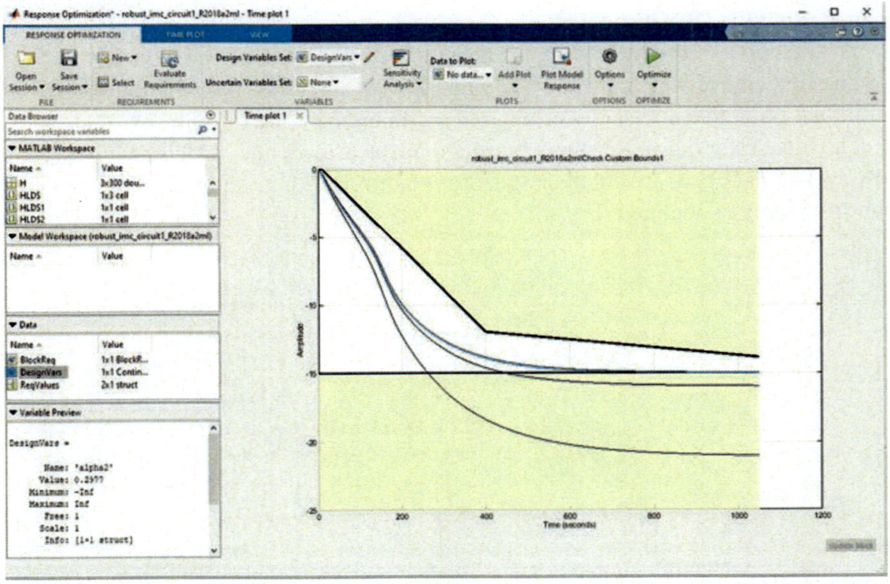

Fig. 10 Optimization process

Table 3 Controller parameter alpha value for every single zone

Zone #1	Zone #2	Zone #3
Alpha1 = 0.7720	Alpha2 = 0.3008	Alpha3 = 0.0444

The controller's parameter alpha for every single zone of secondary cooling of the continuous casting process was obtained at the end of the optimization process, and it shows in Table 3.

5 Simulation and Result

A DPS Blockset library is used in the MATLAB & Simulink environments to set up the distributed parameter system control synthesis of IMC robust control, as shown in Fig. 11.

The distributed reference variable, which was specified as the temperature difference in operating mode, was controlled. The quadratic norm of the distributed control error provides the performance of the control in both the time and space domains.

In a control loop, a robust controller was created by synthesis control based on internal model control as demonstrated above. The block connected with the parameter's alpha identified the transfer function in the form of zeros, poles, and gains representing the nominal transfer functions. In the control space synthesis block the approximation task for the controlled variable Y is solved. HLDS block represents the numerical model of the controlled process of a discrete system with lumped input and distributed output [10], and the desired temperature field is finally displayed using the DPS display block.

The outcome of the IMC robust control process is provided as follows:

The first course which is shown in Fig. 12, shows the fast dynamic of the zones with a set point when we are working only with temperature differences.

The quadratic norm of the controlled error shows the difference between $W(\overline{x},k)-Y(\overline{x},k)$, as much as the quadratic norm has the minimum character the control quality is obtained (Fig. 13).

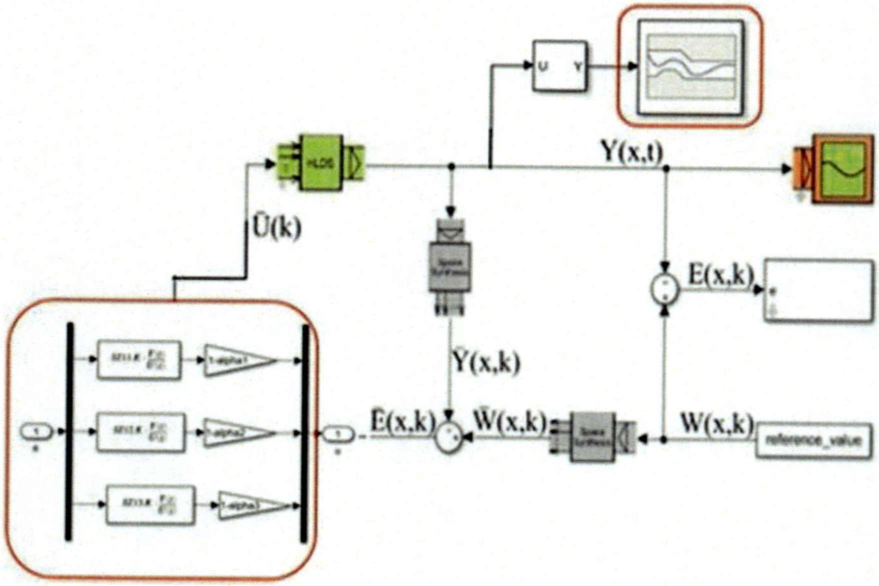

Fig. 11 IMC-based control circuit

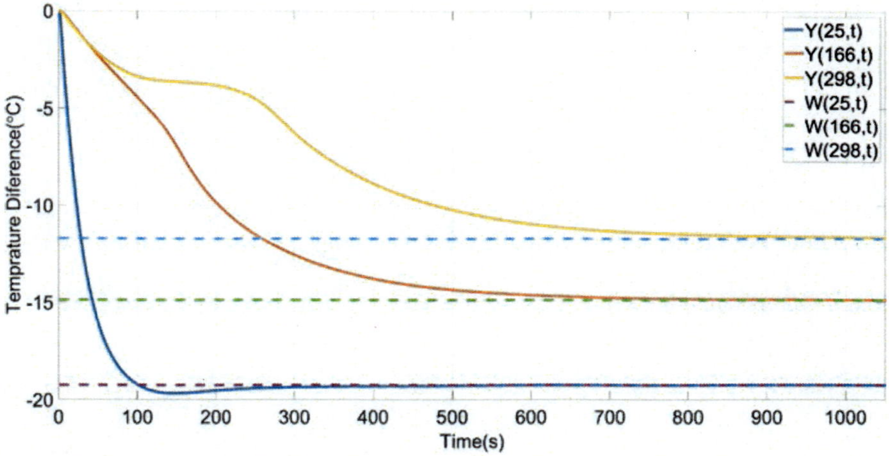

Fig. 12 Set point with controller's output

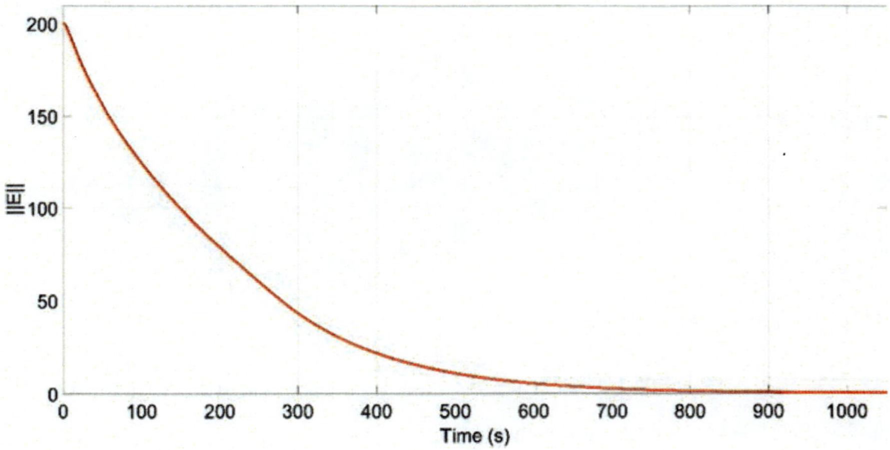

Fig. 13 Minimum error of the controlled process

The produced actuation variable, which is generated by IMC shown in the figure below, shows how fast the controller reacts and keeps the control system in a steady-state operation mode (Fig. 14).

Figure 15 shows the controlled temperature field in the secondary cooling zone of the continuous casting.

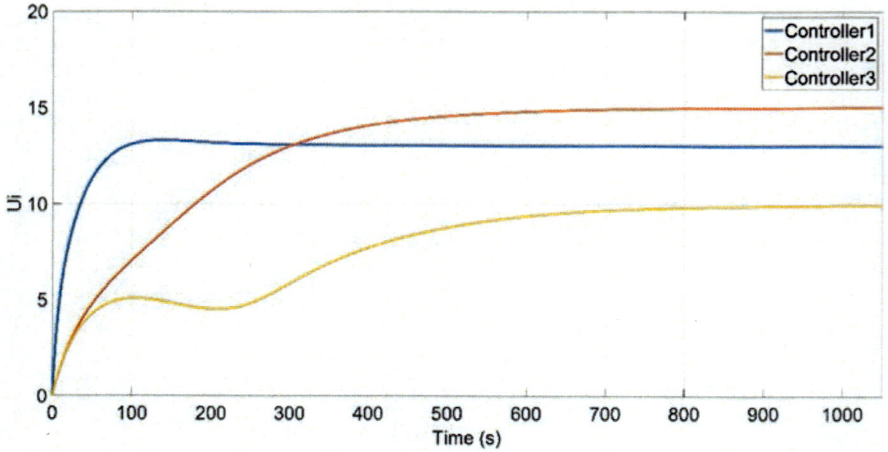

Fig. 14 Acting of the water follow nozzles

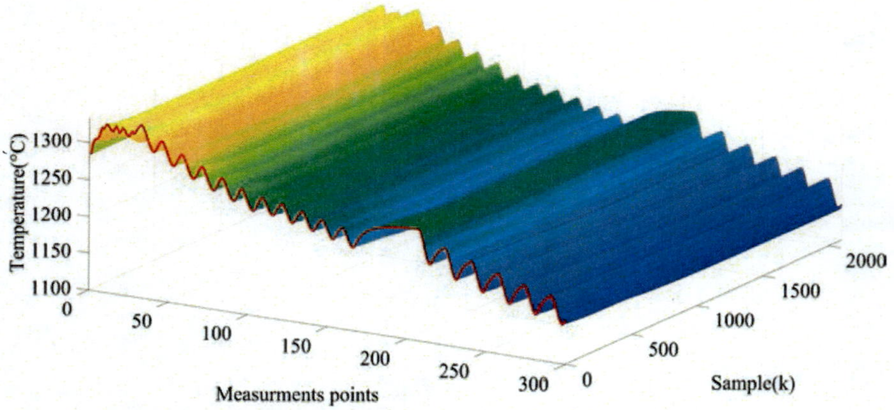

Fig. 15 Steel billet controlled temperature field

6 Conclusion

In the presented article, the task of controlling the temperature field in the secondary cooling zone, which is a typical representative of a system with a distributed parameter, was solved, and the optimized parameters of the IMC controller have been obtained. The goal of the control was to ensure the desired course of technology in the time in the space domain. Hence, a reliable method based on a reliable IMC controller was selected for control synthesis. The uncertainty bounds for the simulation model were established based on the model acquired from the ProCast program, and it is evident from the courses in Fig. 14 that the controllers are operating as intended.

Acknowledgments The authors would like to gratefully thank for the support of the KEGA in the framework of grant project KEGA 032STU-4/2021

References

1. Ramiz, Y.: Optimal control of lumped sources in distributed-parameter systems on classes of impulse and heaviside functions. Cybern. Syst. Anal. **48**(5), 798 (2012). https://doi.org/10.1007/s10559-012-9458-z
2. Belavy, C., Hulko, G., Ondrejkovič, K., Kubiš, M., Bartalský, L.: Robust control of temperature field in continuous casting process as distributed parameter systems. Int. Conf. Proc. Control (PC). **20**, 125–130 (2015)
3. Hulko, G., Belavy, C., Ondrejkovič, K., Bartalský, L.: Control of distributed parameter systems an engineering approach–Tutoria. Int. Conf. Proc. Control. **22**, 120–125 (2019)
4. Hulko, G., Belavy, C., Ondrejkovič, K., Bartalský, L., Bartko, M.: Control of technological and production processes as distributed parameter systems based on advanced numerical modeling. Control. Eng. Pract. **66**, 23–38 (2017)
5. Hulko, G., Belavy, C., Bartalský, L., Ondrejkovič, K., Bartko, M.: Control of the preheating of a casting mold as a distributed parameter system by PLC. Int. Conf. Proc. Control (PC). **20**, 344–349 (2015)
6. Hulko, G., Belavy, C., Noga, P., Ondrejkovič, K.: Control of Technological and Production Processes as Distributed Parameter Systems – Supported by Advanced Numerical Modeling. STU, Bratislava (2014)
7. Ondrejkovič, K., Bartalský, L., Belavy, C., Hulko, H.: Control of Distributed Parameter Systems: Exercise Instructions. Riadenie systémov S rozloženými Parametrami, Návody na cvičenia. STU, Bratislava (2019)
8. Cemernek, D., Cemernek, S., Gursch, H., Pandeshwar, A., Leitner, T., Berger, M., Klosch, G., Kern, R.: Machine learning in continuous casting of steel: a state-of-the-art survey. J. Intell. Manuf. **33**, 1561–1579 (2022). https://doi.org/10.1007/s10845-021-01754-7
9. Bartalský, L., Belavy, C., Bartko, M., Hulko, G., Kubiš, M.: PLC control of casting die preheating process as distributed parameter system. Int. Conf. Proc. Control (PC). **21**, 263–268 (2017)
10. Zarghoon, S., Bartalský, L., Belavy, C.: Robust Control of the Continuous Casting Process for Various Technological Conditions. In: 2022 23rd international Carpathian control conference (ICCC), pp. 250–255. IEEE (2022)
11. Thomas et al.: Online dynamic control of cooling in continuous casting of thin steel slabs. Proceedings of NSF Engineering Research and Innovation Conference, Honolulu (2009)
12. Vilanova, R.: IMC based robust PID design: tuning guidelines and automatic tuning. J. Proc. Control. **18**, 61–70 (2008). https://doi.org/10.1016/j.jprocont.2007.05.004
13. Zhenping, J., Jian, Y.: Online dynamic control of secondary cooling for the continuous casting process. Proceedings of Third international conference on intelligent networks and intelligent systems, IEEE, Shenyang, pp. 262–272 (2010)

Index

M. Balog et al. (eds.), *EAI International Conference on Automation and Control in Theory and Practice*, EAI/Springer Innovations in Communication and Computing, https://doi.org/10.1007/978-3-031-31967-9_1